Latent Structure and Causality

Inference from Data

Latent Structure and Causality

Inference from Data

Qing Zhou

University of California, Los Angeles, USA

World Scientific

NEW JERSEY · LONDON · SINGAPORE · BEIJING · SHANGHAI · TAIPEI · CHENNAI

Published by

World Scientific Publishing Co. Pte. Ltd.

5 Toh Tuck Link, Singapore 596224

USA office: 27 Warren Street, Suite 401-402, Hackensack, NJ 07601

UK office: 57 Shelton Street, Covent Garden, London WC2H 9HE

Library of Congress Control Number: 2024041762

British Library Cataloguing-in-Publication Data
A catalogue record for this book is available from the British Library.

LATENT STRUCTURE AND CAUSALITY
Inference from Data

ISBN 9789811290688 (hardcover)
ISBN 9789811290695 (ebook for institutions)
ISBN 9789811290701 (ebook for individuals)

For any available supplementary material, please visit
https://www.worldscientific.com/worldscibooks/10.1142/13773#t=suppl

Desk Editors: Murali Appadurai/Rosie Williamson

Typeset by Stallion Press
Email: enquiries@stallionpress.com

To my mother, my father, Jane, and Vicky

Preface

Understanding the latent structures behind observable phenomena and identifying causal relationships between variables are among the most profound challenges in contemporary science. This book introduces statistical models and methods for inferring latent structures and causality from data, bridging the gap between abstract theoretical constructs and practical empirical research. It provides readers with the methodology to predict hidden patterns and establish causal connections, thereby enhancing their inferences of the underlying mechanisms that drive observable outcomes.

After an introduction to the EM algorithm for maximum likelihood on incomplete data, the book covers in detail several widely used latent structure models, including mixture models, hidden Markov models, and stochastic block models. We develop EM and variational EM algorithms for parameter estimation under these models and compare them with their Bayesian inference counterparts. Further extensions of these models address related problems, such as clustering, motif discovery, Kalman filtering, and community detection. Conditional independence structures, with graphical representation, are utilized in the inference of the latent structures within these models. This notion naturally generalizes to the second part on graphical models that use graph separation to encode conditional independence. We introduce a variety of graphical models, including undirected graphs, directed acyclic graphs (DAGs), chain graphs, and acyclic directed mixed graphs (ADMGs). A primary motivation for these graphical models is to infer causality from observed data, often referred to as causal discovery, where a directed edge encodes a direct causal relation between two variables. In addition to classical topics in graphical models, we provide a detailed review of recent research in this area, such as continuous optimization for structure learning, identifiability of nonlinear and non-Gaussian DAGs, graphical models for dependent data, divide-and-conquer approaches for learning large graphs, and federated learning across multiple local clients. ADMGs further generalize DAGs to accommodate latent confounders in causal inference. Finally, we briefly introduce the causal bandit problem, which seeks to detect the optimal intervention through a sequential decision process, assuming an underlying causal graphical model for the system.

This book may be used as a textbook for graduate courses in statistical modeling, graphical models, and causal inference. To this end, I have made the topics accessible and suitable for graduate students. Most of the chapters are derived from my lecture notes for two graduate courses at UCLA. Former students enrolled in these courses have provided valuable feedback on the course materials. This book may also serve as a useful resource for researchers interested in graphical models and causal inference, offering a self-contained introduction to these topics and a selective review of recent developments. Readers interested in this area may focus on Chapter 1 and Part 2 (Chapters 6–12).

Some portions of the book are based on joint work with my colleagues, Arash Amini and Oscar Madrid Padilla, and my former PhD students, Fei Fu, Bryon Aragam, Jiaying Gu, Bingling Wang, Qiaoling Ye, Hangjian Li, Zhanhao Peng, Jireh Huang, Dale Kim, and Gabriel Ruiz. My gratitude goes to all of them!

About the Author

Qing Zhou is a professor of statistics and chair of the Department of Statistics and Data Science at the University of California, Los Angeles, USA. His research interests include graphical models, causal inference, high-dimensional statistics, Monte Carlo methods, and bioinformatics. He holds a PhD in statistics from Harvard University. Dr. Zhou has published more than 60 research papers in statistics, machine learning, and computational biology. He has received several NSF research awards, including the Career Award and a Big Data Award.

Contents

Preface   vii

About the Author   ix

1. Introduction and Review   **1**

1.1   Statistical Inference   1
1.2   Bayesian Inference   2
    1.2.1   Main steps   3
    1.2.2   Some basic models   4
1.3   Conditional Independence   10
    1.3.1   Definitions   10
    1.3.2   CI in statistical inference   11

Part 1   Latent Structure Models   **13**

2. Incomplete Data and the EM Algorithm   **15**

2.1   Assumptions   15
    2.1.1   Ignorability   16
    2.1.2   Observed-data likelihood and posterior 17
2.2   The EM Algorithm and Its Properties   18
    2.2.1   The algorithm   18
    2.2.2   EM as an MM algorithm   21
    2.2.3   Properties of the EM   22
    2.2.4   Missing information and convergence rate   24
    2.2.5   Another example   25

2.3   EM for Exponential Families   26
        2.3.1   Exponential families   26
        2.3.2   MLE for complete data   27
        2.3.3   EM for incomplete data   28
2.4   Incomplete Normal Data   30
        2.4.1   The complete-data MLE   31
        2.4.2   EM algorithm for incomplete normal data   32
2.5   Bayesian Inference with Missing Data   34
        2.5.1   Data augmentation   35
        2.5.2   Examples   36
2.6   Problems   39

3.   Mixture Modeling   41

3.1   Mixture Models   41
        3.1.1   Definition   41
        3.1.2   Maximum likelihood estimate using EM   43
3.2   Model-Based Clustering   46
3.3   Motif Discovery   49
        3.3.1   Problem formulation   49
        3.3.2   Maximum likelihood via EM   50
        3.3.3   Bayesian inference via Gibbs sampler   53
3.4   Problems   54

4.   Hidden Markov Models   57

4.1   Elements of an HMM   57
        4.1.1   Example and definition   57
        4.1.2   Graphical representation   58
4.2   Maximum Likelihood Estimate via the EM Algorithm   59
        4.2.1   Problem setup   60
        4.2.2   Forward and backward summations   61
4.3   The Viterbi Algorithm   65
4.4   Extensions   66
        4.4.1   Continuous observation   66
        4.4.2   Kalman filtering   67
4.5   Problems   69

5.   Random Graphs for Modeling Network Data   71

5.1   Network Data   71
5.2   Latent Space Models   72
5.3   Stochastic Block Models   73

5.3.1   Model structure 73

5.3.2   Variational EM algorithm 75

5.3.3   Variational inference for the SBM 77

5.3.4   Community detection 81

5.3.5   Extensions and discussion 83

5.4   Graphon 84

5.5   Problems 87

Part 2   Causal Graphical Models   89

6.   Undirected Graphical Models   91

6.1   Graphoid 91

6.2   Conditional Independence Tests 93

6.3   Undirected Graphs 95

6.4   Markov Properties 96

     6.4.1   Definitions 96

     6.4.2   Examples 97

     6.4.3   Equivalence 99

6.5   Conditional Independence Graphs 99

     6.5.1   Gaussian graphical models 100

     6.5.2   Discrete graphical models 103

6.6   Faithfulness 105

6.7   Markov Blanket 106

6.8   Neighborhood Lattice 108

6.9   Problems 111

7.   Directed Acyclic Graphs   113

7.1   Terminology for DAGs 113

7.2 d-Separation 114

7.3   Markov Properties 116

     7.3.1   Definitions and relations 116

     7.3.2   Moral graphs 118

     7.3.3   Markov equivalence 120

7.4   Parameterizations 122

     7.4.1   Gaussian Bayesian networks 122

     7.4.2   Discrete Bayesian networks 124

7.5   Chain Graphs 125
    7.5.1   Definition and characterization 125
    7.5.2   Markov properties 127
7.6   Faithfulness 128
7.7   Overview of Topics 129
7.8   Problems 131

8.   Causal Inference Based on Directed Acyclic Graphs   133

8.1   Causal DAGs and Intervention 133
8.2   Truncated Factorization 137
8.3   Linear Structural Equation Models 139
8.4   Estimation of Causal Effect 142
    8.4.1   Back-door adjustment 142
    8.4.2   Front-door adjustment 145
    8.4.3   Instrumental variables 147
8.5   Potential Outcome Approach 149
8.6   Problems 152

9.   Structure Learning of Directed Acyclic Graphs   153

9.1   Overview and Assumptions 153
9.2   Equivalence Class and Completed Partially DAGs 155
9.3   Constraint-Based Learning 158
    9.3.1   The PC algorithm 159
    9.3.2 p-value adjacency thresholding 162
9.4   Score-Based Learning 166
    9.4.1   Scoring functions 166
    9.4.2   Consistency 169
    9.4.3   Search strategies 170
9.5   Experimental Data 172
9.6   Continuous Optimization 175
    9.6.1   Maximum penalized likelihood 176
    9.6.2   Differentiable acyclicity constraint 180
9.7   Problems 182

**10.   Learning Generalized Directed Acyclic
     Graphical Models   185**

10.1   DAG Identifiability 185
    10.1.1   Equal error variance 186
    10.1.2   Generalized linear DAGs 187

10.2 Linear Non-Gaussian DAGs 189
    10.2.1 LiNGAM discovery algorithm 189
    10.2.2 Sequential order learning 190
10.3 Nonlinear DAG Models 194
    10.3.1 Additive noise models 194
    10.3.2 Learning with neural networks 196
10.4 Network Data 198
    10.4.1 Gaussian DAG for dependent data 198
    10.4.2 Parameter estimation 200
    10.4.3 Structure learning via de-correlation 202

11. Directed Mixed Graphs for Latent Variables   **203**

11.1 Semi-Markov Causal Models 203
11.2 Acyclic Directed Mixed Graphs 206
11.3 Factorizations on ADMGs 207
    11.3.1 District factorization 207
    11.3.2 Tian factorization 209
    11.3.3 Nested factorization 211
    11.3.4 Finding GCI constraints 213
11.4 Identification of Causal Effects 214
11.5 Linear SEM Associated with ADMG 217
    11.5.1 Parameterization and identifiability 217
    11.5.2 Structure learning 219
11.6 Ancestral Graphs 220
    11.6.1 Maximal ancestral graphs 221
    11.6.2 Partial ancestral graphs 222
    11.6.3 The FCI algorithm 225
11.7 Latent Factor Analysis 226
    11.7.1 Graphical representation 227
    11.7.2 Structure recovery by clique search 228

12. Partitioned, Federated, and Active Learning   **231**

12.1 Divide and Conquer 231
    12.1.1 Partition and estimation 232
    12.1.2 Fusion 234
12.2 Federated Learning 235
    12.2.1 Global objective function 236
    12.2.2 Distributed optimization 237
    12.2.3 Theoretical guarantees 239

12.3   Causal Bandit via Sequential Interventions   240
        12.3.1   Multi-armed bandit   240
        12.3.2   Causal bandit   242
        12.3.3   Bayesian backdoor bandit   244

Bibliography   249

Subject Index   261

Author Index   267

Chapter 1

Introduction and Review

Statistical inference is the process of drawing conclusions about a population from observed data in a sample. This process is crucial in many fields, from scientific research to business analytics, as it provides a systematic framework for making data-driven decisions and for quantifying uncertainty and variability inherent in the data.

1.1 Statistical Inference

In statistics, a data generation process is described by a mathematical model with a set of unknown parameters. Latent variables that are not observed may be involved in some problems as well. Under this framework, the two major tasks of statistical inference are:

(i) to estimate unknown model parameters and predict latent variables from data;
(ii) to quantify the uncertainty in the estimation and prediction.

Let us start with the most basic setting. Suppose we have collected the data

$$y_1, y_2, \ldots, y_n \stackrel{\text{iid}}{\sim} f(y \mid \theta),$$

where $f(y \mid \theta)$ is a probability density function (pdf) of a distribution parameterized by θ. Under this setup, task (i) involves estimating the unknown parameter θ, and task (ii) may be achieved by building a confidence interval for θ.

In general, denote the observed data by $\mathbf{y} = (y_1, y_2, \ldots, y_n)$. A common estimation method is to find the maximum likelihood estimate (MLE) of the unknown parameter. Define the likelihood of θ given data $\mathbf{y}$ as

$$L(\theta \mid \mathbf{y}) := p(y_1, \ldots, y_n \mid \theta) = \prod_{i=1}^{n} f(y_i \mid \theta).$$

The MLE $\widehat{\theta}_{\text{MLE}}$ is defined as the maximizer of $L(\theta \mid \mathbf{y})$ over θ,

$$\widehat{\theta}_{\text{MLE}} := \operatorname*{argmax}_{\theta} L(\theta \mid \mathbf{y}).$$

Moreover, we often estimate the standard error of the MLE, denoted by $\widehat{\text{se}}$, and construct an approximate 95% confidence interval as

$$(\widehat{\theta}_{\text{MLE}} - 2\widehat{\text{se}}, \widehat{\theta}_{\text{MLE}} + 2\widehat{\text{se}})$$

as a way to quantify the uncertainty in our estimate. The interpretation of the interval is

$$\mathbb{P}[\theta \in (\widehat{\theta}_{\text{MLE}} - 2\widehat{\text{se}}, \widehat{\theta}_{\text{MLE}} + 2\widehat{\text{se}})] = 0.95.$$

Here, $\widehat{\theta}_{\text{MLE}}$ is considered a random variable as a function of the random sample $\mathbf{y}$, while θ is an *unknown constant*.

1.2 Bayesian Inference

Bayesian inference relies on posterior distributions to provide solutions to the two inferential tasks (i) and (ii). The unknown parameter θ is regarded as a *random variable*, and thus we need to specify a marginal distribution for θ, denoted by $p(\theta)$, which is called a prior distribution. Here, "prior" means before observing any data, as a prior distribution does not depend on the data. Accordingly, a Bayesian model for our data $\mathbf{y}$ is set up by two distributions:

$$\text{prior: } \theta \sim p(\theta),$$

$$\text{data: } \mathbf{y} = (y_1, \ldots, y_n) \mid \theta \overset{\text{iid}}{\sim} f(y \mid \theta).$$

Together, they define a joint distribution for $(\theta, \mathbf{y})$:

$$p(\theta, \mathbf{y}) = p(\theta)p(\mathbf{y} \mid \theta) = p(\theta) \cdot \prod_{i=1}^{n} f(y_i \mid \theta). \tag{1.1}$$

1.2.1 *Main steps*

Based on (1.1), we find the conditional distribution $[\theta \mid \mathbf{y}]$ to perform inference on θ. This conditional distribution of θ given the data $\mathbf{y}$ is called the posterior distribution, where "posterior" means that the distribution of θ is now updated after observing the data and thus depends on $\mathbf{y}$. Applying Bayes' formula,

$$p(\theta \mid \mathbf{y}) = \frac{p(\theta, \mathbf{y})}{p(\mathbf{y})} = \frac{p(\theta)p(\mathbf{y} \mid \theta)}{p(\mathbf{y})} = \frac{p(\theta) \cdot \prod_{i=1}^{n} f(y_i \mid \theta)}{p(\mathbf{y})},$$

where the marginal density $p(\mathbf{y}) = \int p(\theta, \mathbf{y})d\theta$ does not depend on θ and can be regarded as a normalizing constant. Consequently, it is more convenient to work with an unnormalized posterior density:

$$p(\theta \mid \mathbf{y}) \propto p(\theta)p(\mathbf{y} \mid \theta) = p(\theta) \cdot \prod_{i=1}^{n} f(y_i \mid \theta). \qquad (1.2)$$

We may recognize the posterior distribution given the unnormalized density on the right-hand side if $p(\theta \mid \mathbf{y})$ belongs to a known family of distributions; see Section 1.2.2 for examples.

A Bayesian estimate of θ is usually constructed as the mean of the posterior distribution:

$$\widehat{\theta}_B := \mathbb{E}(\theta \mid \mathbf{y}) = \int \theta \cdot p(\theta \mid \mathbf{y})d\theta. \qquad (1.3)$$

A $(1 - 2\alpha)$ Bayesian credible interval for θ can be constructed by the quantiles of the posterior distribution: $(\theta_{(\alpha)}, \theta_{(1-\alpha)})$, where $\theta_{(\alpha)}$ is the α-quantile for $\alpha \in (0, 1)$. The interpretation of a credible interval is

$$\mathbb{P}(\theta \in (\theta_{(\alpha)}, \theta_{(1-\alpha)}) \mid \mathbf{y}) = 1 - 2\alpha, \qquad (1.4)$$

where θ is a random variable following the posterior distribution $p(\theta \mid \mathbf{y})$. See Figure 1.1 for an illustration.

For complicated problems, the posterior distribution $p(\theta \mid \mathbf{y})$ usually does not belong to any well-studied distribution family. In this case, a Monte Carlo simulation, such as Markov chain Monte Carlo, is applied to draw samples of θ from the posterior distribution $p(\theta \mid \mathbf{y})$, regarding (1.2) as the target density. From the Monte Carlo samples, one can easily calculate the sample mean and sample quantiles to approximate $\widehat{\theta}_B$ and $(\theta_{(\alpha)}, \theta_{(1-\alpha)})$.

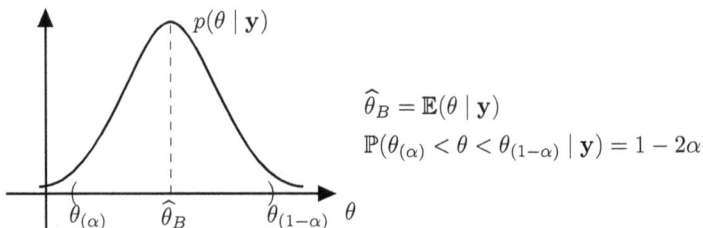

Fig. 1.1. Posterior distribution $p(\theta \mid \mathbf{y})$, posterior mean $\mathbb{E}(\theta \mid \mathbf{y})$, and credible interval $(\theta_{(\alpha)}, \theta_{(1-\alpha)})$.

In summary, the main steps of Bayesian inference are as follows:

(1) Choose a prior distribution $p(\theta)$.
(2) Find the posterior distribution $p(\theta \mid \mathbf{y})$ using (1.2).
(3) Apply a Monte Carlo algorithm to draw samples from $p(\theta \mid \mathbf{y})$.
(4) Construct Bayesian estimates and intervals from the Monte Carlo samples.

1.2.2 *Some basic models*

In this section, we demonstrate the main steps of Bayesian inference with a few simple examples.

Example 1.1 (Binomial distribution). Consider independent coin tossing, with $\theta \in (0,1)$ being the probability of heads. Suppose we toss n times and observe heads x times. How can we estimate θ?

Let the random variable X represent the number of times we observe heads in n coin tosses. The distribution of X given θ is

$$X \mid \theta \sim \mathrm{Bin}(n, \theta).$$

Thus, the likelihood is

$$\mathbb{P}(X = x \mid \theta) = \binom{n}{x} \theta^x (1 - \theta)^{n-x},$$

and the MLE is $\widehat{\theta}_{\mathrm{MLE}} = x/n$.

The main steps of Bayesian inference for this problem are summarized as follows:

(1) Choose a prior distribution for θ: Without any prior knowledge on θ, we usually choose a flat prior,

$$\theta \sim \mathrm{Unif}(0, 1), \quad \text{i.e., } p(\theta) = 1, \ \theta \in (0, 1).$$

(2) Find the posterior distribution:

$$p(\theta \mid X = x) \propto p(\theta) \cdot \mathbb{P}(X = x \mid \theta)$$

$$= \binom{n}{x} \theta^x (1 - \theta)^{n-x}$$

$$\propto \theta^x (1 - \theta)^{n-x}, \tag{1.5}$$

where θ is a random variable.

(3) From (1.5), we recognize that it is an unnormalized Beta density. Therefore, the posterior distribution is

$$\theta \mid x \sim \text{Beta}(x + 1, n - x + 1). \tag{1.6}$$

As a reference, the pdf of $\text{Beta}(\alpha, \beta)$ is

$$f(\theta) = \frac{\Gamma(\alpha + \beta)}{\Gamma(\alpha)\Gamma(\beta)} \theta^{\alpha-1} (1 - \theta)^{\beta-1}, \qquad \theta \in (0, 1),$$

and its mean is $\mathbb{E}(\theta) = \frac{\alpha}{\alpha + \beta}$.

(4) Given (1.6), we find the Bayesian estimate:

$$\widehat{\theta}_B = \mathbb{E}(\theta \mid x) = \frac{x + 1}{n + 2}.$$

To construct a 95% credible interval, we use the 2.5% and 97.5% quantiles of $\text{Beta}(x + 1, n - x + 1)$. For example, if $n = 10$ and $x = 3$, the posterior distribution is $\text{Beta}(4, 8)$, for which the two quantiles are as follows:

```
> qbeta(c(0.025,0.975),4,8)
[1] 0.1092634 0.6097426
```

So, a 95% credible interval is $(0.109, 0.610)$. If $n = 20$ and $x = 6$, the posterior distribution is $\text{Beta}(7, 15)$, with the following quantiles:

```
> qbeta(c(0.025,0.975),7,15)
[1] 0.1458769 0.5217511
```

In this case, a credible interval is $(0.146, 0.522)$, which is shorter than in the first case as the sample size n is larger.

Figure 1.2 shows the shape of the prior (black) and the posterior distributions, in red for $n = 10$ and $x = 3$ and in blue for $n = 20$ and $x = 6$.

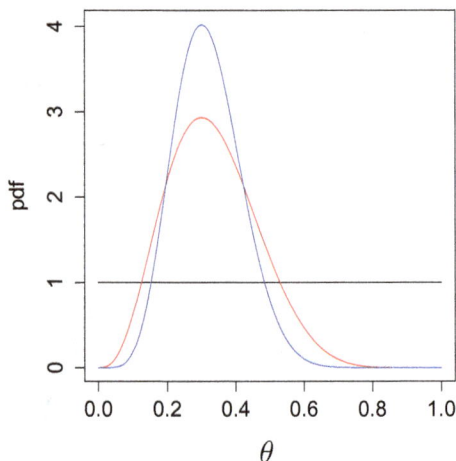

Fig. 1.2. Prior and posterior distributions in the coin toss example.

A credible interval can be used for performing hypothesis testing. Suppose we want to decide whether the coin is fair by testing the null hypothesis

$$H_0 : \theta = 0.5.$$

Based on the data $n = 20$ and $x = 6$, the 95% credible interval $(0.146, 0.522)$ covers 0.5; therefore, we accept the null hypothesis H_0. If we collect more data and observe that $n = 50$ and $x = 15$, then $\theta \mid x \sim \text{Beta}(16, 36)$, and a 95% credible interval will be $(0.191, 0.438)$. Because 0.5 falls outside this interval, we conclude, with 95% probability, that the coin is not fair (rejecting H_0).

The uniform distribution $\text{Unif}(0, 1)$ is equivalent to $\text{Beta}(1, 1)$. We may choose other Beta distributions as the prior for θ:

$$\theta \sim \text{Beta}(\alpha, \beta), \qquad p(\theta) \propto \theta^{\alpha-1}(1 - \theta)^{\beta-1}.$$

Then, the posterior distribution is

$$p(\theta \mid X = x) \propto p(\theta) \cdot \mathbb{P}(X = x \mid \theta)$$

$$\propto \theta^{\alpha-1}(1 - \theta)^{\beta-1} \cdot \binom{n}{x} \theta^x (1 - \theta)^{n-x}$$

$$\propto \theta^{x+\alpha-1}(1 - \theta)^{n-x+\beta-1},$$

and thus

$$\theta \mid x \sim \text{Beta}(x + \alpha, n - x + \beta).$$

We see that the posterior is in the same family of the prior, both Beta distributions, in which case we say that the prior is a *conjugate prior*. That is, the Beta prior is conjugate to the binomial distribution. The Bayesian estimate, or the posterior mean, under this prior is

$$\widehat{\theta}_B = \frac{x + \alpha}{n + \alpha + \beta}. \tag{1.7}$$

Compared to the MLE $\widehat{\theta}_{\text{MLE}} = x/n$, the prior parameters (α, β) may be regarded as pseudo counts added to the two possible outcomes (heads or tails). If there is no prior knowledge about θ, we choose small pseudo counts, $\alpha, \beta \in (0, 1]$. If there is a strong prior for θ, say from historical data, one may choose larger values of α and β to reflect such prior knowledge.

Example 1.2 (Multinomial distribution). We generalize Example 1.1 to multinomial data. Let $\theta = (\theta_1, \theta_2, \ldots, \theta_k)$ be the probabilities of k possible outcomes in a random experiment, $\theta_j > 0$, $\sum_{j=1}^{k} \theta_j = 1$. Suppose we have done this experiment n times independently and observed the jth outcome x_j times. So, the observations follow a multinomial distribution:

$$\mathbf{x} = (x_1, x_2, \ldots, x_k) \mid \theta \sim \text{M}(n, \theta), \quad \sum x_j = n.$$

The likelihood is

$$p(\mathbf{x} \mid \theta) \propto \theta_1^{x_1} \theta_2^{x_2} \cdots \theta_k^{x_k}, \tag{1.8}$$

and the MLE is

$$(\widehat{\theta}_j)_{\text{MLE}} = \frac{x_j}{n}, \qquad j = 1, \ldots, k.$$

To perform Bayesian inference, let us first find a conjugate prior.

Definition 1.1 (Dirichlet distribution). Let $\theta = (\theta_1, \ldots, \theta_k) \in \mathbb{R}^k$ be a random vector such that $\theta_j \geq 0$ for all $j = 1, \ldots, k$ and $\sum_{j=1}^{k} \theta_j = 1$. Then, θ follows the Dirichlet distribution $\text{Dir}(\alpha_1, \ldots, \alpha_k)$, $\alpha_j > 0$ for all j, if the pdf of θ is

$$p(\theta) = \frac{\Gamma(\alpha_1 + \alpha_2 + \cdots + \alpha_k)}{\Gamma(\alpha_1)\Gamma(\alpha_2) \cdots \Gamma(\alpha_k)} \theta_1^{\alpha_1 - 1} \theta_2^{\alpha_2 - 1} \cdots \theta_k^{\alpha_k - 1}.$$

The mean of θ is

$$\mathbb{E}(\theta_j) = \frac{\alpha_j}{\alpha_1 + \alpha_2 + \cdots + \alpha_k}, \qquad j = 1, \ldots, k. \tag{1.9}$$

A Dirichlet distribution $\text{Dir}(\alpha_1, \ldots, \alpha_k)$ can be represented with Gamma random variables, which provides a procedure to draw samples from the Dirichlet distributions:

(i) Draw $v_j \sim \text{Gamma}(\alpha_j, 1)$ independently for $j = 1, \ldots, k$.
(ii) Put $S = \sum_{j=1}^{k} v_j$ and define

$$\theta_j = \frac{v_j}{S} = \frac{v_j}{v_1 + \cdots + v_k}, \qquad j = 1, \ldots, k.$$

Then, $\theta = (\theta_1, \theta_2, \ldots, \theta_k) \sim \text{Dir}(\alpha_1, \alpha_2, \ldots, \alpha_k)$.

It turns out that the Dirichlet distribution is a conjugate prior for multinomial distribution. To see that, let us assume that the prior is

$$\theta \sim \text{Dir}(\alpha_1, \ldots, \alpha_k), \text{ i.e., } p(\theta) \propto \theta_1^{\alpha_1 - 1} \theta_2^{\alpha_2 - 1} \cdots \theta_k^{\alpha_k - 1}. \qquad (1.10)$$

Then, the posterior distribution, by multiplying (1.10) and (1.8), is given by

$$p(\theta \mid \mathbf{x}) \propto p(\theta) p(\mathbf{x} \mid \theta)$$
$$\propto \theta_1^{x_1 + \alpha_1 - 1} \theta_2^{x_2 + \alpha_2 - 1} \cdots \theta_k^{x_k + \alpha_k - 1},$$

which is an unnormalized density of $\text{Dir}(x_1 + \alpha_1, \ldots, x_k + \alpha_k)$. Therefore,

$$\theta \mid \mathbf{x} \sim \text{Dir}(x_1 + \alpha_1, \ldots, x_k + \alpha_k). \qquad (1.11)$$

Put $\alpha_0 = \sum_{j=1}^{k} \alpha_j$. By (1.9), we find the Bayesian estimate of θ using the posterior mean

$$(\widehat{\theta_j})_B = \mathbb{E}(\theta_j \mid \mathbf{x}) = \frac{x_j + \alpha_j}{n + \alpha_0}, \qquad j = 1, \ldots, k.$$

Similar to (1.7), here $\alpha_1, \ldots, \alpha_k$ are also interpreted as pseudo counts for the k possible outcomes. Without any prior knowledge, we choose $\alpha_j \in (0, 1]$. In particular, if $\alpha_j = 1$ for all j, the prior is a uniform distribution as $p(\theta) \propto 1$.

If we wish to build a credible interval for θ_j, we can do so using the quantiles of the posterior distribution $[\theta_j \mid \mathbf{x}]$, which is simply a marginal distribution of the Dirichlet distribution (1.11). By the properties of Dirichlet distributions, the marginal distribution is a Beta distribution:

$$\theta_j \mid \mathbf{x} \sim \text{Beta}(x_j + \alpha_j, n - x_j + \alpha_0 - \alpha_j).$$

Then, we can use the same procedure in Example 1.1 to construct a credible interval for each θ_j.

Example 1.3 (Normal data with known variance). Suppose we have observed

$$y_1, \ldots, y_n \mid \theta \overset{\text{iid}}{\sim} \mathcal{N}(\theta, \sigma^2),$$

where σ^2 is known. Our goal is to make an inference on θ. The likelihood of θ is

$$p(y_1, \ldots, y_n \mid \theta) = \prod_{i=1}^{n} \frac{1}{\sqrt{2\pi}\sigma} \exp\left\{ -\frac{1}{2\sigma^2}(y_i - \theta)^2 \right\}$$

$$\propto \exp\left\{ -\frac{1}{2\sigma^2} \sum_{i=1}^{n}(y_i - \theta)^2 \right\}.$$

The MLE is $\widehat{\theta}_{\text{MLE}} = \bar{y} = \frac{1}{n}\sum_i y_i$. The standard error (standard deviation) of $\bar{y}$ is se $= \sigma/\sqrt{n}$. Thus, we can construct a 95% confidence interval, $(\bar{y} - 2\sigma/\sqrt{n}, \bar{y} + 2\sigma/\sqrt{n})$.

Now, consider the Bayesian inference under a conjugate prior, $\theta \sim \mathcal{N}(\mu_0, \tau_0^2)$. Let us put a flat prior by choosing $\tau_0 \to \infty$:

$$p(\theta) \propto \exp\left\{ -\frac{1}{2\tau_0^2}(\theta - \mu_0)^2 \right\} \to 1, \quad \text{as } \tau_0 \to \infty. \tag{1.12}$$

Then, the posterior distribution $[\theta \mid \mathbf{y} = (y_1, \ldots, y_n)]$ is

$$p(\theta \mid \mathbf{y}) \propto p(\theta)p(\mathbf{y}|\theta) \propto \exp\left\{ -\frac{1}{2\sigma^2} \sum_{i=1}^{n}(\theta - y_i)^2 \right\}.$$

Recall that θ is a random variable and $\mathbf{y}$ is constant. Using the equality

$$\sum_{i=1}^{n}(\theta - y_i)^2 = \sum_{i}(\theta - \bar{y} + \bar{y} - y_i)^2$$

$$= n(\theta - \bar{y})^2 + \sum_{i=1}^{n}(y_i - \bar{y})^2,$$

we get

$$p(\theta \mid \mathbf{y}) \propto \exp\left\{ -\frac{1}{2\sigma^2}n(\theta - \bar{y})^2 \right\} = \exp\left\{ -\frac{(\theta - \bar{y})^2}{2\sigma^2/n} \right\}.$$

This shows that the posterior distribution is

$$\theta \mid \mathbf{y} \sim \mathcal{N}(\bar{y}, \sigma^2/n).$$

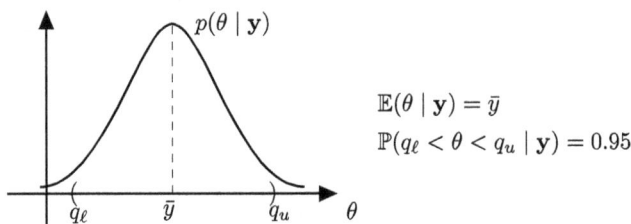

Fig. 1.3. Posterior distribution, posterior mean, and credible interval for normal mean.

Therefore, the Bayesian estimate is $\widehat{\theta}_B = \mathbb{E}(\theta \mid \mathbf{y}) = \bar{y}$, and a 95% credible interval, constructed using the quantiles (q_ℓ, q_u) of $\mathcal{N}(\bar{y}, \sigma^2/n)$, is

$$(\bar{y} - 2\sigma/\sqrt{n}, \bar{y} + 2\sigma/\sqrt{n}).$$

See Figure 1.3 for an illustration.

Again, the interval length $(4\sigma/\sqrt{n})$ shrinks when n increases. For this example, the Bayesian point and interval estimates both coincide with the MLE and the confidence interval.

1.3 Conditional Independence

Conditional independence (CI) is a fundamental concept in statistical modeling and inference. CI relations are frequently used to explore the dependence between two random variables when conditioned on information from a set of other random variables. As demonstrated in the following few chapters, leveraging the structure of CI allows us to construct succinct probabilistic models and devise efficient inference algorithms. Furthermore, representation of CI relationships among a set of random variables is a pivotal motivation for graphical models – the focal point of the latter half of this book.

1.3.1 *Definitions*

We define CI following Lauritzen (1996).

Definition 1.2 (Conditional independence). For three random variables $X, Y,$ and Z, we say that X is conditionally independent of Y given Z, written $X \perp\!\!\!\perp Y \mid Z$, if $\mathbb{P}(X \in A \mid Y, Z)$ is a function of Z only for any measurable set A.

If Z is trivial, then we say that $X \perp\!\!\!\perp Y$. Note that (X, Y, Z) may not have a joint density in the above definition. If X, Y, Z admit a joint density (or mass function) f, then

$$X \perp\!\!\!\perp Y \mid Z \Leftrightarrow f_{XY|Z}(x, y|z) = f_{X|Z}(x|z) f_{Y|Z}(y|z).$$

Using f as a generic symbol for densities, other equivalent conditions for the CI $X \perp\!\!\!\perp Y \mid Z$ include

$$f(x, y, z) = f(x, z) f(y, z) / f(z);$$
$$f(x|y, z) = f(x|z);$$
$$f(x, z|y) = f(x|z) f(z|y);$$
$$f(x, y, z) = h(x, z) k(y, z) \text{ for some } h, k;$$
$$f(x, y, z) = f(x|z) f(y, z).$$

1.3.2 CI in statistical inference

CI is an important and fundamental concept in statistical inference (Dawid, 1979). This is illustrated in the following examples.

Example 1.4 (Sufficient and ancillary statistics). Suppose that X is a random sample from a distribution P_Θ parameterized by Θ, i.e., $X \mid \Theta \sim P_\Theta$. Then:

(i) $T = T(X)$ is a sufficient statistic for Θ if $X \perp\!\!\!\perp \Theta \mid T$;
(ii) $S = S(X)$ is an ancillary statistic if $S \perp\!\!\!\perp \Theta$.

For example, suppose $X = (X_1, \dots, X_n) \mid \mu \overset{\text{iid}}{\sim} \mathcal{N}(\mu, 1)$. Then, $T = \sum_i X_i$ is a sufficient statistic for μ, and $S = \sum_i (X_i - \bar{X})^2 \sim \chi^2_{n-1}$ is an ancillary statistic for μ. To show that T is a sufficient statistic, we verify that the conditional distribution $[X \mid T(X), \mu]$ does not depend on μ. Let $I(\cdot)$ be the indicator function. Assuming $T(x) = t$, the conditional density is

$$p(x \mid t, \mu) \propto f(x \mid \mu) \cdot I(T(x) = t)$$

$$\propto \exp\left\{ -\frac{1}{2}\left(\sum_{i=1}^n x_i^2 - 2\mu t + \mu^2 \right) \right\} \cdot I\left(\sum_i x_i = t \right)$$

$$\propto \exp\left\{ -\frac{1}{2} \sum_{i=1}^n x_i^2 \right\} \cdot I\left(\sum_i x_i = t \right),$$

which is indeed independent of μ. In the last step, we dropped all factors that are not a function of x.

In this example, we do *not* assume a joint density for (X, Θ), as Θ is treated as a constant.

Example 1.5 (Parameter identifiability). Suppose $Y \mid \Theta, \Phi \sim P_{(\Theta, \Phi)}$, where the distribution P is parameterized by Θ and Φ. If $Y \perp\!\!\!\perp \Phi \mid \Theta$, then Φ is not identifiable, since $[Y \mid \Theta, \Phi] = [Y \mid \Theta]$. As an example, consider Gaussian linear regression, $Y = X\beta + \varepsilon$, where $Y, \varepsilon \in \mathbb{R}^n$, X is an $n \times p$ design matrix, and $\beta \in \mathbb{R}^p$. Assume that X does not have a full column rank. Let $\Theta := X\beta \in \mathrm{col}(X)$ (the column space of X) and $\Phi = \beta - X^- X\beta$, where X^- is a generalized inverse of X such that $XX^- X = X$. Then, $X\Phi = 0$, i.e., Φ is in the null space of X, and $Y \perp\!\!\!\perp \Phi \mid \Theta$. This implies that Φ is not identifiable. The linear model is overparameterized because the columns of X are not linearly independent. The data (Y, X) do not contain information for Φ.

CI also plays an essential and important role in statistical modeling. Consider the joint distribution of p random variables $X_1, \ldots, X_p$. Let $\sigma = (\sigma(1), \ldots, \sigma(p))$ be a permutation of the set $[p] := \{1, \ldots, p\}$. Given an ordering σ, the joint density $f(x) = f(x_1, \ldots, x_p)$ can be factorized as

$$f(x) = \prod_{j=1}^{p} f(x_{\sigma(j)} \mid x_{\sigma(1)}, \ldots, x_{\sigma(j-1)})$$

$$= \prod_{j=1}^{p} f(x_{\sigma(j)} \mid x_{A_j}), \qquad (1.13)$$

where $A_j \subseteq \{\sigma(1), \ldots, \sigma(j-1)\}$ is the minimum subset such that (1.13) holds. In other words,

$$X_{\sigma(j)} \perp\!\!\!\perp X_{B_j} \mid X_{A_j}, \qquad B_j = \{\sigma(1), \ldots, \sigma(j-1)\} \setminus A_j.$$

Using such CI relations will lead to a parsimonious model. For example, if $\{X_1, \ldots, X_p\}$ is a Markov chain and $\sigma = (1, \ldots, p)$, then $A_j = \{j - 1\}$.

As another example, consider a model selection problem in linear regression, $Y = X\beta + \varepsilon$. Let $\mathrm{supp}(\beta) := \{i : \beta_i \neq 0\}$. If $\mathrm{supp}(\beta) = S$, then

$$Y \perp\!\!\!\perp \{X_j, j \notin S\} \mid X_S, \qquad (1.14)$$

where X_j is the jth column of X. Given (X, Y), the objective of model selection is to find a subset $\widehat{S}$ such that (1.14) holds.

PART 1
Latent Structure Models

Chapter 2

Incomplete Data and the EM Algorithm

Latent structure models incorporate unobservable or latent variables to capture hidden patterns among observed data. These models are particularly useful to represent underlying structures or relationships that may not be directly measurable. The EM algorithm (Dempster *et al.*, 1977) is a classical method for the estimation of parameters in a latent structure model. It is an iterative optimization algorithm that alternates between an expectation step (E-step), where missing data or latent variables are imputed based on current parameter estimates, and a maximization step (M-step), where model parameters are updated by maximizing the expected log-likelihood. In the context of latent structure models, the EM algorithm enables the estimation of both unknown parameters and latent variables, thereby facilitating a comprehensive inference of the underlying hidden structure given the observed data.

2.1 Assumptions

Let Y be an $n \times p$ data matrix and y_i be the ith row of Y, $i = 1, \ldots, n$. Under the i.i.d. assumption, the probability density of Y is

$$p(Y \mid \theta) = \prod_{i=1}^{n} f(y_i \mid \theta),$$

where θ is the parameter for the data generation model $f(\cdot \mid \theta)$.

Table 2.1.   Example of missing data.

Units	Variable 1	Variable 2	$\cdots$	Variable p
Unit 1				
Unit 2		?		?
$\vdots$				
Unit n	?	?		

Note: Each row shows the data y_i for a unit.

In many applications, missing data occur for various reasons. That is, some components of y_i may not be observed. In the data matrix in Table 2.1, y_1 is fully observed, y_{22} and y_{2p} are missing and marked by question marks, and so on.

We first review some common assumptions for statistical inference with missing data, following the discussion in Chapter 2 of Schafer (1997). Let us denote all observed data and missing data in Y by Y_{obs} and Y_{mis}, respectively. We call $Y = (Y_{\text{obs}}, Y_{\text{mis}})$ the complete data.

2.1.1   *Ignorability*

Missing at random (MAR) is defined in terms of a probability model for the missingness. Let $R = (r_{ij})$ be an $n \times p$ matrix of indicators, where $r_{ij} = 1$ if y_{ij} is observed and $r_{ij} = 0$ otherwise. We put a probability model for R, $p(R \mid Y, \xi)$, where ξ is some parameter. The MAR assumption is that

$$p(R \mid Y_{\text{obs}}, Y_{\text{mis}}, \xi) = p(R \mid Y_{\text{obs}}, \xi), \tag{2.1}$$

i.e., $R \perp\!\!\!\perp Y_{\text{mis}} \mid Y_{\text{obs}}$ (Definition 1.2). Unless otherwise noted, we treat parameters as constants and do not include them in the conditioning set when considering CI relations. A stronger assumption is missing completely at random (MCAR): $R \perp\!\!\!\perp (Y_{\text{mis}}, Y_{\text{obs}})$. If neither holds, then the data are missing not at random (MNAR): R depends on Y_{mis}, written as $Y_{\text{mis}} \to R$.

Consider an example from Mohan and Pearl (2021): a study measured the age (A), gender (G), and obesity (O) of students in a school, with missing values in O since some students failed to reveal their weight. Depending on the reasons for the missingness, we may make different assumptions:

- MCAR: some students accidentally lost questionnaires ($R \perp\!\!\!\perp A, G, O$).
- MAR: some teenagers not reporting weight ($R \perp\!\!\!\perp O \mid A$).
- MNAR: overweight students reluctant to report weight ($O \to R$).

Distinctness of parameters. Let θ denote the parameters of the data model and ξ denote the parameters of the missingness mechanism. Then, θ and ξ are distinct if the following conditions hold:

(a) *Frequentist*: The joint parameter space of (θ, ξ) is the Cartesian product of the individual parameter spaces for $\theta \in \Theta$ and $\xi \in \Gamma$, i.e.,

$$(\theta, \xi) \in \Theta \times \Gamma.$$

(b) *Bayesian*: The joint prior for (θ, ξ) factors into independent marginal priors for θ and ξ, i.e.,

$$\pi(\theta, \xi) = \pi_\theta(\theta)\pi_\xi(\xi).$$

If both MAR and distinctness hold, we say that the missing data mechanism is *ignorable*.

2.1.2 Observed-data likelihood and posterior

Under ignorability, we may ignore R in our inference for θ, the parameter of interest. In other words, we only need to work with observed-data likelihood to draw the inference on θ. To see this, we start with the joint distribution of Y_{obs} and R:

$$\mathbb{P}(R, Y_{\text{obs}} \mid \theta, \xi) = \int \mathbb{P}(R, Y \mid \theta, \xi)dY_{\text{mis}}$$

$$= \int \mathbb{P}(R \mid Y, \xi)\mathbb{P}(Y \mid \theta)dY_{\text{mis}}.$$

If MAR (2.1) holds, then $\mathbb{P}(R \mid Y, \xi) = \mathbb{P}(R \mid Y_{\text{obs}}, \xi)$. Therefore, we have

$$\mathbb{P}(R, Y_{\text{obs}} \mid \theta, \xi) = \mathbb{P}(R \mid Y_{\text{obs}}, \xi) \int \mathbb{P}(Y \mid \theta)dY_{\text{mis}}$$

$$= \mathbb{P}(R \mid Y_{\text{obs}}, \xi)\mathbb{P}(Y_{\text{obs}} \mid \theta).$$

Consider the maximum likelihood estimate (MLE) of (θ, ξ). Under distinctness,

$$\max_{(\theta, \xi) \in \Theta \times \Gamma} \mathbb{P}(R, Y_{\text{obs}} \mid \theta, \xi) = \left\{ \max_{\xi \in \Gamma} \mathbb{P}(R \mid Y_{\text{obs}}, \xi) \right\} \left\{ \max_{\theta \in \Theta} \mathbb{P}(Y_{\text{obs}} \mid \theta) \right\}$$

is separable. Define the observed-data likelihood $L(\theta \mid Y_{\text{obs}}) := \mathbb{P}(Y_{\text{obs}} \mid \theta)$. If both MAR and distinctness hold, the MLE of θ is given by

$$\widehat{\theta}_{\text{MLE}} = \operatorname*{argmax}_{\theta \in \Theta} \mathbb{P}(Y_{\text{obs}} \mid \theta) = \operatorname*{argmax}_{\theta \in \Theta} L(\theta \mid Y_{\text{obs}}),$$

which shows that R is indeed irrelevant to $\widehat{\theta}_{\text{MLE}}$.

For Bayesian inference, consider the joint posterior distribution of all the parameters (θ, ξ):

$$\mathbb{P}(\theta, \xi \mid Y_{\text{obs}}, R) \propto \mathbb{P}(R, Y_{\text{obs}} \mid \theta, \xi)\pi(\theta, \xi)$$

$$= \mathbb{P}(R \mid Y_{\text{obs}}, \xi)\mathbb{P}(Y_{\text{obs}} \mid \theta)\pi(\theta, \xi) \qquad \text{(by MAR)}$$

$$= \mathbb{P}(R \mid Y_{\text{obs}}, \xi)\mathbb{P}(Y_{\text{obs}} \mid \theta)\pi_\theta(\theta)\pi_\xi(\xi). \quad \text{(by distinctness)}$$

Then, we derive the marginal posterior of θ:

$$\mathbb{P}(\theta \mid Y_{\text{obs}}, R) = \int \mathbb{P}(\theta, \xi \mid Y_{\text{obs}}, R)d\xi$$

$$= \mathbb{P}(Y_{\text{obs}} \mid \theta)\pi_\theta(\theta) \int \mathbb{P}(R \mid Y_{\text{obs}}, \xi)\pi_\xi(\xi)d\xi$$

$$\propto \mathbb{P}(Y_{\text{obs}} \mid \theta)\pi_\theta(\theta).$$

This shows that

$$\mathbb{P}(\theta \mid Y_{\text{obs}}, R) = \mathbb{P}(\theta \mid Y_{\text{obs}}) \propto \mathbb{P}(Y_{\text{obs}} \mid \theta)\pi_\theta(\theta),$$

and thus $\theta \perp\!\!\!\perp R \mid Y_{\text{obs}}$ (Definition 1.2). Therefore, the observed-data posterior $\mathbb{P}(\theta \mid Y_{\text{obs}})$ is sufficient for inference on θ.

2.2 The EM Algorithm and Its Properties

Consider the MLE

$$\widehat{\theta}_{\text{MLE}} = \operatorname*{argmax}_{\theta \in \Theta} \mathbb{P}(Y_{\text{obs}} \mid \theta) = \operatorname*{argmax}_{\theta \in \Theta} \int \mathbb{P}(Y_{\text{obs}}, Y_{\text{mis}} \mid \theta)dY_{\text{mis}}.$$

Because the observed-data likelihood is defined by marginalizing out the missing data Y_{mis} through a multi-dimensional integral, the MLE usually cannot be found in closed form. The EM algorithm, developed by Dempster *et al.* (1977), is commonly used to find the MLE for missing data problems.

2.2.1 *The algorithm*

Algorithm 2.1 (EM algorithm). Pick an initial $\theta^{(0)}$. For each iteration $t = 0, 1, 2, \ldots,$

(1) E-step: Calculate the expectation of complete-data log-likelihood:

$$Q(\theta \mid \theta^{(t)}) := \mathbb{E}[\log \mathbb{P}(Y_{\text{obs}}, Y_{\text{mis}} \mid \theta) \mid Y_{\text{obs}}, \theta^{(t)}].$$

(2) M-step: Find $\theta^{(t+1)}$ by maximizing $Q(\theta \mid \theta^{(t)})$:

$$\theta^{(t+1)} = \underset{\theta \in \Theta}{\operatorname{argmax}}\, Q(\theta \mid \theta^{(t)}).$$

Iterate the above two steps until convergence.

Remark 2.1. The expectation in the E-step is taken with respect to $\mathbb{P}(Y_{\text{mis}} \mid Y_{\text{obs}}, \theta^{(t)})$ (conditional distribution), not $\mathbb{P}(Y_{\text{mis}} \mid \theta^{(t)})$ (marginal distribution).

We first go through an example to illustrate the two steps of the EM algorithm.

Example 2.1 (Bivariate binary data). Let Y_1 and Y_2 be two correlated binary variables on $\{1, 2\}$. Their joint distribution is parameterized by $\theta_{ij} = \mathbb{P}(Y_1 = i, Y_2 = j)$, for $i, j \in \{1, 2\}$. Missing values occur on either Y_1 or Y_2 in an i.i.d. sample of n units. We want to estimate $\theta = (\theta_{11}, \theta_{12}, \theta_{21}, \theta_{22})$. Let x_{ij} be the number of units for which $Y_1 = i, Y_2 = j$, for $i, j = 1, 2$. Denote the complete data by $X = (x_{11}, x_{12}, x_{21}, x_{22})$, which is a sufficient statistic for θ.

According to the missingness pattern, we partition the n units into three blocks, as shown in Figure 2.1. The counts from each block are arranged into a 2×2 contingency table. In block A, all four counts $\{x_{ij}^A\}$ are observed;

A: Fully observed

$Y_1 \backslash Y_2$	1	2	
1	x_{11}^A	x_{12}^A	x_{1+}^A
2	x_{21}^A	x_{22}^A	x_{2+}^A
	x_{+1}^A	x_{+2}^A	

B: Y_2 missing

$Y_1 \backslash Y_2$	1	2	
1			x_{1+}^B
2			x_{2+}^B

C: Y_1 missing

$Y_1 \backslash Y_2$	1	2
1		
2		
	x_{+1}^C	x_{+2}^C

Fig. 2.1. Missingness patterns for bivariate binary data.

in block B, only the row sums $x_{i+}^B = x_{i1}^B + x_{i2}^B$, $i = 1, 2$, are observed; in block C, only the column sums x_{+j}^C, $j = 1, 2$, are observed.

The complete-data log-likelihood is

$$\ell(\theta \mid X) = \sum_{i,j=1}^{2} x_{ij} \log \theta_{ij}.$$

Our starting point is to calculate the expectation of $\ell(\theta \mid X)$,

$$Q(\theta \mid \theta^{(t)}) = \mathbb{E}[\ell(\theta \mid X) \mid Y_{\text{obs}}, \theta^{(t)}] = \sum_{i,j} \mathbb{E}(x_{ij} \mid Y_{\text{obs}}, \theta^{(t)}) \log \theta_{ij},$$

which reduces to calculating

$$\mathbb{E}(x_{ij} \mid Y_{\text{obs}}, \theta^{(t)}) = x_{ij}^A + \mathbb{E}(x_{ij}^B \mid Y_{\text{obs}}, \theta^{(t)}) + \mathbb{E}(x_{ij}^C \mid Y_{\text{obs}}, \theta^{(t)}). \quad (2.2)$$

Given $\theta^{(t)} = (\theta_{ij}^{(t)})_{2 \times 2}$, we use the following conditional distributions for blocks B and C to find the expectations in (2.2):

$$(x_{i1}^B, x_{i2}^B) \mid Y_{\text{obs}}, \theta^{(t)} \sim M\left(x_{i+}^B, \left(\frac{\theta_{i1}^{(t)}}{\theta_{i+}^{(t)}}, \frac{\theta_{i2}^{(t)}}{\theta_{i+}^{(t)}}\right)\right), \qquad i = 1, 2,$$

$$(x_{1j}^C, x_{2j}^C) \mid Y_{\text{obs}}, \theta^{(t)} \sim M\left(x_{+j}^C, \left(\frac{\theta_{1j}^{(t)}}{\theta_{+j}^{(t)}}, \frac{\theta_{2j}^{(t)}}{\theta_{+j}^{(t)}}\right)\right), \qquad j = 1, 2,$$

where $\theta_{i+}^{(t)} = \theta_{i1}^{(t)} + \theta_{i2}^{(t)}$ and $\theta_{+j}^{(t)} = \theta_{1j}^{(t)} + \theta_{2j}^{(t)}$. Consequently, we derive the EM algorithm as follows:

(1) E-step: To calculate $\mathbb{E}[\ell(\theta \mid X) \mid Y_{\text{obs}}, \theta^{(t)}]$, let

$$x_{ij}^{(t)} := \mathbb{E}(x_{ij} \mid Y_{\text{obs}}, \theta^{(t)}) = x_{ij}^A + x_{i+}^B \frac{\theta_{ij}^{(t)}}{\theta_{i+}^{(t)}} + x_{+j}^C \frac{\theta_{ij}^{(t)}}{\theta_{+j}^{(t)}}, \quad 1 \le i, j \le 2.$$

Then,

$$Q(\theta \mid \theta^{(t)}) = \mathbb{E}[\ell(\theta \mid X) \mid Y_{\text{obs}}, \theta^{(t)}] = \sum_{i,j} x_{ij}^{(t)} \log \theta_{ij}.$$

(2) M-step: Maximizing $Q(\theta \mid \theta^{(t)})$ subject to $\sum_{i,j} \theta_{ij} = 1$ leads to

$$\theta_{ij}^{(t+1)} = \frac{x_{ij}^{(t)}}{n} = \frac{1}{n}\left[x_{ij}^A + x_{i+}^B \frac{\theta_{ij}^{(t)}}{\theta_{i+}^{(t)}} + x_{+j}^C \frac{\theta_{ij}^{(t)}}{\theta_{+j}^{(t)}}\right], \qquad \text{for all } i, j.$$

2.2.2 EM as an MM algorithm

The EM algorithm can be viewed as an MM algorithm, which is a general method for optimization. See de Leeuw (1994) and Hunter and Lange (2004) for reviews and tutorials on the MM algorithm.

We start with a simple identity:

$$\log \mathbb{P}(Y_{\mathrm{mis}}, Y_{\mathrm{obs}} \mid \theta) = \ell(\theta \mid Y_{\mathrm{obs}}) + \log \mathbb{P}(Y_{\mathrm{mis}} \mid Y_{\mathrm{obs}}, \theta),$$

where $\ell(\theta \mid Y_{\mathrm{obs}}) := \log \mathbb{P}(Y_{\mathrm{obs}} \mid \theta)$ is the observed-data log-likelihood. Denote by F any distribution for Y_{mis}. Then, rearrange the above equation to get

$$\ell(\theta \mid Y_{\mathrm{obs}}) = \log \mathbb{P}(Y_{\mathrm{mis}}, Y_{\mathrm{obs}} \mid \theta) - \log F(Y_{\mathrm{mis}}) + \log \frac{F(Y_{\mathrm{mis}})}{\mathbb{P}(Y_{\mathrm{mis}} \mid Y_{\mathrm{obs}}, \theta)}.$$

Taking expectation on both sides with respect to (w.r.t.) F (note that the left-hand side is a constant, as it does not involve Y_{mis}) leads to

$$\ell(\theta \mid Y_{\mathrm{obs}}) = \mathbb{E}_F[\log \mathbb{P}(Y_{\mathrm{mis}}, Y_{\mathrm{obs}} \mid \theta)] + H(F) + D(F\|\mathbb{P}(Y_{\mathrm{mis}} \mid Y_{\mathrm{obs}}, \theta)),$$

where $H(F) = \mathbb{E}_F[-\log F]$ denotes the entropy of F and $D(\cdot\|\cdot)$ is the Kullback–Leibler divergence. Since $D(\cdot\|\cdot) \geq 0$, we have

$$\ell(\theta \mid Y_{\mathrm{obs}}) \geq \mathbb{E}_F[\log \mathbb{P}(Y_{\mathrm{mis}}, Y_{\mathrm{obs}} \mid \theta)] + H(F) := L(\theta, F), \qquad (2.3)$$

and equality holds when $F = \mathbb{P}(Y_{\mathrm{mis}} \mid Y_{\mathrm{obs}}, \theta)$. This shows that $L(\theta, F)$ is a lower bound for $\ell(\theta \mid Y_{\mathrm{obs}})$ for any F. Let $F^{(t)} = \mathbb{P}(Y_{\mathrm{mis}} \mid Y_{\mathrm{obs}}, \theta^{(t)})$. Then, $L(\theta, F^{(t)})$, called a minorization function of $\ell(\theta \mid Y_{\mathrm{obs}})$, satisfies the following two conditions:

(i) $\ell(\theta \mid Y_{\mathrm{obs}}) \geq L(\theta, F^{(t)})$ for any θ;
(ii) $\ell(\theta^{(t)} \mid Y_{\mathrm{obs}}) = L(\theta^{(t)}, F^{(t)})$.

The MM algorithm iterates between two steps to maximize $\ell(\theta \mid Y_{\mathrm{obs}})$:

(1) Minorization (E-step): Find the minorization function $L(\theta, F^{(t)})$ by calculating

$$\mathbb{E}_{F^{(t)}}[\log \mathbb{P}(Y_{\mathrm{mis}}, Y_{\mathrm{obs}} \mid \theta)] = Q(\theta \mid \theta^{(t)}).$$

Note that $L(\theta, F^{(t)}) = Q(\theta \mid \theta^{(t)}) + H(F^{(t)})$, where $H(F^{(t)})$ is a constant w.r.t. θ and can thus be omitted.

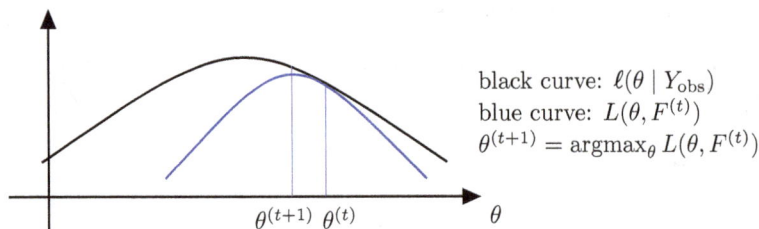

black curve: $\ell(\theta \mid Y_{\text{obs}})$
blue curve: $L(\theta, F^{(t)})$
$\theta^{(t+1)} = \text{argmax}_\theta\, L(\theta, F^{(t)})$

Fig. 2.2. The MM view of the EM algorithm.

(2) Maximization (M-step): Maximize the minorization function to obtain $\theta^{(t+1)}$. This is done by

$$\theta^{(t+1)} = \underset{\theta}{\text{argmax}}\, L(\theta, F^{(t)}) = \underset{\theta}{\text{argmax}}\, Q(\theta \mid \theta^{(t)}).$$

Now, we see that the EM algorithm is recovered by the two steps of this MM algorithm. Furthermore, it is not difficult to show the ascent property (Theorem 2.1) of EM:

$$\begin{aligned}
\ell(\theta^{(t+1)} \mid Y_{\text{obs}}) &\geq L(\theta^{(t+1)}, F^{(t)}) &&\text{(by (i))} \\
&\geq L(\theta^{(t)}, F^{(t)}) &&\text{(M-step)} \\
&= \ell(\theta^{(t)} \mid Y_{\text{obs}}). &&\text{(by (ii))}
\end{aligned}$$

Figure 2.2 illustrates one iteration of the EM algorithm, regarding it as an MM algorithm.

2.2.3 *Properties of the EM*

To establish the ascent property of the EM algorithm, we need the following inequality.

Lemma 2.1 (Jensen's inequality). *Assume that a random variable W is defined in the interval (a, b). If $h(W)$ is convex on (a, b), then*

$$\mathbb{E}[h(W)] \geq h[\mathbb{E}(W)],$$

provided that both expectations exist. For a strictly convex function, equality holds if and only if $W = \mathbb{E}(W)$ almost surely.

Proof. Use the supporting hyperplane theorem. Denote by $g(w)$ the supporting hyperplane of $h(w)$ at point $w_0 = \mathbb{E}(W)$. By convexity, we have $h(w) \geq g(w) \; \forall w \in (a, b)$, and thus

$$\mathbb{E}[h(W)] \geq \mathbb{E}[g(W)] = g[\mathbb{E}(W)] = h[\mathbb{E}(W)].$$

The first equality is due to the linearity of $\mathbb{E}(\cdot)$ and $g(\cdot)$. $\qquad\square$

Theorem 2.1 (Ascent property of the EM). *Let $\ell(\theta \mid Y_{\text{obs}})$ be the observed-data log-likelihood. Then, the EM iterations satisfy*

$$\ell(\theta^{(t+1)} \mid Y_{\text{obs}}) \geq \ell(\theta^{(t)} \mid Y_{\text{obs}}), \qquad \text{for all } t.$$

Proof. There are three steps in this proof. First, write

$$\ell(\theta \mid Y_{\text{obs}}) = \log \mathbb{P}(Y_{\text{obs}} \mid \theta) = Q(\theta \mid \theta^{(t)}) - H(\theta \mid \theta^{(t)}),$$

where

$$H(\theta \mid \theta^{(t)}) = \int [\log \mathbb{P}(Y_{\text{mis}} \mid Y_{\text{obs}}, \theta)] \, \mathbb{P}(Y_{\text{mis}} \mid Y_{\text{obs}}, \theta^{(t)}) dY_{\text{mis}}.$$

Second, we have $Q(\theta^{(t)} \mid \theta^{(t)}) \leq Q(\theta^{(t+1)} \mid \theta^{(t)})$ since $\theta^{(t+1)}$ is a maximizer of $Q(\theta \mid \theta^{(t)})$. Third, by Jensen's inequality and the convexity of $-\log(\cdot)$,

$$H(\theta^{(t)} \mid \theta^{(t)}) - H(\theta^{(t+1)} \mid \theta^{(t)}) = \mathbb{E} \left\{ \log \frac{\mathbb{P}(Y_{\text{mis}} \mid Y_{\text{obs}}, \theta^{(t)})}{\mathbb{P}(Y_{\text{mis}} \mid Y_{\text{obs}}, \theta^{(t+1)})} \, \middle| \, Y_{\text{obs}}, \theta^{(t)} \right\}$$

$$\geq 0.$$

Note that $-H(\theta^{(t)} \mid \theta^{(t)})$ is the entropy of the distribution $[Y_{\text{mis}} \mid Y_{\text{obs}}, \theta^{(t)}]$ and the right-hand side of the above equality is the KL divergence between $\mathbb{P}(Y_{\text{mis}} \mid Y_{\text{obs}}, \theta^{(t)})$ and $\mathbb{P}(Y_{\text{mis}} \mid Y_{\text{obs}}, \theta^{(t+1)})$. Therefore,

$$\ell(\theta^{(t)} \mid Y_{\text{obs}}) = Q(\theta^{(t)} \mid \theta^{(t)}) - H(\theta^{(t)} \mid \theta^{(t)})$$

$$\leq Q(\theta^{(t+1)} \mid \theta^{(t)}) - H(\theta^{(t+1)} \mid \theta^{(t)}) = \ell(\theta^{(t+1)} \mid Y_{\text{obs}}),$$

by combining the last two steps. $\qquad\square$

We provide an informal statement for the convergence of the EM algorithm. See Wu (1983) for the technical conditions for this result.

Theorem 2.2 (Convergence of the EM). *Under some conditions, the sequence $\{\theta^{(t)}\}$ defined by the EM iterations converges to a stationary point of the observed-data log-likelihood $\ell(\theta \mid Y_{\text{obs}})$.*

2.2.4 *Missing information and convergence rate*

Recall that $Q(\theta \mid \theta) = \ell(\theta \mid Y_{\text{obs}}) + H(\theta \mid \theta)$. Taking second derivatives on both sides, we arrive at

$$\underbrace{-\frac{\partial^2}{\partial\theta^2}Q(\theta \mid \theta)}_{\mathcal{I}_C(\theta)} = \underbrace{-\frac{\partial^2}{\partial\theta^2}\ell(\theta \mid Y_{\text{obs}})}_{\mathcal{I}_O(\theta)} + \underbrace{\left[-\frac{\partial^2}{\partial\theta^2}H(\theta \mid \theta)\right]}_{\mathcal{I}_M(\theta)}, \qquad (2.4)$$

where $\mathcal{I}$ stands for the Fisher information. Accordingly, the three terms are called the complete-data Fisher information ($\mathcal{I}_C$), the observed-data Fisher information ($\mathcal{I}_O$), and the missing-data Fisher information ($\mathcal{I}_M$). Equation (2.4) represents a relation among the three quantities:

$$\mathcal{I}_C(\theta) = \mathcal{I}_O(\theta) + \mathcal{I}_M(\theta),$$

which is called the *missing information principle* (Orchard and Woodbury, 1972).

For regular problems where $\theta^{(t+1)}$ is updated by solving the equation $\frac{\partial}{\partial\theta}Q(\theta \mid \theta^{(t)}) = 0$, Dempster *et al.* (1977) showed that

$$\theta^{(t+1)} - \widehat{\theta} \approx D\left(\theta^{(t)} - \widehat{\theta}\right),$$

when $\theta^{(t)}$ is close to the MLE $\widehat{\theta} = \text{argmax}_\theta\, \ell(\theta \mid Y_{\text{obs}})$. Here,

$$D = \mathcal{I}_C(\widehat{\theta})^{-1}\mathcal{I}_M(\widehat{\theta})$$

is called *the fraction of the missing information*. Consequently,

$$\|\theta^{(t+1)} - \widehat{\theta}\| \leq \|D\|_2\|\theta^{(t)} - \widehat{\theta}\| = \lambda_{\max}(D)\|\theta^{(t)} - \widehat{\theta}\|,$$

where $\|D\|_2 = \lambda_{\max}(D) \in (0,1)$ is the operator norm of the matrix D. This shows that the convergence rate of EM is governed by the largest eigenvalue of D. If a dataset contains only a small fraction of missing data, in which case $\lambda_{\max}(D)$ is close to zero, the EM algorithm tends to converge quickly. Conversely, a large amount of missing data often lead to slower convergence. From a geometric point of view, the Fisher information can be interpreted as the curvature of the log-likelihood surface. In the one-dimensional case of Figure 2.2, $\mathcal{I}_C(\widehat{\theta})$ and $\mathcal{I}_O(\widehat{\theta})$ are the respective curvatures of the curves $L(\theta, F^{(t)})$ and $\ell(\theta \mid Y_{\text{obs}})$ at $\theta = \widehat{\theta}$, when $\theta^{(t)}$ is close to $\widehat{\theta}$. As shown in the figure, the curvature of $L(\theta, F^{(t)})$ is larger than the curvature of $\ell(\theta \mid Y_{\text{obs}})$. If the curvature of $L(\theta, F^{(t)})$ increases, we see that $\theta^{(t+1)}$ will be closer to $\theta^{(t)}$, leading to a smaller increase in $\ell(\theta \mid Y_{\text{obs}})$ and slower convergence. Note that $D = 1 - \mathcal{I}_O(\widehat{\theta})/\mathcal{I}_C(\widehat{\theta})$ also increases in this case.

2.2.5 *Another example*

Example 2.2. Suppose the cell probabilities of a multinomial distribution are given by

$$(\pi_1, \pi_2, \pi_3, \pi_4) = \left(\frac{1}{2} + \frac{\theta}{4}, \frac{1-\theta}{4}, \frac{1-\theta}{4}, \frac{\theta}{4} \right),$$

where $\theta \in (0, 1)$ is the only unknown parameter. Given the observations

$$y = (y_1, y_2, y_3, y_4), \qquad \sum_{i=1}^{4} y_i = n,$$

we want to find the MLE of θ.

We could directly maximize the likelihood via numerical optimization. Here, we use the EM algorithm as an alternative tool to find the MLE. To do this, we reformulate this problem by introducing missing data. Split the first category into two sub-categories with cell probabilities π_{11} and π_{12} such that

$$\pi_1 = \pi_{11} + \pi_{12}, \quad \pi_{11} = \frac{1}{2}, \quad \pi_{12} = \frac{\theta}{4}.$$

Under this reformulation, the complete data are $y_{\text{cmp}} = (y_{11}, y_{12}, y_2, y_3, y_4)$, where y_{11} and y_{12} are the counts for the two sub-categories and are unobserved. The complete-data log-likelihood is

$$\ell(\theta \mid y_{\text{cmp}}) = y_{11} \log \frac{1}{2} + (y_{12} + y_4) \log \frac{\theta}{4} + (y_2 + y_3) \log \frac{1-\theta}{4}$$

$$= (y_{12} + y_4) \log \theta + (y_2 + y_3) \log(1 - \theta) + C,$$

where C is a constant w.r.t. θ and can be dropped from the Q function.

The two steps of the EM algorithm for this problem are outlined as follows:

(1) E-step: Calculate

$$\mathbb{E}(y_{12} \mid y, \theta^{(t)}) = y_1 \frac{\theta^{(t)}/4}{1/2 + \theta^{(t)}/4} := y_{12}^{(t)}.$$

Then, up to an additive constant,

$$Q(\theta \mid \theta^{(t)}) = \mathbb{E}[\ell(\theta \mid y_{\text{cmp}}) \mid y, \theta^{(t)}]$$

$$= (y_{12}^{(t)} + y_4) \log \theta + (y_2 + y_3) \log(1 - \theta).$$

(2) M-step: Maximizing $Q(\theta \mid \theta^{(t)})$, which is a binomial log-likelihood, leads to

$$\theta^{(t+1)} = \frac{y_{12}^{(t)} + y_4}{y_{12}^{(t)} + y_4 + y_2 + y_3}.$$

2.3 EM for Exponential Families

2.3.1 *Exponential families*

Definition 2.1. A family of pdfs or probability mass functions (pmfs) is called an exponential family (EF) if it can be expressed as

$$f(x \mid \theta) = h(x)c(\theta) \exp\left[\phi(\theta)^{\mathsf{T}} t(x)\right], \tag{2.5}$$

where $\theta = (\theta_m)_{1:d} \in \mathbb{R}^d$, $\phi(\theta) = (\phi_j(\theta))_{1:k} \in \mathbb{R}^k$, $t(x) = (t_j(x))_{1:k} \in \mathbb{R}^k$, and $d \leq k$. If $d < k$, the family is called a curved EF.

Theorem 2.3. *Suppose that $f(x \mid \theta)$ in (2.5) and its partial derivatives $\partial f(x \mid \theta)/\partial \theta_m$ are continuous in x and θ. If X is a random variable with density $f(x \mid \theta)$, then*

$$\mathbb{E}\left[\sum_{j=1}^{k} \frac{\partial \phi_j(\theta)}{\partial \theta_m} t_j(X)\right] = -\frac{\partial \log c(\theta)}{\partial \theta_m}, \quad \text{for } m = 1, \ldots, d.$$

Theorem 2.4 (Sufficient statistic). *Let $Y_1, \ldots, Y_n$ be an i.i.d. sample of size n from an EF $f(\cdot \mid \theta)$. Then,*

$$T(Y_1, \ldots, Y_n) = \left(\sum_{i=1}^{n} t_1(Y_i), \ldots, \sum_{i=1}^{n} t_k(Y_i)\right) := \sum_{i=1}^{n} t(Y_i)$$

is a sufficient statistic for θ.

Proof. Let $Y = (Y_1, \ldots, Y_n)$ and y_i be the observed value of Y_i. Then,

$$f(y \mid \theta) = f(y_1, \ldots, y_n \mid \theta) = \left[\prod_{i=1}^{n} h(y_i)\right] [c(\theta)]^n \exp\left[\phi(\theta)^{\mathsf{T}} \sum_{i=1}^{n} t(y_i)\right].$$

Suppose $\sum_{i=1}^{n} t(y_i) = t^*$. The conditional distribution $[Y \mid T(Y) = t^*, \theta]$ is given by

$$p(y \mid t^*, \theta) \propto f(y \mid \theta) \cdot I(T(y) = t^*)$$

$$= \prod_{i=1}^{n} h(y_i) \cdot I(T(y) = t^*) \cdot [c(\theta)]^n \exp\left[\phi(\theta)^{\mathsf{T}} t^*\right]$$

$$\propto \prod_{i=1}^{n} h(y_i) \cdot I(T(y) = t^*),$$

which is independent of θ. $\qquad\qquad\qquad\square$

2.3.2 MLE for complete data

Let $T_j(y) = \sum_{i=1}^{n} t_j(y_i)$, $j = 1, \ldots, k$. The log-likelihood given complete data is

$$\ell(\theta \mid y) = n \log c(\theta) + \phi(\theta)^{\mathsf{T}} \sum_{i=1}^{n} t(y_i)$$

$$= n \log c(\theta) + \sum_{j=1}^{k} \phi_j(\theta) T_j(y). \qquad (2.6)$$

The MLE is given by the solution to

$$\frac{\partial \ell(\theta \mid y)}{\partial \theta_m} = n \frac{\partial \log c(\theta)}{\partial \theta_m} + \sum_{j=1}^{k} \frac{\partial \phi_j(\theta)}{\partial \theta_m} T_j(y) = 0, \quad m = 1, \ldots, d.$$

From Theorem 2.3 and that $Y_i \sim f(\cdot \mid \theta)$, we have

$$n \frac{\partial \log c(\theta)}{\partial \theta_m} = -n\mathbb{E}\left[\sum_{j=1}^{k} \frac{\partial \phi_j(\theta)}{\partial \theta_m} t_j(Y_1)\right];$$

therefore, the MLE is given by the solution to

$$\sum_{j=1}^{k} \frac{\partial \phi_j(\theta)}{\partial \theta_m} T_j(y) = n \sum_{j=1}^{k} \frac{\partial \phi_j(\theta)}{\partial \theta_m} \mathbb{E}\left[t_j(Y_1)\right], \quad m = 1, \ldots, d.$$

Assume that $d = k$ and the matrix

$$\frac{\partial \phi}{\partial \theta} = \left(\frac{\partial \phi_j(\theta)}{\partial \theta_m}\right)_{k \times k}$$

is invertible, where $\partial \phi_j(\theta)/\partial \theta_m$ is the (m,j)th element. Then, the MLE $\widehat{\theta} = (\widehat{\theta}_1, \ldots, \widehat{\theta}_k)$ is the solution to the following system of equations:

$$\left(\frac{\partial \phi}{\partial \theta}\right)\begin{pmatrix} T_1(y) \\ \vdots \\ T_k(y) \end{pmatrix} = n \left(\frac{\partial \phi}{\partial \theta}\right)\begin{pmatrix} \mathbb{E}t_1(Y_1) \\ \vdots \\ \mathbb{E}t_k(Y_1) \end{pmatrix}$$

$$\Longleftrightarrow \qquad T_j(y) = n\mathbb{E}_\theta[t_j(Y_1)], \quad j = 1, \ldots, k.$$

That is,

$$\sum_{i=1}^n t_j(y_i) = n\mathbb{E}_\theta[t_j(Y_1)] = \mathbb{E}_\theta\left[\sum_{i=1}^n t_j(Y_i)\right], \quad j = 1, \ldots, k. \tag{2.7}$$

Note that the left-hand side is the observed value of the sufficient statistic, while the right-hand side is the expectation, which depends on θ.

Example 2.3. Given $y_1, \ldots, y_n \overset{\text{iid}}{\sim} \mathcal{N}(\mu, \sigma^2)$, we may use (2.7) to find the MLE of (μ, σ^2). The sufficient statistic is $(T_1, T_2) = (\sum_i y_i, \sum_i y_i^2)$. The MLE $(\widehat{\mu}, \widehat{\sigma}^2)$ is the solution to

$$\sum_i y_i = \mathbb{E}[T_1] = n\mu,$$

$$\sum_i y_i^2 = \mathbb{E}[T_2] = n\mathbb{E}(Y_1^2) = n(\mu^2 + \sigma^2).$$

This leads to

$$\widehat{\mu} = \frac{1}{n}\sum_i y_i = \bar{y},$$

$$\widehat{\sigma}^2 = \frac{1}{n}\sum_i y_i^2 - \bar{y}^2 = \frac{1}{n}\sum_i (y_i - \bar{y})^2,$$

indeed consistent with the maximizer of the log-likelihood.

2.3.3   *EM for incomplete data*

Let y_{obs} be the observed data. We outline one iteration of an EM algorithm for data from a regular EF:

(1) E-step: We calculate

$$Q(\theta \mid \theta^{(t)}) = \mathbb{E}\left[\ell(\theta \mid Y) \mid y_{\text{obs}}, \theta^{(t)}\right]$$

$$= n \log c(\theta) + \sum_{j=1}^{k} \phi_j(\theta) \mathbb{E}\left[T_j(Y) \mid y_{\text{obs}}, \theta^{(t)}\right] \quad \text{due to (2.6)}$$

$$= n \log c(\theta) + \sum_{j=1}^{k} \phi_j(\theta) \mathbb{E}\left[\sum_{i=1}^{n} t_j(Y_i) \,\middle|\, y_{\text{obs}}, \theta^{(t)}\right].$$

This shows that we need to compute $\mathbb{E}\left[t_j(Y_i) \mid y_{\text{obs}}, \theta^{(t)}\right]$ for each Y_i that contains a missing value.

(2) M-step: By essentially the same derivation for (2.7), we see that $\theta^{(t+1)}$ is the solution to the following system of equations:

$$\mathbb{E}\left[\sum_{i=1}^{n} t_j(Y_i) \,\middle|\, y_{\text{obs}}, \theta^{(t)}\right] = n\mathbb{E}_\theta[t_j(Y_1)], \quad j = 1, \ldots, k,$$

where the left-hand side is given by the E-step.

Example 2.4. Let $y_1, \ldots, y_n$ be i.i.d. observations from $\mathcal{N}(\mu, 1)$, but only $\text{sgn}(y_i)$ are observed for $i = 1, \ldots, k$. Find the MLE of μ.

Let $\phi(\cdot)$ and $\Phi(\cdot)$ be the pdf and cdf of $\mathcal{N}(0, 1)$, respectively. Suppose that $\text{sgn}(y_i) = 1$, for $i = 1, \ldots, k_1$, and $\text{sgn}(y_i) = -1$, for $i = k_1 + 1, \ldots, k$, as shown in the following:

$$(\underbrace{\overbrace{+\cdots+}^{k_1} \mid \overbrace{-\cdots-}^{k_2}}_{k} \mid y_{k+1}, \ldots, y_n),$$

where $k_1 + k_2 = k$. We compare two different approaches.

(1) *By EM:* Regard $y_1, \ldots, y_k$ as missing. A sufficient statistic for μ is $T = \sum_{i=1}^{n} Y_i$. We compute $\mathbb{E}(T \mid y_{\text{obs}}, \mu^{(t)}) = \sum_i \mathbb{E}(Y_i \mid y_{\text{obs}}, \mu^{(t)})$ in the E-step as follows.

(a) For $i > k$, $\mathbb{E}(Y_i \mid y_{\text{obs}}, \mu^{(t)}) = y_i$, as they are observed.
(b) For $i = 1, \ldots, k_1$,

$$\mathbb{E}(Y_i \mid y_{\text{obs}}, \mu^{(t)}) = \mathbb{E}(Y_i \mid Y_i > 0, \mu^{(t)}) = \mu^{(t)} + \frac{\phi(\mu^{(t)})}{\Phi(\mu^{(t)})}.$$

(c) For $i = k_1 + 1, \ldots, k$,

$$\mathbb{E}(Y_i \mid y_{\text{obs}}, \mu^{(t)}) = \mathbb{E}(Y_i \mid Y_i < 0, \mu^{(t)}) = \mu^{(t)} - \frac{\phi(\mu^{(t)})}{1 - \Phi(\mu^{(t)})}.$$

In the M-step, solve $\mathbb{E}(T \mid y_{\text{obs}}, \mu^{(t)}) = n\mu \ (= \mathbb{E}_\mu(T))$ to update

$$\mu^{(t+1)} = \frac{1}{n} \left[\sum_{i>k} y_i + k\mu^{(t)} + \left(\frac{k_1}{\Phi(\mu^{(t)})} - \frac{k_2}{1 - \Phi(\mu^{(t)})} \right) \phi(\mu^{(t)}) \right]. \qquad (2.8)$$

(2) *Direct approach*: Since $\mathbb{P}(Y_i > 0) = \Phi(\mu)$ and $\mathbb{P}(Y_i < 0) = 1 - \Phi(\mu)$,

$$p(y_{\text{obs}} \mid \mu) \propto [\Phi(\mu)]^{k_1} [1 - \Phi(\mu)]^{k_2} \exp\left[-\frac{1}{2} \sum_{i>k} (y_i - \mu)^2 \right].$$

Thus, the observed-data log-likelihood is

$$\ell(\mu \mid y_{\text{obs}}) = k_1 \log \Phi(\mu) + k_2 \log[1 - \Phi(\mu)] - \frac{1}{2} \sum_{i>k} (\mu - y_i)^2.$$

Therefore, setting

$$\frac{\partial \ell(\mu \mid y_{\text{obs}})}{\partial \mu} = \frac{k_1 \phi(\mu)}{\Phi(\mu)} - \frac{k_2 \phi(\mu)}{1 - \Phi(\mu)} - (n - k)\mu + \sum_{i>k} y_i = 0$$

shows that the MLE $\hat{\mu}$ satisfies the equation

$$\hat{\mu} = \frac{1}{n} \left[\sum_{i>k} y_i + k\hat{\mu} + \left(\frac{k_1}{\Phi(\hat{\mu})} - \frac{k_2}{1 - \Phi(\hat{\mu})} \right) \phi(\hat{\mu}) \right]. \qquad (2.9)$$

Comparing (2.8) and (2.9) shows that the MLE $\hat{\mu}$ is indeed a fixed point of the EM iteration. More specifically, if we write the EM iteration (2.8) as a map $\mu^{(t+1)} = M(\mu^{(t)})$, then the MLE satisfies $\hat{\mu} = M(\hat{\mu})$.

2.4   Incomplete Normal Data

In this section, we develop an EM algorithm for incomplete multivariate normal data. The complete data are $Y = (y_{ij})_{n \times p}$, where each row $y_i = (y_{i1}, y_{i2}, \ldots, y_{ip}) \in \mathbb{R}^p$ is a sample from a p-variate normal distribution. That is, we assume that

$$y_i = (y_{i1}, y_{i2}, \ldots, y_{ip}) \overset{\text{i.i.d.}}{\sim} \mathcal{N}_p(\mu, \Sigma), \quad i = 1, \ldots, n,$$

where $\mu \in \mathbb{R}^p$ and Σ is a $p \times p$ positive definite matrix.

2.4.1   *The complete-data MLE*

Put $\theta = (\mu, \Sigma)$. The complete-data likelihood is

$$L(\theta \mid Y) \propto |\Sigma|^{-n/2} \exp\left\{ -\frac{1}{2} \sum_{i=1}^{n} (y_i - \mu)^{\mathsf{T}} \Sigma^{-1} (y_i - \mu) \right\}.$$

Let $\bar{y} = n^{-1} \sum_i y_i$ and $S = \sum_{i=1}^{n} (y_i - \bar{y})(y_i - \bar{y})^{\mathsf{T}} \in \mathbb{R}^{p \times p}$. The quadratic form in the exponent is

$$\sum_{i=1}^{n} (y_i - \mu)^{\mathsf{T}} \Sigma^{-1} (y_i - \mu) = \operatorname{tr}\left[\sum_i \Sigma^{-1} (y_i - \mu)(y_i - \mu)^{\mathsf{T}} \right]$$

$$= \operatorname{tr}(\Sigma^{-1} S) + \operatorname{tr}[\Sigma^{-1} n (\bar{y} - \mu)(\bar{y} - \mu)^{\mathsf{T}}]$$

$$= \operatorname{tr}(\Sigma^{-1} S) + n(\bar{y} - \mu)^{\mathsf{T}} \Sigma^{-1} (\bar{y} - \mu). \qquad (2.10)$$

In the above derivation, we used $\operatorname{tr}(AB) = \operatorname{tr}(BA)$ for the first equality and

$$\sum_i (y_i - \mu)(y_i - \mu)^{\mathsf{T}} = S + n(\bar{y} - \mu)(\bar{y} - \mu)^{\mathsf{T}}$$

for the second equality. Based on (2.10), the log-likelihood of θ is

$$\ell(\theta \mid Y) = -\frac{n}{2} \log |\Sigma| - \frac{1}{2} \operatorname{tr}(\Sigma^{-1} S) - \frac{1}{2} n(\bar{y} - \mu)^{\mathsf{T}} \Sigma^{-1} (\bar{y} - \mu).$$

Maximizing $\ell(\theta \mid Y)$ over (μ, Σ) gives us the MLE:

$$\widehat{\mu}_{\mathrm{MLE}} = \bar{y}, \quad \widehat{\Sigma}_{\mathrm{MLE}} = \frac{1}{n} S.$$

An alternative approach is to find the sufficient statistic and solve Equation (2.7) for the EF to find the MLE. We start with the log-likelihood up to a constant independent of (θ, Y):

$$\ell(\theta \mid Y) = -\frac{n}{2} \log |\Sigma| - \frac{1}{2} \sum_{i=1}^{n} \left(\mu^{\mathsf{T}} \Sigma^{-1} \mu - 2\mu^{\mathsf{T}} \Sigma^{-1} y_i + y_i^{\mathsf{T}} \Sigma^{-1} y_i \right)$$

$$= -\frac{n}{2} \log |\Sigma| - \frac{n}{2} \mu^{\mathsf{T}} \Sigma^{-1} \mu + \mu^{\mathsf{T}} \Sigma^{-1} \sum_i y_i - \frac{1}{2} \sum_i y_i^{\mathsf{T}} \Sigma^{-1} y_i.$$

Using the properties of trace,

$$\sum_i y_i^{\mathsf{T}} \Sigma^{-1} y_i = \sum_i \operatorname{tr}\left(y_i^{\mathsf{T}} \Sigma^{-1} y_i \right) = \sum_i \operatorname{tr}\left(\Sigma^{-1} y_i y_i^{\mathsf{T}} \right) = \operatorname{tr}\left(\Sigma^{-1} \sum_i y_i y_i^{\mathsf{T}} \right).$$

Letting

$$T_1 := \sum_{i=1}^{n} y_i = n\bar{y}, \qquad T_2 := \sum_{i=1}^{n} y_i y_i^{\mathsf{T}} = Y^{\mathsf{T}} Y,$$

we arrive at

$$\ell(\theta \mid Y) = -\frac{n}{2} \log |\Sigma| - \frac{n}{2} \mu^{\mathsf{T}} \Sigma^{-1} \mu + \mu^{\mathsf{T}} \Sigma^{-1} T_1 - \frac{1}{2} \operatorname{tr}(\Sigma^{-1} T_2). \qquad (2.11)$$

Let $\langle u, v \rangle := u^{\mathsf{T}} v$ for the two vectors u and v, and let $\operatorname{vec}(\cdot)$ be the vectorization operator that stacks the columns of a matrix into a vector. Note that

$$\mu^{\mathsf{T}} \Sigma^{-1} T_1 = \langle \Sigma^{-\mathsf{T}} \mu, T_1 \rangle = \langle \Sigma^{-1} \mu, T_1 \rangle,$$

$$\operatorname{tr}(\Sigma^{-1} T_2) = \langle \operatorname{vec}(\Sigma^{-\mathsf{T}}), \operatorname{vec}(T_2) \rangle = \langle \operatorname{vec}(\Sigma^{-1}), \operatorname{vec}(T_2) \rangle,$$

using the fact that Σ is symmetric. We see that $\mathcal{N}_p(\mu, \Sigma)$ is an EF and (T_1, T_2) is a sufficient statistic for $\theta = (\mu, \Sigma)$. Moreover, the expectation of (T_1, T_2) is given by

$$\mathbb{E}_\theta(T_1) = n\mu, \qquad \mathbb{E}_\theta(T_2) = n(\Sigma + \mu\mu^{\mathsf{T}}).$$

Now, we find the MLE of (μ, Σ) by solving the system of equations

$$\sum_{i=1}^{n} y_i = n\mu,$$

$$\sum_{i=1}^{n} y_i y_i^{\mathsf{T}} = n(\Sigma + \mu\mu^{\mathsf{T}}),$$

which leads to

$$\hat{\mu}_{\mathrm{MLE}} = \frac{1}{n} \sum_{i=1}^{n} y_i = \bar{y}, \qquad \hat{\Sigma}_{\mathrm{MLE}} = \frac{1}{n} \sum_{i=1}^{n} y_i y_i^{\mathsf{T}} - \bar{y}\bar{y}^{\mathsf{T}} = \frac{1}{n} S. \qquad (2.12)$$

2.4.2 *EM algorithm for incomplete normal data*

Let $O(i) \subseteq [p] := \{1, \dots, p\}$ index the observed components of y_i and $M(i)$ index the missing components. For the example data point in Figure 2.3, $O(i) = \{2, 4, \dots, p\}$ and $M(i) = \{1, 3\}$.

Variables	1	2	3	4	...	p
y_i	?	✓	?	✓	✓	✓

Fig. 2.3. Example multivariate data point with missing values.

Our goal is to derive an EM algorithm to find the MLE of (μ, Σ). In the E-step, for each y_i with missing data, we make use of the conditional distribution $[y_{i,M(i)} \mid y_{i,O(i)}]$, which is given by the following general result.

Theorem 2.5 (Conditional distributions). *Let a random vector* $\mathbf{x}$ *be partitioned into two sub-vectors* $\mathbf{x}_1$ *and* $\mathbf{x}_2$. *If*

$$\begin{bmatrix} \mathbf{x}_1 \\ \mathbf{x}_2 \end{bmatrix} \sim \mathcal{N}\left(\begin{bmatrix} \mu_1 \\ \mu_2 \end{bmatrix}, \begin{bmatrix} \Sigma_{11} & \Sigma_{12} \\ \Sigma_{21} & \Sigma_{22} \end{bmatrix} \right),$$

then $\mathbf{x}_1 | \mathbf{x}_2 \sim \mathcal{N}(\mu_{1|2}(\mathbf{x}_2), \Sigma_{1|2})$, *where*

$$\mu_{1|2}(\mathbf{x}_2) := \mu_1 + \Sigma_{12}\Sigma_{22}^{-1}(\mathbf{x}_2 - \mu_2), \qquad \Sigma_{1|2} := \Sigma_{11} - \Sigma_{12}\Sigma_{22}^{-1}\Sigma_{21}.$$

By Theorem 2.5,

$$y_{i,M(i)} \mid y_{i,O(i)} \sim \mathcal{N}\left(\mu_{M(i)|O(i)}(y_{i,O(i)}), \Sigma_{M(i)|O(i)} \right),$$

which will be used in the E-step to find the expectation of the sufficient statistic.

Now, we derive the two steps of the EM algorithm for incomplete normal data. To simplify notation, let $y_{i,O} \equiv y_{i,O(i)}$ and $y_{i,M} \equiv y_{i,M(i)}$, for all i.

E-step: The expectation of the complete-data log-likelihood,

$$\mathbb{E}[\ell(\theta \mid Y)|Y_{\text{obs}}, \theta^{(t)}] = \mu^{\mathsf{T}}\Sigma^{-1}\underbrace{\mathbb{E}(T_1|Y_{\text{obs}}, \theta^{(t)})}_{*} - \frac{1}{2}\text{tr}[\Sigma^{-1}\underbrace{\mathbb{E}(T_2|Y_{\text{obs}}, \theta^{(t)})}_{\boxtimes}]$$

$$- \frac{n}{2}\log|\Sigma| - \frac{n}{2}\mu^{\mathsf{T}}\Sigma^{-1}\mu, \qquad (2.13)$$

depends on the conditional expectations of T_1 and T_2, which are computed as follows:

(1) $* = \sum_i \mathbb{E}(y_i|Y_{\text{obs}}, \theta^{(t)})$: Let

$$y^*_{i,M(i)} := \mathbb{E}(y_{i,M}|y_{i,O}, \theta^{(t)}) = \mu^{(t)}_{M(i)|O(i)}(y_{i,O}).$$

Then, it is easy to find

$$\mathbb{E}(y_{ij}|Y_{\text{obs}}, \theta^{(t)}) = \mathbb{E}(y_{ij}|y_{i,O}, \theta^{(t)}) = \begin{cases} y_{ij} & \text{if } j \in O(i), \\ y^*_{ij} & \text{if } j \in M(i). \end{cases}$$

(2) $\boxtimes = \sum_i \mathbb{E}(y_i y_i^{\mathsf{T}} | Y_{\text{obs}}, \theta^{(t)})$: Note that

$$\mathbb{E}(y_i y_i^{\mathsf{T}} | Y_{\text{obs}}, \theta^{(t)}) = [\mathbb{E}(y_{ij} y_{ik} | y_{i,O}, \theta^{(t)})]_{p \times p}$$

is a matrix. We compute the expectation of the (j, k)th element,

$$\mathbb{E}(y_{ij} y_{ik} | y_{i,O}, \theta^{(t)}) = \begin{cases} y_{ij} y_{ik} & \text{if } j, k \in O(i), \\ y_{ij} y_{ik}^* & \text{if } j \in O(i), k \in M(i), \\ y_{ij}^* y_{ik} & \text{if } j \in M(i), k \in O(i), \\ \left(\Sigma_{M(i)|O(i)}^{(t)} \right)_{jk} + y_{ij}^* y_{ik}^* & \text{if } j, k \in M(i). \end{cases}$$

The last case, i.e., $j, k \in M(i)$, is due to

$$\text{Cov}(y_{ij}, y_{ik} | y_{i,O}, \theta^{(t)}) = \mathbb{E}(y_{ij} y_{ik} | y_{i,O}, \theta^{(t)}) - y_{ij}^* y_{ik}^*.$$

<u>M-step</u>: Let $T_1^{(t)} := \mathbb{E}(T_1 | Y_{\text{obs}}, \theta^{(t)})$ and $T_2^{(t)} := \mathbb{E}(T_2 | Y_{\text{obs}}, \theta^{(t)})$, which have been computed in the E-step. We either maximize (2.13) over $\theta = (\mu, \Sigma)$ or solve the system of equations

$$T_1^{(t)} = \mathbb{E}_\theta(T_1) = n\mu,$$
$$T_2^{(t)} = \mathbb{E}_\theta(T_2) = n(\Sigma + \mu\mu^{\mathsf{T}})$$

for (μ, Σ) to update the parameter estimate

$$\mu^{(t+1)} = \frac{1}{n} T_1^{(t)}, \qquad \Sigma^{(t+1)} = \frac{1}{n} T_2^{(t)} - (\mu^{(t+1)})(\mu^{(t+1)})^{\mathsf{T}}.$$

The update here replaces the observed value of the sufficient statistic in the MLE (2.12) by its current expectation $(T_1^{(t)}, T_2^{(t)})$.

2.5   Bayesian Inference with Missing Data

Assuming the missing-data mechanism is ignorable (Section 2.1.1), the Bayesian inference on the model parameter θ can be performed based on the observed-data posterior $p(\theta \mid y_{\text{obs}})$.

2.5.1 Data augmentation

Bayesian inference for missing data problems follows the simple principle of inferring unknown variables (θ, Y_{mis}) conditioning on the observed data Y_{obs}. Thus, we start with the joint posterior distribution of (θ, Y_{mis}):

$$p(\theta, Y_{\text{mis}} | Y_{\text{obs}}) \propto p(\theta) p(Y_{\text{obs}}, Y_{\text{mis}} | \theta), \qquad (2.14)$$

where $p(\theta)$ is the prior for θ and

$$p(Y_{\text{obs}}, Y_{\text{mis}} | \theta) = p(Y | \theta) = \prod_i f(\mathbf{y}_i | \theta)$$

is the complete-date likelihood. Accordingly, the posterior distribution of θ is a marginal of the joint posterior (2.14):

$$p(\theta | Y_{\text{obs}}) = \int p(\theta, Y_{\text{mis}} | Y_{\text{obs}}) dY_{\text{mis}}$$

$$\propto p(\theta) \int p(Y_{\text{obs}}, Y_{\text{mis}} | \theta) dY_{\text{mis}},$$

which involves marginalization over the missing data Y_{mis}.

Usually, there are no closed-form formulas for the mean or quantiles of the posterior distribution of θ. Instead, we may draw samples of (θ, Y_{mis}) from the joint posterior distribution $[\theta, Y_{\text{mis}} | Y_{\text{obs}}]$ to perform Bayesian inference. Once we have samples of (θ, Y_{mis}), the marginal posterior distribution $p(\theta | Y_{\text{obs}})$ can be approximated using the samples of θ. Simulation from the joint posterior $p(\theta, Y_{\text{mis}} | Y_{\text{obs}})$ can be implemented with a Gibbs sampler (Geman and Geman, 1984).

Algorithm 2.2. Pick an initial $\theta^{(0)}$. For each iteration $t = 0, 1, 2, \ldots$:

(1) given $\theta^{(t)}$, draw $Y_{\text{mis}}^{(t+1)} \sim p(Y_{\text{mis}} | Y_{\text{obs}}, \theta^{(t)})$;
(2) given $Y_{\text{mis}}^{(t+1)}$, draw $\theta^{(t+1)} \sim p(\theta | Y_{\text{obs}}, Y_{\text{mis}}^{(t+1)}) = p(\theta | Y^{(t+1)})$, where $Y^{(t+1)} = (Y_{\text{obs}}, Y_{\text{mis}}^{(t+1)})$ is a complete-data matrix with missing values imputed as $Y_{\text{mis}}^{(t+1)}$.

As illustrated by the diagram in Figure 2.4, this Gibbs sampler cycles through two conditional sampling steps. It is essentially an iterative simulation of θ and Y_{mis}. Step (1) imputes the missing data from its conditional distribution given θ, while step (2) draws θ from its conditional distribution given Y_{mis}.

$$\theta^{(0)} \qquad \theta^{(1)} \qquad \theta^{(2)} \qquad \theta^{(t)} \qquad \theta^{(t+1)}$$

$$Y_{\text{mis}}^{(1)} \qquad Y_{\text{mis}}^{(2)} \qquad \cdots \qquad Y_{\text{mis}}^{(t+1)} \qquad \cdots$$

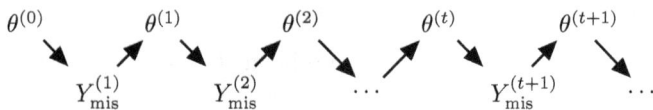

Fig. 2.4.   Diagram for a two-block Gibbs sampler.

	1	2	3
x_1	×	?	?
x_2	?	×	?
x_3	?	?	×
x_4	✓		
$\vdots$			
x_n		✓	

× not in this category
? possible categories
✓ observed category

Fig. 2.5.   An example of coarsened discrete data.

Remark 2.2. This two-block Gibbs sampler can be viewed as a stochastic version of the EM algorithm and was developed under the name of "data augmentation" by Tanner and Wong (1987).

For many commonly used models, both conditional sampling steps are easy to implement, as shown by the following examples.

2.5.2   *Examples*

Example 2.5 (Coarsened discrete data). Suppose $(x_1, x_2, \ldots, x_n)$ is an i.i.d. sample from a discrete distribution with three possible categories such that

$$\mathbb{P}(x_i = k) = \theta_k, \quad k = 1, 2, 3, \ i = 1, \ldots, n.$$

As shown in Figure 2.5, some of the data are coarsened, where x_1, x_2, and x_3 are only partially classified: $x_1 \in \{2, 3\}$, $x_2 \in \{1, 3\}$, and $x_3 \in \{1, 2\}$. The other data points are fully observed, e.g., $x_4 = 1, \ldots, x_n = 2$.

We put a conjugate prior $\theta \sim \text{Dir}(\alpha_1, \alpha_2, \alpha_3)$:

$$p(\theta_1, \theta_2, \theta_3) \propto \theta_1^{\alpha_1 - 1} \theta_2^{\alpha_2 - 1} \theta_3^{\alpha_3 - 1}.$$

There are missing data in x_1, x_2, and x_3, as they have been coarsened. What we observe is

$$Y_{\text{obs}} = \{x_1 \neq 1, x_2 \neq 2, x_3 \neq 3, x_4, \ldots, x_n\}.$$

Let $C_j^{\text{obs}} = \sum_{i=4}^{n} I(x_i = j)$ be the observed count for the jth category from the fully observed units $x_4, \ldots, x_n$. In what follows, we derive the two conditional sampling steps in data augmentation (Algorithm 2.2):

(1) Given θ, we impute the coarsened data according to the conditional distribution $[x_i \mid Y_{\text{obs}}, \theta]$, for $i = 1, 2, 3$. Note that

$$\mathbb{P}(x_1 = j \mid Y_{\text{obs}}, \theta) = \mathbb{P}(x_1 = j \mid x_1 \neq 1, \theta) = \frac{\theta_j}{1 - \theta_1}, \quad \text{for } j = 2, 3.$$

Similarly,

$$\mathbb{P}(x_2 = j \mid x_2 \neq 2, \theta) = \frac{\theta_j}{1 - \theta_2}, \quad j = 1, 3.$$

$$\mathbb{P}(x_3 = j \mid x_3 \neq 3, \theta) = \frac{\theta_j}{1 - \theta_3}, \quad j = 1, 2.$$

Then, we draw x_1, x_2, and x_3 independently according to the above conditional probabilities.

(2) Given the imputed value of (x_1, x_2, x_3), we compute the counts for the coarsened data $C_j^{(\text{mis})} = \sum_{i=1}^{3} I(x_i = j)$, for all j. This gives the complete-data posterior distribution, as in Example 1.2. We draw

$$\theta \mid \mathbf{x} \sim \text{Dir}(C_1^{(\text{obs})} + C_1^{(\text{mis})} + \alpha_1, \; C_2^{(\text{obs})} + C_2^{(\text{mis})} + \alpha_2, \; C_3^{(\text{obs})} + C_3^{(\text{mis})} + \alpha_3),$$

where $\mathbf{x} = (x_1, \ldots, x_n)$ is the complete data.

We iterate between the two steps to generate $(\theta^{(t)}, x_{1,2,3}^{(t)})$, for $t = 1, \ldots, m$. After we generate a large number of samples, the Bayesian estimate is

$$\widehat{\theta}_B = \mathbb{E}(\theta \mid Y_{\text{obs}}) \approx \frac{1}{m} \sum_{t=1}^{m} \theta^{(t)},$$

and the posterior distribution $[\theta \mid Y_{\text{obs}}]$ can be approximated using the empirical distribution of $\{\theta^{(t)} : t = 1, \ldots, m\}$.

Example 2.6 (Incomplete Gaussian data). Suppose that $\{y_1, y_2, \ldots, y_n\}$ is an i.i.d. sample from $\mathcal{N}_2(\mu, \Sigma)$, where $y_i = (y_{i1}, \; y_{i2})$ and

$$\mu = \begin{pmatrix} \mu_1 \\ \mu_2 \end{pmatrix}, \quad \Sigma = \begin{pmatrix} \sigma_1^2 & \rho\sigma_1\sigma_2 \\ \rho\sigma_1\sigma_2 & \sigma_2^2 \end{pmatrix}.$$

As shown in Figure 2.6, the dataset contains missing values, where the missing data $Y_{\text{mis}} = \{y_{11}, y_{22}\}$ and the observed data $Y_{\text{obs}} = \{y_{12}, y_{21}, y_3, \ldots, y_n\}$. We assume that Σ is known and specify an improper flat prior on μ, $p(\mu) \propto 1$, similar to (1.12).

	Y_1	Y_2
y_1	?	✓
y_2	✓	?
y_3	✓	✓
y_4	✓	✓
⋮	⋮	⋮
y_n	✓	✓

? missing value,
✓ observed value.

Fig. 2.6. Bivariate Gaussian data with missing values.

Now, let us derive the two conditional sampling steps in one iteration of the data augmentation for this problem.

(1) $[Y_{\text{mis}} \mid Y_{\text{obs}}, \mu]$: Given μ, draw y_{11} and y_{22} from their conditional distribution given the observed data. This step is similar to the E-step of an EM algorithm, but we draw random samples of Y_{mis} instead of filling in its expectation. We could use Theorem 2.5 to work out the conditional distributions $[y_{11}|y_{12}, \mu]$ and $[y_{22}|y_{21}, \mu]$. Alternatively, we directly derive the conditional densities from the joint distribution. Recall $y_1 = (y_{11}, y_{12})$. The conditional density is

$$p(y_{11}|y_{12}, \mu) \propto p(y_{11}, y_{12}|\mu)$$

$$\propto \exp\left\{ -\frac{1}{2(1-\rho^2)\sigma_1^2}\left[(y_{11} - \mu_1)^2 - \frac{2\rho\sigma_1}{\sigma_2}(y_{12} - \mu_2)(y_{11} - \mu_1) \right] \right\}$$

$$= \exp\left\{ -\frac{1}{2(1-\rho^2)\sigma_1^2}\left[y_{11} - \mu_1 - \frac{\rho\sigma_1}{\sigma_2}(y_{12} - \mu_2) \right]^2 + C \right\},$$

where C is constant w.r.t. y_{11}. This shows that

$$y_{11} \mid y_{12}, \mu \sim \mathcal{N}\left(\mu_1 + \frac{\rho\sigma_1}{\sigma_2}(y_{12} - \mu_2), (1-\rho^2)\sigma_1^2 \right). \qquad (2.15)$$

Similarly,

$$y_{22}|y_{21}, \mu \sim \mathcal{N}\left(\mu_2 + \frac{\rho\sigma_2}{\sigma_1}(y_{21} - \mu_1), (1-\rho^2)\sigma_2^2 \right). \qquad (2.16)$$

Given the current μ, we draw y_{11} and y_{22} independently from the above two normal distributions.

(2) $[\mu \mid Y_{\text{obs}}, Y_{\text{mis}}]$: Given the imputed values of y_{11} and y_{22}, we have the complete data $y = (y_1, \ldots, y_n)$, which allows us to sample μ from its complete-data posterior:

$$p(\mu \mid y_1, y_2, \ldots, y_n) \propto p(y_1, \ldots, y_n \mid \mu) p(\mu)$$

$$\propto |2\pi\Sigma|^{-\frac{n}{2}} \exp\left[-\frac{1}{2} \sum_{i=1}^{n} (y_i - \mu)^{\mathsf{T}} \Sigma^{-1} (y_i - \mu) \right]$$

$$\propto \exp\left[-\frac{1}{2} \sum_{i=1}^{n} (y_i - \mu)^{\mathsf{T}} \Sigma^{-1} (y_i - \mu) \right].$$

Let $\bar{y} = \sum_i y_i / n$. It follows from (2.10) that

$$\sum_i (\mu - y_i)^{\mathsf{T}} \Sigma^{-1} (\mu - y_i) = n(\mu - \bar{y})^{\mathsf{T}} \Sigma^{-1} (\mu - \bar{y}) + C,$$

where C does not depend on μ and thus can be regarded as a constant. This shows that

$$\mu \mid y_1, \ldots, y_n \sim \mathcal{N}_2 \left(\bar{y}, \Sigma/n \right). \tag{2.17}$$

In summary, the data augmentation algorithm iterates between drawing (y_{11}, y_{22}) according to (2.15) and (2.16) and drawing μ according to (2.17).

2.6 Problems

(1) Prove the following statements:

(a) Let $f(x)$ and $g(x)$ be probability densities defined on $\mathbb{R}^n$. Suppose $f(x) > 0$ and $g(x) > 0$, for all x. Show that $\mathbb{E}_f(\log f) \geq \mathbb{E}_f(\log g)$ using Jensen's inequality, where $\mathbb{E}_f(h) = \int h(x)f(x)dx$.

(b) The entropy of a probability distribution $p(x)$ on $\mathbb{R}^n$ is

$$H(p) := -\mathbb{E}_p(\log p) = -\int p(x) \log p(x)dx.$$

Among all distributions with mean $\mu = \int xp(x)dx$ and covariance matrix $\Sigma = \int (x - \mu)(x - \mu)^{\mathsf{T}} p(x)dx$, prove that the multivariate normal distribution has the maximum entropy.

Hint: In fact, (b) is a special case of the following more general result. Consider the Boltzmann distribution

$$p_\beta(x) \propto \exp[-\beta h(x)]$$

with energy function $h(x)$ at inverse temperature $\beta > 0$. Define the average energy of a distribution $q(x)$ by $\mathbb{E}_q(h) = \int h(x)q(x)dx$. Let $U(\beta)$ be the average energy of p_β. Then, among all distributions with average energy $U(\beta)$, the Boltzmann distribution p_β has the maximum entropy.

Proof outline: First, show that the cross-entropy $-\mathbb{E}_q(\log p_\beta)$ is a constant depending on β for any q with average energy $U(\beta)$. Then, apply (a).

(2) In a genetic linkage experiment, 197 animals are randomly assigned to four categories according to the multinomial distribution with cell probabilities $\pi_1 = \frac{1}{2} + \frac{\theta}{4}$, $\pi_2 = \frac{1-\theta}{4}$, $\pi_3 = \frac{1-\theta}{4}$, and $\pi_4 = \frac{\theta}{4}$. The corresponding observations are $y = (y_1, y_2, y_3, y_4) = (125, 18, 20, 34)$.

(a) Derive and implement an EM algorithm to estimate θ.

(b) Plot the observed-data log-likelihood function $\ell(\theta \mid y)$, for $\theta \in (0, 1)$. Compare the maximizer of this function with your EM estimate.

(3) Consider an i.i.d. sample drawn from a bivariate normal distribution with mean $\mu = (\mu_1, \mu_2)$ and covariance matrix

$$\Sigma = \begin{pmatrix} \sigma_1^2 & \sigma_{12} \\ \sigma_{12} & \sigma_2^2 \end{pmatrix}.$$

Suppose that the first k observations are missing their first component, the next m observations are missing their second component, and the last r observations are complete. Derive an EM algorithm for estimating the mean, assuming that the covariance matrix Σ is known.

(4) Prove the following results:

(a) If $Y \sim \mathcal{N}(\mu, 1)$, then $\mathbb{E}(Y \mid Y > 0) = \mu + \phi(\mu)/\Phi(\mu)$.

(b) Theorem 2.3. Hint: Start from the equality $\int f(x \mid \theta)dx = 1$ and differentiate both sides w.r.t. θ_m.

Chapter 3

Mixture Modeling

Mixture modeling offers a powerful framework for understanding complex data by assuming that the data are generated from a mixture of underlying probability distributions. Each component in the mixture corresponds to a latent class, and the task is to estimate both the parameters of these distributions and the assignment of data points to these latent classes. The EM algorithm plays a pivotal role in this process, iteratively updating the parameters and latent class assignments until convergence. From an application perspective, mixture modeling provides a statistical approach to clustering, where data points are probabilistically assigned to clusters rather than definitively assigned. This flexibility allows mixture models to capture complex data structures more effectively, accommodating situations where multiple clusters have substantial overlaps or where the boundaries between clusters are blurred. Thus, mixture modeling serves as a principled tool in both understanding latent structures within data and solving clustering problems in a probabilistic manner.

3.1 Mixture Models

3.1.1 *Definition*

Let $y = (y_1, y_2, \ldots, y_n)$ be a dataset composed of n independent and identically distributed (i.i.d.) units. Under a mixture model, the distribution of each y_i is specified as a mixture of K components,

$$\mathbb{P}(y_i|\theta, \lambda) = \sum_{m=1}^{K} \lambda_m f_m(y_i|\theta_m), \qquad (3.1)$$

where λ_m is the proportion of the mth component, $\sum_{m=1}^{K} \lambda_m = 1$, and $f_m(y_i|\theta_m)$ is the distribution of the mth component with the parameter θ_m.

To better understand the model structure, let us introduce missing indicator variables, $z_i = (z_{i1}, \ldots, z_{iK})$, such that

$$z_{im} = \begin{cases} 1 & \text{if } y_i \text{ is drawn from the } m\text{th mixture component,} \\ 0 & \text{otherwise.} \end{cases}$$

We construct the following two-layer model for (z_i, y_i):

$$z_i \sim M(1, (\lambda_1, \ldots, \lambda_K)),$$

$$y_i|z_i \sim f_m(y_i|\theta_m) \text{ if } z_{im} = 1.$$

It is easy to see that the marginal distribution of y_i under this two-layer model is the mixture distribution (3.1),

$$\mathbb{P}(y_i|\theta, \lambda) = \sum_{z_i} \mathbb{P}(y_i, z_i \mid \lambda, \theta) = \sum_{m=1}^{K} \lambda_m f_m(y_i|\theta_m), \qquad (3.2)$$

by summing over the range of $z_i \in \{e_1, \ldots, e_K\}$, where e_m are the standard basis vectors in $\mathbb{R}^K$, e.g., $e_1 = (1, 0, \ldots, 0)$. Based on (3.2), we may reformulate the mixture model as a latent structure model with two sets of variables:

- $y = (y_1, \ldots, y_n)^{\mathsf{T}}$ is an $n \times p$ matrix of observed data (p is the dimension of y_i);
- $z = (z_1, z_2, \ldots, z_n)^{\mathsf{T}}$ is an $n \times K$ matrix of latent or missing data.

Hereafter, we write $f(\cdot \mid \theta_m)$ instead of f_m to simplify our notation. Using the indicator variables z_{im}, the joint probability density of (z_i, y_i) is written as $\prod_{m=1}^{K} (\lambda_m f(y_i|\theta_m))^{z_{im}}$. Accordingly, the complete-data likelihood is

$$\mathbb{P}(y, z|\theta, \lambda) = \prod_{i=1}^{n} \prod_{m=1}^{K} (\lambda_m f(y_i|\theta_m))^{z_{im}}.$$

Remark 3.1. For a distribution $\mathbb{P}_\theta = \mathbb{P}(x|\theta)$, the parameter θ is *identifiable* if the mapping $\theta \mapsto \mathbb{P}_\theta$ is one-to-one. For the mixture model defined in (3.1), the parameters (θ, λ) are *not* identifiable due to permutation of the group labels $\{1, \ldots, K\}$. However, the non-identifiability of a mixture model is

usually not an issue in practice since most methods will produce an estimate
of the parameters defined by an arbitrary permutation of the group labels.

3.1.2   *Maximum likelihood estimate using EM*

Regarding the class indicators z as missing data, we may use the EM
algorithm to find the maximum likelihood estimate (MLE) of (θ, λ). The
complete-data log-likelihood is

$$\log(\mathbb{P}(y, z | \theta, \lambda)) = \sum_{i=1}^{n} \sum_{m=1}^{K} z_{im} [\log \lambda_m + \log f(y_i | \theta_m)].$$

Taking expectation with respect to (w.r.t.) $[z \mid y, \theta^{(t)}, \lambda^{(t)}]$,

$$\mathbb{E}\left[\log(\mathbb{P}(y, z | \theta, \lambda)) | y, \theta^{(t)}, \lambda^{(t)}\right]$$

$$= \sum_{i=1}^{n} \sum_{m=1}^{K} \mathbb{E}(z_{im} | y_i, \theta^{(t)}, \lambda^{(t)}) [\log \lambda_m + \log f(y_i | \theta_m)].$$

This shows that, in the E-step, we need to calculate the conditional expectation

$$\mathbb{E}(z_{im} | y_i, \theta^{(t)}, \lambda^{(t)}) = \mathbb{P}(z_{im} = 1 | y_i, \theta^{(t)}, \lambda^{(t)})$$

$$= \frac{\mathbb{P}(y_i | z_{im} = 1, \theta_m^{(t)}) \mathbb{P}(z_{im=1} | \lambda^{(t)})}{\sum_{j=1}^{K} \mathbb{P}(y_i | z_{ij} = 1, \theta_j^{(t)}) \mathbb{P}(z_{ij=1} | \lambda^{(t)})}$$

$$= \frac{\lambda_m^{(t)} f(y_i | \theta_m^{(t)})}{\sum_{j=1}^{K} \lambda_j^{(t)} f(y_i | \theta_j^{(t)})}$$

$$\triangleq w_{im}^{(t)} : \text{weight of } y_i \text{ from } f(\cdot | \theta_m). \qquad (3.3)$$

Note that $\sum_m w_{im}^{(t)} = 1$. The weight $w_{im} = \mathbb{P}(z_{im} = 1 \mid y_i)$ is the *posterior*
probability of $z_{im} = 1$ given the data point y_i, while $\lambda_m = \mathbb{P}(z_{im} = 1)$ is
the *prior* probability before observing any data.

Given the current parameters $(\lambda^{(t)}, \theta^{(t)})$, one iteration of the EM algorithm is described as follows:

(1) E-step: calculate the weights $w_{im}^{(t)}$ using (3.3), for $m = 1, \ldots, K$ and
$i = 1, \ldots, n$. Then, the Q-function is

$$Q(\theta, \lambda | \theta^{(t)}, \lambda^{(t)}) = \mathbb{E}\left[\log(\mathbb{P}(y, z | \theta, \lambda)) | y, \theta^{(t)}, \lambda^{(t)}\right]$$

$$= \sum_{m=1}^{K} \left\{ \underbrace{\left(\sum_{i=1}^{n} w_{im}^{(t)}\right)}_{\triangleq w_{\cdot m}^{(t)}} \log \lambda_m + \sum_{i=1}^{n} w_{im}^{(t)} \log f(y_i | \theta_m) \right\}$$

$$= \sum_{m=1}^{K} w_{\cdot m}^{(t)} \log \lambda_m + \sum_{m=1}^{K} \left[\sum_{i=1}^{n} w_{im}^{(t)} \log f(y_i | \theta_m)\right].$$

$$(3.4)$$

(2) M-step: in (3.4), λ and θ_m's are separable, which allows us to maximize them independently:

$$\max_{\lambda} \sum_{m=1}^{K} w_{\cdot m}^{(t)} \log \lambda_m, \qquad \text{subject to } \sum_{m} \lambda_m = 1,$$

$$\max_{\theta_m} \left\{ Q_m(\theta_m | \theta^{(t)}, \lambda^{(t)}) := \sum_{i=1}^{n} w_{im}^{(t)} \log f(y_i | \theta_m) \right\}.$$

Then, for $m = 1, \ldots, K$, we update the parameters

$$\lambda_m^{(t+1)} = \frac{w_{\cdot m}^{(t)}}{\sum_k w_{\cdot k}^{(t)}} = \frac{w_{\cdot m}^{(t)}}{n}, \tag{3.5}$$

$$\theta_m^{(t+1)} = \underset{\theta}{\operatorname{argmax}} \, Q_m(\theta_m | \theta^{(t)}, \lambda^{(t)}). \tag{3.6}$$

The update of λ by (3.5) is the same for all models. In the following examples, we show how to update θ_m.

Example 3.1 (Mixture exponential). Let $\mathcal{E}(\theta)$ be the exponential distribution with a mean of $\theta > 0$. Assume that

$$y_i \mid z_{im} = 1 \sim \mathcal{E}(\theta_m),$$

i.e., the conditional density

$$p(y_i \mid z_{im} = 1) = f(y_i \mid \theta_m) = \frac{1}{\theta_m} \exp\left(-\frac{y_i}{\theta_m}\right)$$

and $\mathbb{E}(y_i \mid z_{im} = 1) = \theta_m$.

For this problem, we find the Q_m function:

$$Q_m(\theta_m \mid \theta^{(t)}, \lambda^{(t)}) = \sum_{i=1}^{n} w_{im}^{(t)} \log \left[\frac{1}{\theta_m} \exp \left(-\frac{y_i}{\theta_m} \right) \right]$$

$$= -w_{.m}^{(t)} \log \theta_m - \frac{\sum_{i=1}^{n} w_{im}^{(t)} y_i}{\theta_m}.$$

Taking the derivative and setting it to zero, we find the maximizer $\theta_m^{(t+1)}$:

$$\frac{\partial Q_m}{\partial \theta_m} = 0 \Rightarrow \theta_m^{(t+1)} = \frac{\sum_{i=1}^{n} w_{im}^{(t)} y_i}{w_{.m}^{(t)}},$$

which is a *weighted average* of y_i. As a comparison, if all the data $y_1, \ldots, y_n$ were from $\mathcal{E}(\theta_m)$, the MLE would be $\widehat{\theta}_m = \bar{y}$, which corresponds to $\theta_m^{(t+1)}$ if the weights $w_{im}^{(t)} = 1$ for $i \in [n]$.

Example 3.2 (Exponential family). Suppose each component distribution is an exponential family (Definition 2.1) with pdf

$$f(y_i \mid \theta_m) = h(y_i) c(\theta_m) \exp[\phi(\theta_m)^\mathsf{T} t(y_i)], \qquad m = 1, \ldots, K.$$

Let us work out the Q_m function and the update of θ_m. Based on the above pdf, it is straightforward to find

$$Q_m(\theta_m \mid \theta^{(t)}, \lambda^{(t)}) = \sum_{i=1}^{n} w_{im}^{(t)} \left[\log h(y_i) + \log c(\theta_m) + \phi(\theta_m)^\mathsf{T} t(y_i) \right]$$

$$= w_{.m}^{(t)} \log c(\theta_m) + \phi(\theta_m)^\mathsf{T} \left(\sum_{i=1}^{n} w_{im}^{(t)} t(y_i) \right) + \text{const},$$

where const is a constant w.r.t. θ_m. Following an essentially identical derivation as in Section 2.3.2, the updated parameter $\theta_m^{(t+1)}$ is the solution (for θ_m) to the equation

$$\sum_{i=1}^{n} w_{im}^{(t)} t(y_i) = \mathbb{E}_{\theta_m} \left[\sum_{i=1}^{n} w_{im}^{(t)} t(y_i) \right] = w_{.m}^{(t)} \mathbb{E}_{\theta_m}[t(y_1)]. \qquad (3.7)$$

This equation can be viewed as a weighted version, assigning y_i the weight $w_{im}^{(t)}$, of the complete-data case, where $\widehat{\theta}_{\text{MLE}}$ satisfies

$$\sum_{i=1}^{n} t(y_i) = n \mathbb{E}_{\theta_m}[t(y_1)].$$

3.2 Model-Based Clustering

Suppose we have observed feature vectors, $y_1, \ldots, y_n$, for n units (individuals), where each $y_i \in \mathbb{R}^p$. These units come from K groups; however, the group labels are unknown. We want to group them into K clusters according to the observed feature vectors. This is the so-called clustering problem. Figure 3.1 shows a two-dimensional dataset with $K = 3$ clusters, represented by three different point symbols.

There are many clustering methods, such as hierarchical clustering and k-means clustering. In this section, we formulate the clustering problem as a mixture model and utilize the EM algorithm to infer the model parameters and cluster labels. This approach is called EM clustering or model-base clustering. Denote by z_i the cluster label of y_i, which is a latent variable. We assume the following two-layer mixture model for a clustering problem in a Euclidean feature space:

$$z_i \sim \mathrm{M}(1, \lambda), \quad \lambda = (\lambda_1, \ldots, \lambda_K),$$

$$y_i | z_{im} = 1 \sim \mathcal{N}_p(\mu_m, \Sigma_m),$$

where $\mu_m \in \mathbb{R}^p$ is the mean and Σ_m is the covariance matrix of a p-variate Gaussian distribution.

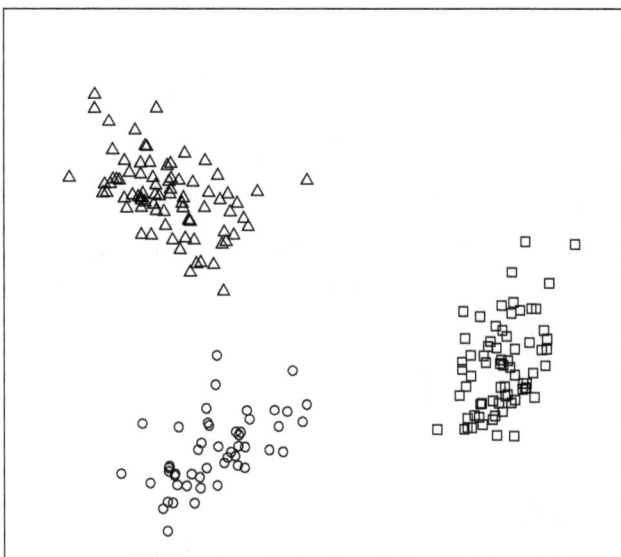

Fig. 3.1. Scatter plot of three clusters of data points in a two-dimensional feature space.

The EM clustering approach proceeds in two stages. First, we find the MLE $\widehat{\theta}$ of the parameters $\theta = \{\lambda, \mu_m, \Sigma_m, m = 1, \ldots, K\}$ using an EM algorithm. Then, we predict the cluster labels according to the posterior probability $\mathbb{P}(z_{im} = 1 | y_i, \widehat{\theta})$.

As the multivariate Gaussian is an exponential family, the EM algorithm follows the outline in Example 3.2:

(1) E-step: For $i = 1, \ldots, n$ and $m = 1, \ldots, K$, calculate

$$w_{im}^{(t)} = \frac{\lambda_m^{(t)} \phi_p(y_i; \mu_m^{(t)}, \Sigma_m^{(t)})}{\sum_{j=1}^K \lambda_j^{(t)} \phi_p(y_i, \mu_j^{(t)}, \Sigma_j^{(t)})},$$

which is the weight in (3.3). The weights $\{w_{im}^{(t)} : m \in [K]\}$ define a probabilistic assignment of y_i to the K clusters.

(2) M-step: Update $\lambda^{(t+1)}$ using (3.5). For $m = 1, \ldots, K$, solve the system of equations

$$\sum_i w_{im}^{(t)} y_i = w_{\cdot m}^{(t)} \mu_m,$$

$$\sum_i w_{im}^{(t)} y_i y_i^{\mathsf{T}} = w_{\cdot m}^{(t)} \left(\Sigma_m + \mu_m \mu_m^{\mathsf{T}} \right)$$

for μ_m and Σ_m to update

$$\mu_m^{(t+1)} = \frac{\sum_i w_{im}^{(t)} y_i}{w_{\cdot m}^{(t)}},$$

$$\Sigma_m^{(t+1)} = \frac{\sum_i w_{im}^{(t)} y_i y_i^{\mathsf{T}}}{w_{\cdot m}^{(t)}} - \mu_m^{(t+1)} (\mu_m^{(t+1)})^{\mathsf{T}}.$$

This step follows from the equation in (3.7) with the sufficient statistic of $\mathcal{N}_p(\mu_m, \Sigma_m)$ (Section 2.4.1).

Suppose we have run the EM for T iterations and estimated $\widehat{\theta} = \theta^{(T)}$. The predicted cluster label is

$$\widehat{z}_i = \operatorname*{argmax}_{1 \leq m \leq K} \mathbb{P}(z_{im} = 1 | y_i, \widehat{\theta}) = \operatorname*{argmax}_{1 \leq m \leq K} w_{im}^{(T)},$$

i.e., we assign unit i to the most likely cluster according to posterior probability evaluated with the MLE $\widehat{\theta}$. This is the *maximum a posteriori* (MAP) estimate.

When p is large, the $p \times p$ covariance matrix Σ_m has too many parameters. We usually simplify the model by assuming that Σ_m is a diagonal matrix, or even $\Sigma_m = \sigma_m^2 I_p$, i.e., each cluster has a spherical shape. There is an interesting connection between EM clustering and k-means clustering when the clusters are spherical.

Theorem 3.1. *Assume that $\Sigma_1 = \cdots = \Sigma_K = \sigma^2 I_p$ and $\sigma^2 > 0$ is known. If $\sigma^2 \to 0^+$ and $\lambda_m^{(t)} > 0$ for all $m \in [K]$, then the above EM algorithm is equivalent to k-means clustering.*

Proof. If $\Sigma_m = \sigma^2 I_p$, the E-step simplifies to

$$w_{im}^{(t)} = \frac{\lambda_m^{(t)} \exp\left(-\frac{\|y_i - \mu_m^{(t)}\|_2^2}{2\sigma^2}\right)}{\sum_{j=1}^{K} \lambda_j^{(t)} \exp\left(-\frac{\|y_i - \mu_j^{(t)}\|_2^2}{2\sigma^2}\right)}.$$

As $\sigma^2 \to 0^+$,

$$w_{im}^{(t)} = \begin{cases} 1 & \text{if } m = \text{argmin}_j \|y_i - \mu_j^{(t)}\|_2^2, \\ 0 & \text{otherwise}, \end{cases}$$

which assigns y_i to the closest center. Let $\mathcal{C}_m^{(t)} = \{i : w_{im}^{(t)} = 1\}$ be the mth cluster and $|\mathcal{C}_m^{(t)}|$ be its size in the current iteration. Then, parameter update in the M-step is given by

$$\mu_m^{(t+1)} = \frac{\sum_{i \in \mathcal{C}_m^{(t)}} y_i}{|\mathcal{C}_m^{(t)}|},$$

i.e., update μ_m using the sample mean of $\mathcal{C}_m^{(t)}$. □

This result shows that the k-means can be regarded as an approximate EM algorithm when the clusters are well separated, that is, the within-cluster variance σ^2 is small relative to the distances between the cluster centers. It suggests that we may apply the k-means clustering for such cases. However, when the clusters have substantial overlaps, the two algorithms may produce quite different clustering results, and the EM clustering, derived under the maximum likelihood principle, is expected to be more accurate.

3.3 Motif Discovery

In genomics and molecular biology, a sequence motif is a nucleotide or amino-acid sequence pattern that is widespread and has, or is conjectured to have, a biological significance. Figure 3.2(a) illustrates a DNA sequence motif that is recognized by a transcription factor (TF). After the TF binds to the DNA sequence, the downstream gene will be activated or suppressed. The binding sites or motif sequences show similarity in their nucleotide compositions, as shown in Figure 3.2(b). For example, the third nucleotide of this motif is uniformly a "T" across all the binding sites. Such similarity is also seen from the count matrix of the binding sites, whose column reports the observed counts of $\{A, C, G, T\}$ for one position of this motif. A motif pattern may be visualized with a logo plot (Schneider and Stephens, 1990) shown in Figure 3.2(d).

3.3.1 *Problem formulation*

Given a set of sequences, we want to identify the motif sites in these sequences. This is the so-called motif discovery problem. Here, we formulate the simplest case of this problem using a mixture model with two component distributions, one for the motif and one for background letters:

(1) *Motif model.* Let $X = (x_1, \ldots, x_w)$ be a motif of length w, where each $x_i \in \{A, C, G, T\}$. We assume $x_i \perp\!\!\!\perp x_j$, and each x_i follows a discrete distribution with cell probabilities $\theta_i = (\theta_{iA}, \theta_{iC}, \theta_{iG}, \theta_{iT})$, which we write as $x_i \sim \theta_i$ for brevity. Set $\Theta = [\theta_1 \mid \cdots \mid \theta_w]$ as a $4 \times w$ matrix,

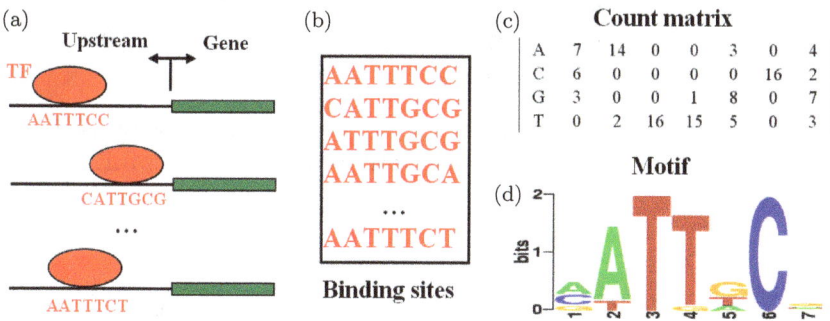

Fig. 3.2. Sequence motif: (a) upstream sequences of genes that share a common motif recognized by a TF; (b) examples of the TF binding sites (motif sequences); (c) count matrix from the motif sequences; (d) logo plot for the motif.

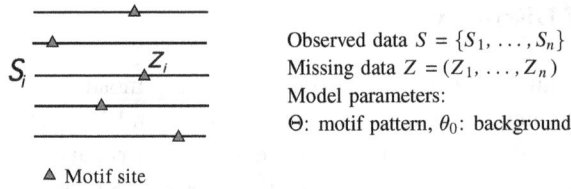

Observed data $S = \{S_1, \ldots, S_n\}$
Missing data $Z = (Z_1, \ldots, Z_n)$
Model parameters:
Θ: motif pattern, θ_0: background

▲ Motif site

Fig. 3.3. Motif-finding problem, with one motif site (at Z_i) in each sequence S_i.

which is called a position-specific weight matrix in the bioinformatics literature. The distribution of X is a *product multinomial* distribution with parameter Θ. As an example,

$$\mathbb{P}\{X = AATGC \mid \Theta\} = \theta_{1A}\theta_{2A}\theta_{3T}\theta_{4G}\theta_{5C}$$

for a motif site of length $w = 5$.

(2) *Background model.* We assume that a background letter $\tilde{x}$ follows a discrete distribution, $\theta_0 = (\theta_{0A}, \theta_{0C}, \theta_{0G}, \theta_{0T})$ independently. That is,

$$\mathbb{P}(\tilde{x} = j \mid \theta_0) = \theta_{0j}, \quad j \in \{A, C, G, T\}.$$

Under this mixture model formulation, the observed data are a set of sequences on the alphabets $\{A, C, G, T\}$, denoted by $S = \{S_1, S_2, \ldots, S_n\}$. The hidden variables are the start locations of the motif sites. We consider the simple scenario that each sequence contains one motif site. Let Z_i denote the beginning location of the motif site in S_i, and $Z = (Z_1, Z_2, \ldots, Z_n)$ are regarded as missing data. For simplicity, we assume that θ_0 is known. Figure 3.3 summarizes the key ingredients of this problem.

3.3.2   *Maximum likelihood via EM*

Using the maximum likelihood approach, we first estimate Θ using the EM algorithm and then predict the motif locations Z. The MLE for Θ is given by

$$\widehat{\Theta} = \operatorname*{argmax}_{\Theta} \underbrace{\mathbb{P}(S \mid \Theta)}_{\text{obs-data lik}} = \operatorname*{argmax}_{\Theta} \sum_{Z} \underbrace{\mathbb{P}(S, Z \mid \Theta)}_{\text{comp-data lik}}.$$

Lawrence and Reilly (1990) developed an EM algorithm to find motifs in biopolymer sequences.

Let us first look at one sequence and find the conditional distribution of Z_i given the observed sequence S_i. Define

$S_i(j, w) :=$ the segment of S_i starting at its jth position with length w.

Note that j ranges from 1 to $\ell_i := L_i - w + 1$, where $L_i = |S_i|$ is the length of the ith sequence. Assuming Z_i is uniform *in priori*, i.e., $\mathbb{P}(Z_i = j) = 1/\ell_i$, we have, for $j \in [\ell_i]$,

$$\mathbb{P}(S_i, Z_i = j \mid \Theta) = \mathbb{P}(Z_i = j)\mathbb{P}(S_i \mid Z_i = j, \Theta)$$

$$= \frac{1}{\ell_i}\mathbb{P}(S_i(j, w) \mid \Theta)\mathbb{P}(S_i \setminus S_i(j, w) \mid \theta_0)$$

$$= \frac{1}{\ell_i}\frac{\mathbb{P}(S_i(j, w) \mid \Theta)}{\mathbb{P}(S_i(j, w) \mid \theta_0)}\mathbb{P}(S_i \mid \theta_0)$$

$$\propto \frac{\mathbb{P}(S_i(j, w) \mid \Theta)}{\mathbb{P}(S_i(j, w) \mid \theta_0)} \equiv r_{ij}(\Theta) \quad \text{(likelihood ratio).} \quad (3.8)$$

Therefore, the posterior probability of $[Z_i = j \mid S_i]$ is

$$w_{ij}(\Theta) := \mathbb{P}(Z_i = j \mid S_i, \Theta) = \frac{\mathbb{P}(S_i, Z_i = j \mid \Theta)}{\sum_{k=1}^{\ell_i} \mathbb{P}(S_i, Z_i = k \mid \Theta)}$$

$$= \frac{r_{ij}(\Theta)}{\sum_{k=1}^{\ell_i} r_{ik}(\Theta)}, \quad (3.9)$$

which can be viewed as a weight for the motif site starting at the jth position of S_i.

Since $[S_i \mid Z_i]$ is an exponential family (product multinomial), we only need to derive the sufficient statistic for Θ, which will be used in the EM algorithm. Recall the EM algorithm for exponential families in Section 2.3.

It is easy to see that the count matrix of motif sites, such as that in Figure 3.2(c), is a sufficient statistic for Θ. Let us define the count matrix for a single sequence segment. For example, for $S_i(j, w) = ACCTG$, its count matrix is

$$C(S_i(j, w)) := \begin{bmatrix} 1 & 0 & 0 & 0 & 0 \\ 0 & 1 & 1 & 0 & 0 \\ 0 & 0 & 0 & 0 & 1 \\ 0 & 0 & 0 & 1 & 0 \end{bmatrix} \begin{matrix} A \\ C \\ G \\ T \end{matrix},$$

where an indicator vector of length four is used for each letter in the segment. If $X_1, \ldots, X_n$ are the count matrices of the n motif sequences in S, then the sufficient statistic is $X := \sum_i X_i$ and the MLE of Θ is

$$\widehat{\Theta} = \frac{1}{n}\sum_{i=1}^{n} X_i = \frac{1}{n}X. \tag{3.10}$$

An example count matrix of $n = 15$ motif sequences is given as follows:

$$X = \sum_{i=1}^{n} X_i = \begin{bmatrix} 10 & 1 & \cdots & 1 \\ 1 & 10 & \cdots & 2 \\ 1 & 3 & \cdots & 7 \\ 3 & 1 & \cdots & 5 \end{bmatrix}_{4 \times w}.$$

Now, we are ready to derive the EM algorithm for the motif-finding problem:

(1) E-Step: Given $\Theta^{(t)}$, find $\mathbb{E}(X \mid S, \Theta^{(t)})$, the expectation of the sufficient statistic X. To do that, we compute

$$\mathbb{E}(X_i \mid S_i, \Theta^{(t)}) = \sum_{j=1}^{\ell_i} C[S_i(j, w)]\mathbb{P}(Z_i = j \mid S_i, \Theta^{(t)})$$

$$= \sum_{j=1}^{\ell_i} w_{ij}(\Theta^{(t)})C[S_i(j, w)],$$

for $i = 1, \ldots, n$. Then, let

$$X^{(t)} := \mathbb{E}(X \mid S, \Theta^{(t)}) = \sum_{i=1}^{n} \mathbb{E}(X_i \mid S_i, \Theta^{(t)})$$

$$= \sum_{i=1}^{n}\sum_{j=1}^{\ell_i} w_{ij}(\Theta^{(t)})C[S_i(j, w)].$$

(2) M-Step: Regarding $X^{(t)}$ as the sufficient statistic, find the MLE as in (3.10) to update

$$\Theta^{(t+1)} = \frac{X^{(t)}}{n}.$$

After we find $\widehat{\Theta}$ using the above EM algorithm, we predict Z through the MAP criterion

$$\widehat{Z}_i = \underset{j\in[\ell_i]}{\operatorname{argmax}}\,\mathbb{P}(Z_i = j \mid S_i, \widehat{\Theta}) = \underset{j\in[\ell_i]}{\operatorname{argmax}}\,w_{ij}(\widehat{\Theta})$$

according to (3.9).

3.3.3 *Bayesian inference via Gibbs sampler*

We may also take a Bayesian approach to motif discovery. Assume a conjugate prior, $\theta_j \sim \text{Dir}(\alpha, \ldots, \alpha)$, independently for $j = 1, \ldots, w$. In short, we say that the prior of Θ is product-Dirichlet:

$$\Theta \sim \text{Prod-Dir}(\alpha). \tag{3.11}$$

Given the motif locations Z, let $X_{\bullet j}$ be the jth column of the count matrix X. We have $X_{\bullet j} \mid \theta_j \sim \text{M}(n, \theta_j)$. Then, the posterior distribution is

$$\theta_j \mid X_{\bullet j} \sim \text{Dir}(X_{\bullet j} + \alpha), \, j = 1, \ldots, w \iff \Theta \mid X \sim \text{Prod-Dir}(X + \alpha). \tag{3.12}$$

The posterior mean is

$$\mathbb{E}(\Theta \mid X) = \frac{X + \alpha}{n + 4\alpha}.$$

As Z is unknown, we work with the joint posterior $[Z, \Theta \mid S]$. Depending on the application, if our primary interest is to predict the motif locations Z, we may want to draw Z from its posterior distribution

$$\mathbb{P}(Z_1, \ldots, Z_n \mid S) = \int \mathbb{P}(Z_1, \ldots, Z_n, \Theta \mid S) d\Theta$$

$$\propto \int \mathbb{P}(Z, S \mid \Theta) \pi(\Theta) d\Theta,$$

where π is the product-Dirichlet prior for Θ. We develop a Gibbs sampler to draw $[Z_1, \ldots, Z_n \mid S]$. The Gibbs sampler cycles through the conditional distributions $[Z_i \mid Z_{-i}, S]$, for $i = 1, \ldots, n$, in each iteration. The key is to calculate

$$\mathbb{P}(Z_i = j \mid Z_{-i}, S) \propto \mathbb{P}(S_i, Z_i = j \mid Z_{-i}, S_{-i})$$

$$= \int \mathbb{P}(S_i, Z_i = j \mid \Theta) \, p(\Theta \mid X_{-i}) d\Theta, \tag{3.13}$$

where the count matrix X_{-i} is computed from (S_{-i}, Z_{-i}) and the posterior distribution $\Theta \mid X_{-i} \sim \text{Prod-Dir}(X_{-i} + \alpha)$, as in (3.12). Regarding

(S_{-i}, Z_{-i}) as the observed data, (3.13) is the posterior predictive distribution for the "new" data (S_i, Z_i). Using (3.8),

$$\mathbb{P}(Z_i = j \mid Z_{-i}, S) \propto \int \frac{\mathbb{P}(S_i(j, w) \mid \Theta)}{\mathbb{P}(S_i(j, w) \mid \theta_0)} p(\Theta \mid X_{-i}) d\Theta$$

$$= \frac{1}{\mathbb{P}(S_i(j, w) \mid \theta_0)} \int \mathbb{P}(S_i(j, w) \mid \Theta) p(\Theta \mid X_{-i}) d\Theta$$

$$= \frac{\mathbb{P}(S_i(j, w) \mid \widehat{\Theta}_{-i})}{\mathbb{P}(S_i(j, w) \mid \theta_0)} = r_{ij}(\widehat{\Theta}_{-i}),$$

where $\widehat{\Theta}_{-i}$ is the posterior mean given the count matrix X_{-i},

$$\widehat{\Theta}_{-i} = \mathbb{E}(\Theta \mid X_{-i}) = \frac{X_{-i} + \alpha}{n - 1 + 4\alpha}. \tag{3.14}$$

After normalization,

$$\mathbb{P}(Z_i = j \mid Z_{-i}, S) = \frac{r_{ij}(\widehat{\Theta}_{-i})}{\sum_{k=1}^{\ell_i} r_{ik}(\widehat{\Theta}_{-i})} = w_{ij}(\widehat{\Theta}_{-i}), \quad j = 1, \ldots, \ell_i. \tag{3.15}$$

In summary, each iteration of this Gibbs sampler consists of a loop that cycles through n posterior predictive distributions. For $i = 1, \ldots, n$:

(1) compute the posterior mean $\widehat{\Theta}_{-i} = \mathbb{E}(\Theta \mid X_{-i})$ using (3.14);
(2) draw $[Z_i \mid Z_{-i}, S]$ according to (3.15).

More detailed development of the Gibbs motif sampler can be found in the work of Liu *et al.* (1995). An interesting application to a protein sequence set was reported by Lawrence *et al.* (1993). As a generalization, Zhou and Wong (2004) proposed a hierarchical mixture model for a *cis*-regulatory module, which is a set of multiple motifs located closely, and developed a Gibbs sampler for the *de novo* discovery of *cis*-regulatory modules from sequence data. A review of sequence motifs and motif-finding methods has been presented by Jensen *et al.* (2004).

3.4   Problems

(1) Suppose that X follows a two-component mixture distribution with mixture proportions λ_1 and λ_2 ($\lambda_1 + \lambda_2 = 1$). The mean and the variance of the mth component distribution are μ_m and σ_m^2, respectively, for $m = 1, 2$. Find $\mathbb{E}(X)$ and $\text{Var}(X)$.

(2) Simulate a dataset consisting of three clusters of sizes $n_1 = 30$, $n_2 = 20$, and $n_3 = 50$. The data points in the mth ($m = 1, 2, 3$) cluster are i.i.d. from $\mathcal{N}_p(\mu_m, \sigma_m^2 I_p)$, where $p = 2$ is the dimension of the data and $\mu_m \in \mathbb{R}^2$.

   (a) Derive an EM algorithm to find the MLE of the unknown parameters.

   (b) Implement the EM algorithm to cluster the data points into three groups. Compare the estimated parameters with their true values. Make a scatter plot of the data points with the predicted cluster labels.

   (c) Vary the values of μ_m and σ_m^2 and repeat (b).

(3) We have observed $n = 10$ sites of a motif, summarized into a count matrix, X_{obs}, shown in Table 3.1. A position-specific weight matrix, Θ, is used as the model for the motif sites. Assume an i.i.d. background model with $\theta_0 = (0.24, 0.26, 0.26, 0.24)$ for $\{A, C, G, T\}$. In addition to X_{obs}, we know that the sequence

$$S = \text{ACCATTATCCCTGT}$$

contains another site of this motif, and let $Z \in \{1, \ldots, 10\}$ be its start position. Assume that the marginal probability $\mathbb{P}(Z = i)$ is identical for all possible i.

   (a) Let $\widehat{\Theta}_{\text{obs}} = \frac{1}{n+4\alpha}(X_{\text{obs}} + \alpha)$, where $\alpha = 1$ is a pseudo count. Find the most likely start position $\widehat{z}$ of the site in S, i.e.,

$$\widehat{z} = \operatorname*{argmax}_{1 \le i \le 10} \mathbb{P}(Z = i \mid S, \widehat{\Theta}_{\text{obs}}).$$

Table 3.1.   Count matrix X_{obs}.

Position	1	2	3	4	5
A	1	9	0	0	8
C	3	0	0	0	0
G	6	1	0	0	1
T	0	0	10	10	1

(b) Regarding both X_{obs} and S as our data, develop a method to find the MLE of Θ, i.e.,

$$\widehat{\Theta}_{\mathrm{MLE}} = \underset{\Theta}{\mathrm{argmax}}\,\mathbb{P}(S, X_{\mathrm{obs}} \mid \Theta).$$

Implementation is not required. Write down only the main steps.

(c) Here, we consider this problem in a Bayesian way, assuming a product-Dirichlet prior for Θ, as in (3.11), with $\alpha = 1$. Find the posterior distribution of Θ given the observed count matrix, $[\Theta \mid X_{\mathrm{obs}}]$.

(d) Implement a Monte Carlo method to draw 2,000 samples of Θ from the posterior distribution $[\Theta \mid X_{\mathrm{obs}}, S]$. Use the samples to approximate the posterior mean $\widehat{\Theta}_B = \mathbb{E}[\Theta \mid X_{\mathrm{obs}}, S]$ and the posterior probability $\mathbb{P}(\theta_{jk} > 0.5 \mid X_{\mathrm{obs}}, S)$ for $j = 1, \ldots, 5$ and $k \in \{A, C, G, T\}$.

Hint:

$$p(\Theta \mid X_{\mathrm{obs}}, S) = \sum_i p(\Theta \mid X_{\mathrm{obs}}, S, Z = i)\mathbb{P}(Z = i \mid X_{\mathrm{obs}}, S).$$

Chapter 4

Hidden Markov Models

Hidden Markov models (HMMs) are widely utilized in various fields due to their capability of modeling sequential data with latent structures. At its core, an HMM consists of a sequence of observed variables depending on underlying, unobserved states, making it a latent structure model. This inherent flexibility enables its application across diverse domains, ranging from signal processing to computational biology. In the realm of speech recognition, HMMs excel in modeling the temporal dependencies of speech signals, aiding in accurate transcription and comprehension (Rabiner, 1989). In biological sequence modeling, HMMs play a pivotal role in deciphering genetic sequences, enabling researchers to infer hidden biological phenomena and unravel complex genomic structures (Krogh et al., 1994). With their versatility and effectiveness in capturing hidden dynamics, hidden Markov models continue to be a cornerstone in data analysis and pattern recognition.

4.1 Elements of an HMM

4.1.1 *Example and definition*

Example 4.1 (Coin toss). We toss two coins. Let P_H and P_T be the probabilities of observing heads and tails, respectively. Coin 1 is unbiased with $P_H = P_T = 0.5$, while coin 2 is biased with $P_H = 0.9$ and $P_T = 0.1$. Let $Z_t \in \{1, 2\}$ be the coin used at the tth toss. The switch between the two coins is modeled by a Markov chain with transition matrix

$$A = \begin{bmatrix} a_{11} & a_{12} \\ a_{21} & a_{22} \end{bmatrix},$$

where $a_{ij} = \mathbb{P}(Z_{t+1} = j \mid Z_t = i)$. We observe the outcome $Y_t \in \{H, T\}$ for each toss but not the coin used, i.e., Z_t is hidden. Example sequences of the outcomes Y_t and the hidden coin labels Z_t are shown as follows:

$$\begin{array}{lccccccccccc} Y \text{ (observed):} & H & T & T & T & H & H & H & T & H & H & \cdots \\ Z \text{ (hidden):} & 1 & 1 & 1 & 2 & 2 & 2 & 2 & 1 & 1 & 2 & \cdots \end{array}$$

We want to predict the coin labels Z from the observed outcomes Y.

The above is an example of an HMM in which Z_t are the hidden states and Y_t the observed variables. In general, the elements of an HMM with discrete states and discrete observations include the following:

(1) hidden states, $\{1, \ldots, N\}$, i.e., the state space for the Markov chain Z_t;
(2) observed symbols, $\{1, \ldots, M\}$, i.e., the space for Y_t.
(3) a state transition matrix, $A = (a_{ij})_{N \times N}$, where

$$a_{ij} = \mathbb{P}(Z_{t+1} = j \mid Z_t = i); \tag{4.1}$$

(4) emission probabilities $B = (b_j(k))$, where

$$b_j(k) = \mathbb{P}(Y_t = k \mid Z_t = j); \tag{4.2}$$

(5) an initial state distribution, $\pi = (\pi_1, \ldots, \pi_N)$, for Z_1, i.e., $\mathbb{P}(Z_1 = j) = \pi_j$, for $j \in [N] := \{1, \ldots, N\}$.

4.1.2   *Graphical representation*

Based on this model, it is not hard to work out the joint probability of $Y = \{Y_t\}$ and $Z = \{Z_t\}$:

$$\mathbb{P}(Y, Z) = \mathbb{P}(Z_1)\mathbb{P}(Y_1 \mid Z_1)\mathbb{P}(Z_2 \mid Z_1)\mathbb{P}(Y_2 \mid Z_2)$$

$$\cdots \mathbb{P}(Z_n \mid Z_{n-1})\mathbb{P}(Y_n \mid Z_n)$$

$$= \mathbb{P}(Z_1)\mathbb{P}(Y_1 \mid Z_1) \prod_{t=2}^{n} \mathbb{P}(Z_t \mid Z_{t-1})\mathbb{P}(Y_t \mid Z_t)$$

$$= f_1(Z_1, Y_1) \prod_{t=2}^{n} g_t(Z_{t-1}, Z_t) f_t(Z_t, Y_t). \tag{4.3}$$

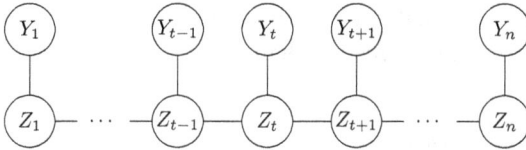

Fig. 4.1. HMM represented as an undirected graph.

In the last line, we have defined functions f_t and g_t in the factorization of the joint probability mass function $\mathbb{P}(Y, Z)$, which allows us to represent the joint distribution by an undirected graph. Each node of the graph represents a random variable V_i. If two random variables, V_i and V_j, appear in the same factor function (e.g., f_t and g_t in (4.3)), then the two nodes are connected by an undirected edge, $V_i - V_j$. See Section 6.4 for more rigorous definitions. Following this rule, the graph in Figure 4.1 is constructed for $\{(Z_t, Y_t) : t = 1, \ldots, n\}$.

One can easily identify conditional independence (CI) relations from the graph. We say that a subset of nodes S separates the nodes V_i and V_k if all paths from V_i to V_k in the graph intersect some node(s) in S. If S separates the nodes i and k, then

$$V_i \perp\!\!\!\perp V_k \mid S.$$

This is the global Markov property, which is implied by the fact that the graph is constructed according to factorization of the joint density. For an HMM, through graph separation, we see that V_{t-i}, Y_t and V_{t+j} are conditionally independent given Z_t. Here, V_k can be either Y_k or Z_k.

The two basic problems that we should solve for an HMM are as follows:

(i) Given Y, how do we estimate the model parameters $\theta = (A, B)$?
(ii) Given Y and the model parameter θ (or its estimate $\widehat{\theta}$), how do we predict the hidden states Z?

4.2 Maximum Likelihood Estimate via the EM Algorithm

We consider problem (i), the estimation of the parameters $\theta = (A, B)$, where $A = (a_{ij})$ is the transition matrix (4.1) and $B = (b_j(k))$ is the emission probabilities (4.2). The Baum–Welch algorithm (Baum *et al.*, 1970; Baum, 1972) is commonly used to find the maximum likelihood estimate (MLE) of the parameters through an iterative approach, which is equivalent to an EM algorithm applied to this problem.

4.2.1 *Problem setup*

By regarding the hidden states Z as missing data, we maximize the observed-data likelihood $\mathbb{P}(Y \mid \theta)$ to find

$$\hat{\theta} = \underset{\theta}{\mathrm{argmax}}\, \mathbb{P}(Y \mid \theta) = \underset{\theta}{\mathrm{argmax}} \sum_{Z_1} \cdots \sum_{Z_n} \mathbb{P}(Y, Z_1, \ldots, Z_n \mid \theta). \qquad (4.4)$$

Because Z is missing, the observed-data likelihood is defined by marginalizing out $Z_1, \ldots, Z_n$. We may develop an EM algorithm for this problem, which uses the conditional distribution $[Z \mid Y, \theta]$ in the E-step. In an HMM, $(Z_1, \ldots, Z_n)$ given Y is a Markov chain. Regarding Y_t as a constant in (4.3),

$$\mathbb{P}(Z \mid Y) \propto \mathbb{P}(Z_1)\mathbb{P}(Y_1 \mid Z_1) \prod_{t=2}^{n} \mathbb{P}(Z_t \mid Z_{t-1})\mathbb{P}(Y_t \mid Z_t)$$

$$= g_1(Z_1) \prod_{t=2}^{n} g_t(Z_{t-1}, Z_t),$$

where we have omitted θ in the conditioning for notational brevity. The above factorization of $\mathbb{P}(Z \mid Y)$ shows that $[Z \mid Y]$ is indeed a Markov chain. This fact will be utilized to simplify the E-step.

We assume that the initial distribution π is known. For example, we may simply assume $\mathbb{P}(Z_1 = j) = 1/N$ for all $j \in [N]$. Let us first consider the complete-data likelihood $\mathbb{P}(Y, Z \mid \theta)$. It is convenient to introduce indicators, $Z_{tj} = I(Z_t = j)$ for all t and j. Then, the complete-data likelihood and log-likelihood are given by

$$\mathbb{P}(Y, Z \mid \theta) \propto \prod_{j=1}^{N}\prod_{k=1}^{M} \prod_{t:Y_t=k} \{b_j(k)\}^{Z_{tj}} \times \prod_{i=1}^{N}\prod_{j=1}^{N}\prod_{t=2}^{n}(a_{ij})^{Z_{(t-1)i}Z_{tj}}$$

and

$$\log \mathbb{P}(Y, Z \mid \theta) = \sum_{j,k} \underbrace{\sum_{t:Y_t=k} Z_{tj}}_{D_{jk}} \log b_j(k) + \sum_{i,j} \underbrace{\sum_{t=2}^{n} Z_{(t-1)i}Z_{tj}}_{C_{ij}} \log a_{ij}$$

$$= \sum_{j}\left[\sum_{k=1}^{M} D_{jk}\log b_j(k)\right] + \sum_{i}\left[\sum_{j=1}^{N} C_{ij}\log a_{ij}\right]. \qquad (4.5)$$

We have defined two sets of counts in (4.5):

$$D_{jk} = \sum_{t:Y_t=k} Z_{tj}$$

is the number of emissions of symbol k from state j, while

$$C_{ij} = \sum_{t=2}^{n} Z_{(t-1)i} Z_{tj}$$

is the number of transitions from state i to state j. They form a sufficient statistic for $\theta = (A, B)$. Subject to the normalization constraints, $\sum_k b_j(k) = 1$ for all j and $\sum_j a_{ij} = 1$ for all i, we maximize (4.5) to find the complete-data MLE:

$$\widehat{b}_j(k) = \frac{D_{jk}}{D_{j\bullet}}, \quad j = 1, \ldots, N, \quad k = 1, \ldots, M, \tag{4.6}$$

$$\widehat{a}_{ij} = \frac{C_{ij}}{C_{i\bullet}}, \quad i, j = 1, \ldots, N, \tag{4.7}$$

where $D_{j\bullet} = \sum_k D_{jk}$ and $C_{i\bullet} = \sum_j C_{ij}$.

4.2.2 Forward and backward summations

As Z is unobserved, we derive an EM algorithm to find the MLE $\widehat{\theta}$, as in (4.4). The two steps of the EM algorithm are outlined as follows.

(1) E-step: Given the current parameters $\theta^{(m)}$, calculate the expectation of the complete-data log-likelihood (4.5) by

$$\mathbb{E}\{\log \mathbb{P}(Y, Z \mid \theta) \mid Y, \theta^{(m)}\}$$
$$= \sum_{j,k} \underbrace{\mathbb{E}(D_{jk} \mid Y, \theta^{(m)})}_{D_{jk}^{(m)}} \log b_j(k) + \sum_{i,j} \underbrace{\mathbb{E}(C_{ij} \mid Y, \theta^{(m)})}_{C_{ij}^{(m)}} \log a_{ij}.$$

(2) M-step: Given $\{D_{jk}^{(m)}\}$ and $\{C_{ij}^{(m)}\}$, update the parameters

$$b_j(k)^{(m+1)} = \frac{D_{jk}^{(m)}}{D_{j\bullet}^{(m)}}, \quad a_{ij}^{(m+1)} = \frac{C_{ij}^{(m)}}{C_{i\bullet}^{(m)}}$$

to obtain $\theta^{(m+1)}$, where $D_{j\bullet}^{(m)} = \sum_k D_{jk}^{(m)}$, and similarly for $C_{i\bullet}^{(m)}$. The update in the M-step is identical to the MLE calculation in (4.6) and (4.7) regarding $\{D_{jk}^{(m)}\}$ and $\{C_{ij}^{(m)}\}$ as the sufficient statistic.

We see that the main task in the E-step is to calculate the expectations $D_{jk}^{(m)}$ and $C_{ij}^{(m)}$. By definition,

$$D_{jk}^{(m)} = \mathbb{E}(D_{jk} \mid Y, \theta^{(m)}) = \sum_{t:Y_t=k} \mathbb{E}(Z_{tj} \mid Y, \theta^{(m)}),$$

$$C_{ij}^{(m)} = \mathbb{E}(C_{ij} \mid Y, \theta^{(m)}) = \sum_{t=2}^{n} \mathbb{E}(Z_{(t-1)i} Z_{tj} \mid Y, \theta^{(m)}).$$

Thus, given any value of the model parameter θ, we need to calculate

$$\mathbb{E}(Z_{tj} \mid Y, \theta) = \mathbb{P}(Z_t = j \mid Y, \theta),$$

$$\mathbb{E}(Z_{(t-1)i} Z_{tj} \mid Y, \theta) = \mathbb{P}(Z_{t-1} = i, Z_t = j \mid Y, \theta),$$

for each t and all i, j. To simplify notation, we drop θ in the derivation of algorithms to calculate the above two conditional probabilities.

The key is to use CI in an HMM. First, consider

$$\mathbb{P}(Z_t = j \mid Y) \propto \mathbb{P}(Y, Z_t = j)$$

$$= \underbrace{\mathbb{P}(Y_{1:t}, Z_t = j)}_{:= \alpha_t(j)} \cdot \underbrace{\mathbb{P}(Y_{(t+1):n} \mid Z_t = j)}_{:= \beta_t(j)} \qquad (4.8)$$

$$= \alpha_t(j)\beta_t(j), \quad \text{for each } j \in [N],$$

where we have used the CI relation $Y_{(t+1):n} \perp\!\!\!\perp Y_{1:t} \mid Z_t$. Therefore, through normalization, we have

$$\mathbb{P}(Z_t = j \mid Y) = \frac{\alpha_t(j)\beta_t(j)}{\sum_{i=1}^{N} \alpha_t(i)\beta_t(i)} := u_t(j), \quad j = 1, \ldots, N. \qquad (4.9)$$

In a similar way,

$$\mathbb{P}(Z_{t-1} = i, Z_t = j \mid Y) \propto \mathbb{P}(Y, Z_{t-1} = i, Z_t = j)$$

$$= \underbrace{\mathbb{P}(Y_{1:(t-1)}, Z_{t-1} = i)}_{\alpha_{t-1}(i)} \cdot \underbrace{\mathbb{P}(Z_t = j \mid Z_{t-1} = i)}_{a_{ij}}$$

$$\times \underbrace{\mathbb{P}(Y_t \mid Z_t = j)}_{b_j(Y_t)} \cdot \underbrace{\mathbb{P}(Y_{(t+1):n} \mid Z_t = j)}_{\beta_t(j)}$$

$$= a_{ij} b_j(Y_t) \alpha_{t-1}(i) \beta_t(j), \quad \text{for all } (i, j) \in [N] \times [N].$$

Through normalization, for all i and j,

$$\mathbb{P}(Z_{t-1} = i, Z_t = j \mid Y) = \frac{a_{ij} b_j(Y_t) \alpha_{t-1}(i) \beta_t(j)}{\sum_k \sum_\ell a_{k\ell} b_\ell(Y_t) \alpha_{t-1}(k) \beta_t(\ell)}$$

$$:= w_t(i, j). \tag{4.10}$$

By (4.9) and (4.10), it suffices to calculate $\alpha_t(i)$ and $\beta_t(j)$, for all t, i, and j to complete the E-step. This can be done by the forward and backward summations. Consequently, $\alpha_t(i)$ and $\beta_t(j)$ are sometimes called the forward and the backward functions, respectively.

Recall that $\alpha_t(i) = \mathbb{P}(Y_{1:t}, Z_t = i)$. We may derive a recursion to find $\alpha_{t+1}(j)$ given all $\alpha_t(i)$ for $i \in [N]$:

$$\alpha_{t+1}(j) = \mathbb{P}(Y_{1:(t+1)}, Z_{t+1} = j)$$

$$= \sum_{i=1}^{N} \mathbb{P}(Y_{1:(t+1)}, Z_t = i, Z_{t+1} = j) \quad \text{(marginalizing out } Z_t\text{).}$$

Next, using the CI relations of an HMM to factorize the joint probability $\mathbb{P}(Y_{1:(t+1)}, Z_t = i, Z_{t+1} = j)$, we have

$$\alpha_{t+1}(j) = \sum_{i=1}^{N} \mathbb{P}(Y_{t+1} \mid Z_{t+1} = j) \mathbb{P}(Z_{t+1} = j \mid Z_t = i) \mathbb{P}(Y_{1:t}, Z_t = i)$$

$$= b_j(Y_{t+1}) \sum_{i=1}^{N} a_{ij} \alpha_t(i).$$

This leads to the *forward summation* algorithm to calculate $\alpha_t(j)$, for all $j \in [N]$ and all $t \in [n]$:

Algorithm 4.1 (Forward summation).

(1) Initialization: $\alpha_1(i) = \pi_i b_i(Y_1)$, for $i = 1, \ldots, N$.
(2) Recursion: For $t = 1, \ldots, n-1$,

$$\alpha_{t+1}(j) = b_j(Y_{t+1}) \sum_{i=1}^{N} a_{ij} \alpha_t(i), \quad j = 1, \ldots, N. \tag{4.11}$$

Similarly, the following *backward summation* algorithm is used to calculate $\beta_t(i)$, for all i and t:

Algorithm 4.2 (Backward summation).

(1) Initialization: $\beta_n(i) = 1$, for $i = 1, \ldots, N$.
(2) Recursion: For $t = n - 1, \ldots, 1$,

$$\beta_t(i) = \sum_{j=1}^{N} a_{ij} b_j(Y_{t+1}) \beta_{t+1}(j), \quad i = 1, \ldots, N. \tag{4.12}$$

Verification of the backward summation algorithm is left as an exercise. Note that both $\alpha_t(i)$ and $\beta_t(i)$ are calculated given the current parameter value $\theta^{(m)}$ at the mth iteration of the EM algorithm. The recursions in both algorithms make use of the fact that $[Z \mid Y]$ is a Markov chain.

Now, we are ready to present the full EM algorithm for HMMs.

Algorithm 4.3 (EM for HMMs). Pick an initial $\theta^{(0)}$. For $m = 0, 1, \ldots$, iterate between the following two steps until convergence:

(1) E-step: Given $\theta^{(m)}$,

    (i) apply the forward and backward summations to calculate $\alpha_t(i)$ and $\beta_t(i)$ for all t and i;
    (ii) calculate $u_t(j)$ and $w_t(i, j)$ using (4.9) and (4.10), for all t, i, and j;
    (iii) $D_{jk}^{(m)} = \sum_{t:Y_t=k} u_t(j)$ and $C_{ij}^{(m)} = \sum_{t=2}^{n} w_t(i, j)$, for all i, j, and k.

(2) M-step:

$$b_j(k)^{(m+1)} = \frac{D_{jk}^{(m)}}{D_{j\bullet}^{(m)}}, \quad j \in [N], k \in [M];$$

$$a_{ij}^{(m+1)} = \frac{C_{ij}^{(m)}}{C_{i\bullet}^{(m)}}, \quad i \in [N], j \in [N].$$

It is recommended to monitor the observed-data likelihood, which should be non-decreasing, as a simple diagnosis for algorithm implementation. The observed-data likelihood can be calculated using the forward functions

$$\mathbb{P}(Y \mid \theta^{(m)}) = \sum_{i} \mathbb{P}(Y_{1:n}, Z_n = i \mid \theta^{(m)}) = \sum_{i} \alpha_n(i),$$

where $\alpha_n(i)$ is evaluated given $\theta^{(m)}$. Another common practice is to store and calculate $\log \alpha_t(i)$ and $\log \beta_t(i)$ to avoid underflow.

4.3 The Viterbi Algorithm

In this section, we consider problem (ii) posed at the end of Section 4.1: how to predict the hidden states Z given the model parameter $\theta = \widehat{\theta}$? Throughout this section, we assume that θ is fixed at a given or estimated value. Thus, we drop it in the conditioning to simplify our notations.

The basic principle is quite straightforward. We predict Z using maximum *a posteriori* (MAP) given the observed symbols Y,

$$\widehat{z} = \operatorname*{argmax}_{z} \mathbb{P}(Z = z \mid Y) = \operatorname*{argmax}_{z} \mathbb{P}(Y, Z = z), \tag{4.13}$$

since the marginal probability $\mathbb{P}(Y)$ is a constant.

The maximization in (4.13) is over all $z = (z_1, \ldots, z_n) \in [N]^n$. Thus, in general, one cannot enumerate all possible values of z to find the maximizer. Using the Markovian structure of $[Z \mid Y]$, however, we may derive a recursion to maximize $\mathbb{P}(Y, z_1, \ldots, z_n)$ via dynamic programming. To do that, let us define the maximum of a partial problem:

$$\delta_{t+1}(j) := \max_{z_1, \ldots, z_t} \mathbb{P}(Y_{1:(t+1)}, z_{1:t}, Z_{t+1} = j)$$

for $t = 0, 1, \ldots$ and $j \in [N]$. By definition, we see that

$$\max_{z_1, \ldots, z_n} \mathbb{P}(Y, z_1, \ldots, z_n) = \max_{1 \leq i \leq N} \delta_n(i).$$

Therefore, we can find the maximum if we have a way to compute $\delta_n(i)$, for all $i \in [N]$.

Using CI, we factorize the probability

$$\mathbb{P}(Y_{1:(t+1)}, z_{1:t}, Z_{t+1} = j)$$
$$= \mathbb{P}(Y_{1:t}, z_{1:(t-1)}, Z_t = z_t)\mathbb{P}(Z_{t+1} = j \mid Z_t = z_t)b_j(Y_{t+1}).$$

With this factorization, we arrive at the following recursion for $\delta_{t+1}(j)$:

$$\delta_{t+1}(j) = \max_{z_1, \ldots, z_t} \mathbb{P}(Y_{1:(t+1)}, z_{1:t}, Z_{t+1} = j)$$

$$= \max_{1 \leq i \leq N} \underbrace{\max_{z_1, \ldots, z_{t-1}} \mathbb{P}(Y_{1:t}, z_{1:(t-1)}, Z_t = i)}_{\delta_t(i)} a_{ij} b_j(Y_{t+1})$$

$$= \max_{1 \leq i \leq N} \{\delta_t(i) a_{ij}\} b_j(Y_{t+1}). \tag{4.14}$$

The recursion (4.14) is the core of the Viterbi (1967) algorithm, or dynamic programming, for finding the optimal hidden states defined by the MAP criterion (4.13). The Viterbi algorithm consists of two key steps: forward maximization to calculate $\delta_t(i)$, for $t = 1, \ldots, n$, and backward tracking to find $\widehat{z}_t$, for $t = n, n - 1, \ldots, 1$.

Algorithm 4.4 (The Viterbi algorithm).

(1) Initialization: $\delta_1(i) = \pi_i b_i(Y_1)$ for $i = 1, \ldots, N$.
(2) Forward maximization: for $t = 1, \ldots, n - 1$,

$$\delta_{t+1}(j) = \max_{1 \leq i \leq N} \{\delta_t(i) a_{ij}\} b_j(Y_{t+1}), \quad j = 1, \ldots, N,$$

$$\gamma_{t+1}(j) = \underset{1 \leq i \leq N}{\operatorname{argmax}} \{\delta_t(i) a_{ij}\}.$$

(3) Backward tracking to find $\widehat{z}$: put $\widehat{z}_n = \operatorname{argmax}_i \delta_n(i)$.
    For $t = n - 1, \ldots, 1$,

$$\widehat{z}_t = \gamma_{t+1}(\widehat{z}_{t+1}).$$

Let us go through the main steps of this algorithm, assuming there are $N = 3$ hidden states. In the forward maximization, we find

$$\delta_{t+1}(1) = \max\{\delta_t(1) a_{11}, \delta_t(2) a_{21}, \delta_t(3) a_{31}\} b_1(Y_{t+1}).$$

Suppose the maximum is attained with $\delta_t(2) a_{21}$. Then, we record

$$\gamma_{t+1}(1) = 2,$$

which is represented by a pointer from $\delta_{t+1}(1)$ to $\delta_t(2)$ in the diagram in Figure 4.2. This means that if $\widehat{z}_{t+1} = 1$ in the backward tracking, then we assign $\widehat{z}_t = \gamma_{t+1}(\widehat{z}_{t+1}) = 2$ and subsequently $\widehat{z}_{t-1} = 1$, following the arrows $\delta_{t+1}(1) \to \delta_t(2) \to \delta_{t-1}(1)$ in the diagram.

4.4 Extensions

We consider a few extensions to the HMM.

4.4.1 *Continuous observation*

First, we may generalize the model for continuous observations, $y_t \in \mathbb{R}$. Given $Z_t = j$, the emission model is then defined by a continuous distribution with density $f(y_t \mid \gamma_j)$, where γ_j is the parameter for the emission distribution.

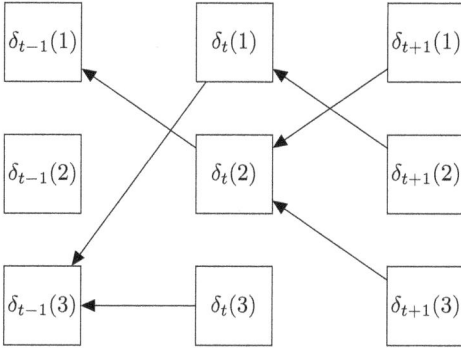

Fig. 4.2. Backward tracking in the Viterbi algorithm for a three-state HMM.

Accordingly, we modify the forward summation (4.11) to

$$\alpha_{t+1}(j) = f(y_{t+1} \mid \gamma_j) \sum_{i=1}^{N} \alpha_t(i) a_{ij}$$

and similarly replace $b_j(Y_{t+1})$ by $f(y_{t+1} \mid \gamma_j)$ in the backward summation (4.12). In the M-step, the update of γ_j depends on the parametric family f and its sufficient statistic.

4.4.2 *Kalman filtering*

Next, we consider both continuous observations y_t and continuous states z_t. In this case, state transition and emission are described by the following state-space model:

$$z_{t+1} = a z_t + \varepsilon_t, \quad \varepsilon_t \overset{\text{iid}}{\sim} \mathcal{N}(0, \tau^2), \tag{4.15}$$

$$y_t = z_t + \eta_t, \quad \eta_t \overset{\text{iid}}{\sim} \mathcal{N}(0, \xi^2), \tag{4.16}$$

where a, τ^2, and ξ^2 are parameters. The model structure is represented graphically in Figure 4.3.

For simplicity, let us assume that the parameters are given. Our goal is to make an online prediction of the hidden state z_t given the observations $\{y_1, \ldots, y_t\}$ collected up to time t. We complete this by updating the conditional distribution $p(z_t \mid y_1, \ldots, y_t)$ recursively. In signal processing, y_t is a noisy version of the underlying signal z_t. The process of predicting z_t from y_t is called filtering. The method we discuss in the following is called Kalman (1960) filtering.

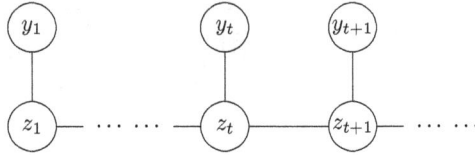

Fig. 4.3. An HMM with continuous states and observations.

We review two lemmas pertaining to the Gaussian distribution, which will be used in the derivation for the Kalman filter.

Lemma 4.1. *If* $X \sim \mathcal{N}(\mu_1, \sigma_1^2)$ *and* $Y \mid X \sim \mathcal{N}(aX, \sigma_2^2)$, *then*

$$Y \sim \mathcal{N}(a\mu_1, a^2\sigma_1^2 + \sigma_2^2).$$

Proof. We may represent Y as $Y = aX + \varepsilon$, where $\varepsilon \sim \mathcal{N}(0, \sigma_2^2)$ and $\varepsilon \perp\!\!\!\perp X$. Then, the marginal distribution of Y follows immediately from the sum of the two independent normal random variables aX and ε. □

Lemma 4.2. *If* $X \sim \mathcal{N}(\mu_1, \sigma_1^2)$ *and* $Y \mid X \sim \mathcal{N}(X, \sigma_2^2)$, *then* $X \mid Y \sim \mathcal{N}(\mu, \sigma^2)$, *where*

$$\mu = \frac{\sigma_1^2 Y + \sigma_2^2 \mu_1}{\sigma_1^2 + \sigma_2^2},$$

$$1/\sigma^2 = 1/\sigma_1^2 + 1/\sigma_2^2.$$

Proof. Denote the density of $\mathcal{N}(\mu, \sigma^2)$ by $\phi(\cdot; \mu, \sigma^2)$. By Bayes' theorem,

$$\begin{aligned} p(x \mid y) &\propto p(x)p(y \mid x) \\ &= \phi(x; \mu_1, \sigma_1^2)\phi(y; x, \sigma_2^2) \\ &= \phi(x; \mu_1, \sigma_1^2)\phi(x; y, \sigma_2^2). \end{aligned}$$

Now, we are working with the product of two Gaussian densities. After normalization, this leads to another normal density with a mean of μ and a variance of σ^2, as in the lemma. □

Remark 4.1. Lemma 4.2, in fact, shows that the normal is a conjugate prior for normal distributions: Suppose $y \mid \theta \sim \mathcal{N}(\theta, \sigma^2)$. If we assume a normal prior $\theta \sim \mathcal{N}(\mu, \tau^2)$, then Lemma 4.2 implies that the posterior $[\theta \mid y]$ is also a normal distribution.

We are now ready to find $[z_t \mid y_1, \ldots, y_t]$, for any $t \geq 1$, by induction:

(i) For $t = 1$, we have $y_1 \mid z_1 \sim \mathcal{N}(z_1, \xi^2)$ by (4.16). Since there is no information at this time about z_1, we simply use a flat prior, $p(z_1) \propto 1$. By Lemma 4.2, the posterior of z_1 given y_1 is

$$z_1 \mid y_1 \sim \mathcal{N}(y_1, \xi^2) := \mathcal{N}(\mu_1, \sigma_1^2).$$

(ii) Assume that

$$z_t \mid y_1, \ldots, y_t \sim \mathcal{N}(\mu_t, \sigma_t^2), \qquad (4.17)$$

we show that $[z_{t+1} \mid y_1, \ldots, y_{t+1}]$ is also a Gaussian distribution.

Since $z_{t+1} \mid z_t \sim \mathcal{N}(az_t, \tau^2)$ by transition model (4.15), together with hypothesis (4.17), we have

$$z_{t+1} \mid y_1, \ldots, y_t \sim \mathcal{N}(a\mu_t, \tau^2 + a^2\sigma_t^2), \qquad (4.18)$$

by Lemma 4.1. From emission model (4.16), $y_{t+1} \mid z_{t+1} \sim \mathcal{N}(z_{t+1}, \xi^2)$. Because $y_{t+1} \perp\!\!\!\perp \{y_i : i \leq t\} \mid z_{t+1}$, we have

$$y_{t+1} \mid z_{t+1}, y_1, \ldots, y_t \sim \mathcal{N}(z_{t+1}, \xi^2). \qquad (4.19)$$

Applying Lemma 4.2 with (4.18) and (4.19), we see that

$$z_{t+1} \mid y_1, \ldots, y_t, y_{t+1} \sim \mathcal{N}(\mu_{t+1}, \sigma_{t+1}^2),$$

where the mean and the variance are given by

$$\mu_{t+1} = \frac{w_1^{(t)} a\mu_t + w_2 y_{t+1}}{w_1^{(t)} + w_2}, \qquad (4.20)$$

$$1/\sigma_{t+1}^2 = w_1^{(t)} + w_2, \qquad (4.21)$$

$$w_1^{(t)} = \left(\tau^2 + a^2\sigma_t^2\right)^{-1}, \quad w_2 = 1/\xi^2.$$

Therefore, in Kalman filtering, we only need to update μ_t and σ_t^2 by (4.20) and (4.21), respectively, after the current observation y_t is collected.

4.5 Problems

(1) The transition probability matrix of a Markov chain, $\{X_n\}$, on the state space $\{1, 2\}$ is

$$A = \begin{bmatrix} 1 - \alpha & \alpha \\ \beta & 1 - \beta \end{bmatrix},$$

where $0 < \alpha, \beta < 1$.

(a) Determine $A^{(\infty)} := \lim_{n \to \infty} A^{(n)}$.

(b) Find the following probabilities as $n \to \infty$:

(i) $\mathbb{P}(X_n = 2, X_{n+1} = 1 \mid X_1 = 1)$; (ii) $\mathbb{P}(X_{n-1} = 2 \mid X_n = 1)$.

(2) Consider the state-space model $y_t = Ay_{t-1} + B\epsilon_t$ and $x_t = Cy_t + D\eta_t$, for $t = 1, \ldots, n$, where ϵ_t and η_t are i.i.d. $\mathcal{N}(0,1)$ and A, B, C, D are known constants. Assume that $y_0 = 0$ in the following questions.

(a) Find the conditional distribution of y_1 given x_1.

(b) Find the joint distribution of x_1 and y_2.

(c) Derive a recursive algorithm to determine the conditional distribution $[y_t \mid x_1, \ldots, x_t]$.

(3) Let $X_i, i = 1, \ldots, n$, be random variables assuming the values of 1 or -1. Suppose that the joint distribution of $X_1, \ldots, X_n$ is given by

$$\mathbb{P}(X_i = x_i, i = 1, \ldots, n) \propto \exp\left(\theta \sum_{i=1}^{n-1} x_i x_{i+1}\right) \equiv h(x_1, \ldots, x_n; \theta),$$

for $x_i \in \{1, -1\}$, where $\theta \in \mathbb{R}$ is a parameter.

(a) Find the normalizing constant

$$Z_n(\theta) = \sum_{x_1, \ldots, x_n} h(x_1, \ldots, x_n; \theta).$$

(b) Suppose that we have observed $X_i = x_i$ for $i = 1, \ldots, n$. Find the MLE of θ. Is there any potential problem with the MLE?

(c) Is X_t a Markov chain? If so, determine its transition matrix.

(4) Recall the definitions of $\alpha_t(i)$ and $\beta_t(i)$ in (4.8).

(a) Show that recursion (4.12) in the backward summation is valid.

(b) Show that $\sum_{i=1}^{N} \alpha_n(i) = \sum_{j=1}^{N} \alpha_t(j)\beta_t(j)$, for $t = 1, \ldots, n-1$.

Chapter 5

Random Graphs for Modeling Network Data

5.1 Network Data

In recent years, network data have become popular in many domains, including biology, sociology, and communication (Albert and Barabási, 2002; Fienberg, 2012). Some examples of network data include social networks, protein–protein interaction networks, and biomedical data with family history. In a social network, individuals with certain social connections are linked by an edge, while in a protein–protein interaction network, two proteins that may form physical interactions are connected.

In general, network data, consisting of relations among a set of individuals, are modeled by a random graph. Each individual in the network corresponds to a vertex or node in the graph, while their relations are represented by edges between the vertices. Thus, the observed data are a network (graph) among n nodes. Let $V = \{1, \ldots, n\}$ be the vertex set of a random graph. Each vertex corresponds to an individual $i \in V$ in the network. Connections among the individuals are represented by a (symmetric) adjacency matrix $A = (Y_{ij})_{n \times n}$ such that

$$Y_{ij} = 0 \quad \text{if there is no edge between } i \text{ and } j;$$
$$Y_{ij} = 1 \quad \text{if there is an edge between } i \text{ and } j.$$

Since each edge is modeled by a binary variable, the graph is called a binary graph. For some data, a real value may be associated with an edge so that $Y_{ij} \in \mathbb{R} \setminus \{0\}$ when there is an edge. Such a graph is called a weighted graph. Unless noted otherwise, we always assume that there is no self-loop

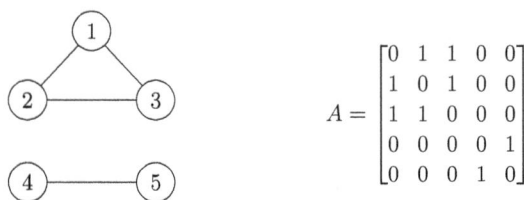

Fig. 5.1. A friendship network and its adjacency matrix.

in the graph, i.e., $Y_{ii} = 0$ for all i. Figure 5.1 shows an example friendship network, where each vertex represents an individual. If two individuals are friends, there is an edge between them in the graph. The adjacency matrix of this network is shown in the right panel.

To simplify notation, we often identify a graph with its adjacency matrix. To model network data, we build a probabilistic model for the random graph A. That is, A is a random matrix following some probability distribution. Equivalently, the edge variables $\{Y_{ij} : i \in [n], j \in [n]\}$ follow a joint distribution. An observed network $(y_{ij})_{n \times n}$ is a realization of A.

A focus of this chapter is to model heterogeneity in network data (Matias and Robin, 2014), say nodes that share a large number of connections form a community. The purpose is similar to the clustering problem discussed in Section 3.2; however, here, a community, or a cluster of nodes, is defined by observed network connections instead of the similarity among the features of individuals. In the network in Figure 5.1, there are two communities, $\{1, 2, 3\}$ and $\{4, 5\}$, which give rise to a block-diagonal structure in the adjacency matrix. We start with a general latent space model and then discuss in more detail the stochastic block model and the graphon.

5.2   Latent Space Models

Under the latent space model (Hoff *et al.*, 2002), each node $i \in V$ is associated with an independent latent variable $Z_i \in \mathbb{R}^q$. The Euclidean space $\mathbb{R}^q$ for Z_i is called the latent space. The dimension q is usually chosen to be a small number, say $q \leq 10$. The distribution of the edge Y_{ij} between the nodes i and j depends on $\|Z_i - Z_j\|$, the distance between Z_i and Z_j in the latent space. As an example, the conditional distribution $[Y_{ij}|Z_i, Z_j]$ may be specified as

$$Y_{ij} = Y_{ji} \mid Z_i, Z_j \sim \text{Bern}(\gamma_{ij}),$$

$$\text{logit}\{\gamma_{ij}\} = \alpha - \|Z_i - Z_j\|,$$

where $\gamma_{ij} \in (0,1)$ and $\alpha \in \mathbb{R}$. If $\|Z_i - Z_j\|$ is small, then $\mathbb{P}(Y_{ij} = 1 \mid Z_i, Z_j)$ is large, so that the nodes i and j are more likely to connect. In this sense, we may regard Z_i as the social position of individual i in the latent space. If two individuals are close with respect to (w.r.t.) their social positions, it is more likely to observe a connection between them in the network. Given the network (Y_{ij}), we may predict the hidden social positions $\{Z_i\}$ and then cluster individuals according to $\{Z_i\}$ to detect communities. In this procedure, we essentially map each node to a point $Z_i \in \mathbb{R}^q$ and, consequently, translate the community detection problem into a clustering problem in the Euclidean space.

The latent space model provides a very general framework for modeling network data. We focus on two related models that may be understood as special models under this general framework. The first one is the graphon model, in which the latent variables $U_i \overset{\text{iid}}{\sim} \text{Unif}(0,1)$, $i = 1, \ldots, n$. Then, we assume that

$$Y_{ij} \mid U_i, U_j \sim \text{Bern}(\gamma_{ij}),$$

where $\gamma_{ij} = g(U_i, U_j)$ and g is a symmetric function, called a *graphon*. Nonparametric estimation methods can be used to estimate the graphon g. The second model is the stochastic block model (SBM), in which the latent variables reduce to community (or cluster) labels, $Z_i \in \{1, \ldots, K\}$, where K is the number of communities in the network.

5.3 Stochastic Block Models

In this section, we give a detailed treatment of the SBM. We first go over the model structure and formulate it as a latent variable model. Then, we discuss parameter estimation for the SBM and review methods for community detection. This section is concluded with a brief discussion on various extensions to the SBM. See Abbe (2018) and Lee and Wilkinson (2019) for reviews of SBMs.

5.3.1 *Model structure*

Assume that there are K communities (clusters) among the n nodes of a network. Encode the latent cluster label for node i by the vector

$$Z_i = (Z_{i1}, \ldots, Z_{iK}) \in \{e_1, \ldots, e_K\},$$

where e_j is the jth standard basis vector in $\mathbb{R}^K$. Let $\pi = (\pi_1, \ldots, \pi_K)$ be a vector of cell probabilities, i.e., $\sum_m \pi_m = 1, \pi_m > 0$, for all $m \in [K]$. Under the SBM, we assume a multinomial distribution for Z_i:

$$Z_i = (Z_{i1}, \ldots, Z_{iK}) \overset{\text{iid}}{\sim} M(1, \pi),$$

i.e., $\mathbb{P}(Z_{im} = 1) = \pi_m$. Given Z_i and Z_j, the edge $Y_{ij} = Y_{ji}$ is drawn independently:

$$Y_{ij} \mid Z_{im} = 1, Z_{j\ell} = 1 \sim f(\cdot; \gamma_{m\ell}),$$

where $f(\cdot; \gamma_{m\ell})$ is a distribution for the edge variable parameterized by $\gamma_{m\ell}$. The matrix $\gamma = (\gamma_{m\ell})_{K \times K}$ contains all parameters for the connection probabilities among the K communities. As a concrete example, for most parts of this section, we assume that

$$Y_{ij} \mid Z_{im} = 1, Z_{j\ell} = 1 \sim \text{Bern}(\gamma_{m\ell}), \tag{5.1}$$

for which

$$f(y; \gamma_{m\ell}) = \gamma_{m\ell}^y (1 - \gamma_{m\ell})^{1-y}, \quad y \in \{0, 1\}.$$

In this case, we are dealing with the binary edges $Y_{ij} \in \{0, 1\}$.

We formulate the SBM as a latent variable model. The observed data Y form a graph over n nodes, represented by its adjacency matrix $A = (Y_{ij})_{n \times n}$. The latent variables are the community labels $Z = (Z_1, \ldots, Z_n)$. Denote all the model parameters by $\theta = (\pi, \gamma)$. The maximum likelihood estimate (MLE) $\widehat{\theta}$ is defined by maximizing the observed-data likelihood:

$$\widehat{\theta} = \underset{\theta}{\text{argmax}} \left\{ \mathbb{P}(Y \mid \theta) = \sum_{Z_1} \cdots \sum_{Z_n} \mathbb{P}(Y, Z_1, \ldots, Z_n \mid \theta) \right\}.$$

Let us consider developing an EM algorithm to find the MLE. The complete data are given by (Y, Z). Based on the model structure, the complete-data log-likelihood is

$$\ell(\theta \mid Y, Z) = \log \mathbb{P}(Y, Z \mid \theta)$$

$$= \sum_{i=1}^{n} \sum_{m=1}^{K} Z_{im} \log \pi_m + \frac{1}{2} \sum_{i \neq j} \sum_{m,\ell} Z_{im} Z_{j\ell} \log f(Y_{ij} \mid \gamma_{m\ell}). \tag{5.2}$$

It is seen that in the E-step, one needs to calculate $\mathbb{E}(Z_{im} \mid Y, \theta^{(t)})$ and $\mathbb{E}(Z_{im} Z_{j\ell} \mid Y, \theta^{(t)})$. For any $i \neq j$, Z_i and Z_j are dependent given Y_{ij};

therefore, $Z_1, \ldots, Z_n$ are all dependent given $Y = (Y_{ij})$. This means that $\mathbb{P}(Z_1, \ldots, Z_n \mid Y, \theta^{(t)})$ does not factorize according to any simple structure. Consequently, the E-step is computationally intractable.

Remark 5.1. For a mixture model, due to the i.i.d. assumption among $\{Z_i, Y_i\}$ for $i = 1, \ldots, n$, we have $Z_i \perp\!\!\!\perp Z_j \mid Y$. For an HMM, we demonstrate that $[Z_1, \ldots, Z_n \mid Y]$ is a Markov chain (Section 4.2.1), which allows us to derive efficient forward and backward summations for the E-step. However, for the SBM, the dependence among Z_i's given Y complicates the E-step of the EM algorithm. In such scenarios, we may consider variational inference or variational Bayes approach to approximate $[Z \mid Y, \theta]$, which will lead to the so-called variational EM algorithm.

5.3.2 *Variational EM algorithm*

Variational inference (Jordan *et al.*, 1999), a powerful technique in statistics and machine learning, is used to approximate complex posterior distributions. Unlike traditional methods, which often struggle with intractable integrals or computationally expensive sampling techniques such as Markov chain Monte Carlo, variational inference offers a scalable and efficient approach. The key idea behind this technique is to frame the problem as an optimization task, where a simpler distribution, known as the variational distribution, is iteratively adjusted to approximate the true posterior. By minimizing the difference between the variational distribution and the true posterior, variational inference provides a computationally feasible means to approximate complex probabilistic models, enabling practitioners to perform maximum likelihood and Bayesian inference in a wide range of applications. A review paper by Blei *et al.* (2017) provides an in-depth overview of variational inference, covering its theoretical foundations, practical implementation, and various applications across different domains.

Variational inference may be motivated from an iterative maximization view of the EM algorithm, which leads to the variational EM algorithm. We start with the observed-data log-likelihood

$$\ell(\theta \mid Y) = \log \mathbb{P}(Y \mid \theta) = \log \mathbb{P}(Y, Z \mid \theta) - \log \mathbb{P}(Z \mid Y, \theta).$$

Taking expectation w.r.t. a distribution F over Z on both sides, we arrive at

$$\ell(\theta \mid Y) = \mathbb{E}_F \left\{ \log \mathbb{P}(Y, Z \mid \theta) \right\} + H(F) + KL(F \,\|\, \mathbb{P}(Z \mid Y, \theta)), \qquad (5.3)$$

where $H(F) = \mathbb{E}_F\{-\log F(Z)\}$ is the entropy of F and $KL(\cdot\|\cdot)$ is the Kullback–Leibler divergence. As $KL(F\,\|\,\mathbb{P}(Z\mid Y,\theta)) \geq 0$ for any F, we have

$$\ell(\theta\mid Y) \geq \mathbb{E}_F\{\log\mathbb{P}(Y,Z;\theta)\} + H(F) := L(\theta,F).$$

This shows that the $L(\theta,F)$ defined above is a lower bound of the log-likelihood $\ell(\theta\mid Y)$ for any chosen distribution $F(Z)$. An important special case is that $L(\theta,F) = \ell(\theta\mid Y)$ if $F = \mathbb{P}(Z\mid Y,\theta)$. In the variational inference literature, $L(\theta,F)$ is called the evidence lower bound (ELBO) and F the variational distribution.

At a conceptual level, the EM algorithm iterates between two maximization steps to maximize the ELBO $L(\theta,F)$ over both F and θ:

$$\max_{F,\theta}\{L(\theta,F) = \mathbb{E}_F\{\log\mathbb{P}(Y,Z\mid\theta)\} + H(F)\}.$$

(1) E-step: Given $\theta^{(t)}$, we maximize $L(\theta^{(t)},F)$ over F. Due to (5.3), this is equivalent to

$$\min_F KL(F\|\mathbb{P}(Z\mid Y,\theta^{(t)})) \Rightarrow F^{(t)} = \mathbb{P}(Z\mid Y,\theta^{(t)}).$$

(2) M-step: Given $F^{(t)}$, we maximize $L(\theta,F^{(t)})$ over θ. Since $H(F^{(t)})$ is a constant w.r.t. θ, the maximization is equivalent to

$$\max_\theta \mathbb{E}_{F^{(t)}}\{\log\mathbb{P}(Y,Z\mid\theta)\} = \max_\theta \mathbb{E}\left\{\log\mathbb{P}(Y,Z\mid\theta)\mid Y,\theta^{(t)}\right\}$$
$$= \max_\theta Q(\theta\mid\theta^{(t)}),$$

where $Q(\theta\mid\theta^{(t)})$ is the Q-function in the EM algorithm. Consequently, we update

$$\theta^{(t+1)} = \operatorname*{argmax}_\theta Q(\theta\mid\theta^{(t)}).$$

Clearly, this recovers exactly the same EM iteration.

Note that $L(\theta,F^{(t)})$ is the minorization function (2.3) in the MM view of the EM algorithm.

When the E-step is intractable, as for the SBM, we may maximize $L(\theta,F)$ over a restricted class of $F \in \mathcal{F}$ so that the E-step is tractable, which is the key idea behind the variational EM. This will change the range of the maximization in the E-step in the above formulation. The two steps in one iteration of the variational EM are as follows:

(1) E-step: Given $\theta^{(t)}$, we maximize $L(\theta^{(t)}, F)$ over $F \in \mathcal{F}$:

$$\max_{F \in \mathcal{F}} \mathbb{E}_F \left\{ \log \mathbb{P}(Y, Z \mid \theta^{(t)}) \right\} + H(F). \qquad (5.4)$$

Let $F^{(t)} \in \mathcal{F}$ be a maximizer of the above optimization program.

(2) M-step: Given $F^{(t)}$, we maximize $L(\theta, F^{(t)})$ over θ to obtain $\theta^{(t+1)}$, which is equivalent to

$$\theta^{(t+1)} = \operatorname*{argmax}_{\theta} \mathbb{E}_{F^{(t)}} \left\{ \log \mathbb{P}(Y, Z \mid \theta) \right\}.$$

Figure 5.2 illustrates one iteration of the variational EM. Note that $L(\theta, F^{(t)})$ is always a lower bound of $\ell(\theta \mid Y)$ for any $F^{(t)}$. The distance between $L(\theta^{(t)}, F^{(t)})$ and $\ell(\theta^{(t)} \mid Y)$ is the KL divergence between $F^{(t)}$ and $\mathbb{P}(Z \mid Y, \theta^{(t)})$. For the EM algorithm, the two curves would coincide at $\theta^{(t)}$ since $F^{(t)} = \mathbb{P}(Z \mid Y, \theta^{(t)})$; see Figure 2.2 for a comparison. The maximization in (5.4) approximates the posterior distribution $[Z \mid Y, \theta^{(t)}]$ by minimizing the KL divergence over the restricted class $\mathcal{F}$.

5.3.3 *Variational inference for the SBM*

We apply the variational EM to the SBM, following the work by Daudin *et al.* (2008). We assume $Z_i \sim M(1, \tau_i)$ independently for each i under F so that $F(Z) = \prod_{i=1}^n h(Z_i; \tau_i)$, for some h. Note that the multinomial parameter $\tau_i = (\tau_{i1}, \ldots, \tau_{iK})$ is node-specific. Under this choice of F,

$$\mathbb{E}_F(Z_{im}) = \tau_{im}, \qquad \mathbb{E}_F(Z_{im} Z_{j\ell}) = \mathbb{E}_F(Z_{im})\mathbb{E}_F(Z_{j\ell}) = \tau_{im}\tau_{j\ell}. \qquad (5.5)$$

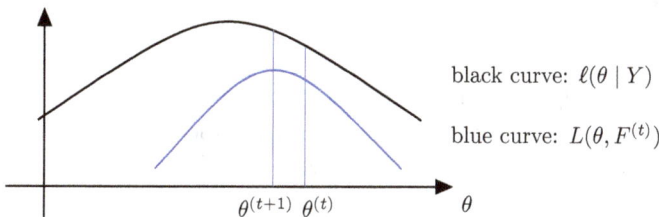

black curve: $\ell(\theta \mid Y)$

blue curve: $L(\theta, F^{(t)})$

Fig. 5.2. Illustration of the variational EM algorithm.

Given (5.5), we find the expectation of the complete-data log-likelihood, as in (5.2), to calculate the ELBO

$$L(\theta, F) = \sum_{i=1}^{n} \sum_{m} \tau_{im} \log \pi_m + \frac{1}{2} \sum_{i \neq j} \sum_{m,\ell} \tau_{im} \tau_{j\ell} \log f(Y_{ij} \mid \gamma_{m\ell})$$

$$- \sum_{i=1}^{n} \sum_{m} \tau_{im} \log \tau_{im} := L(\theta, \tau), \qquad (5.6)$$

where the last term is the entropy $H(F)$ and $\tau = (\tau_1, \ldots, \tau_n)$. The variational EM will maximize $L(\theta, \tau)$ over τ (E-step) and θ (M-step) iteratively.

Let us first work out the E-step. Given $\theta^{(t)} = (\pi^{(t)}, \gamma^{(t)})$, we solve the following constraint optimization problem to find $\tau^{(t)}$:

$$\max_{\tau} L(\theta^{(t)}, \tau), \qquad \text{subject to} \sum_{m=1}^{K} \tau_{im} = 1, \text{ for all } i = 1, \ldots, n. \qquad (5.7)$$

Using the Lagrange multipliers λ_i, we maximize the Lagrange function

$$\max_{\tau, \{\lambda_i\}} L(\theta^{(t)}, \tau) + \sum_{i=1}^{n} \lambda_i \left(1 - \sum_{m} \tau_{im} \right).$$

Taking derivative w.r.t. τ_{im} and setting it to zero, we arrive at the equations

$$\log \pi_m^{(t)} - \log \tau_{im} + \sum_{j \neq i} \sum_{\ell=1}^{K} \tau_{j\ell} \log f(Y_{ij} \mid \gamma_{m\ell}^{(t)}) = \lambda_i + 1,$$

for $m \in [K]$ and $i \in [n]$. Although there is no closed-form solution, it is seen that $\tau^{(t)}$ must satisfy

$$\tau_{im} \propto \pi_m^{(t)} \prod_{j \neq i} \prod_{\ell=1}^{K} \left\{ f(Y_{ij} \mid \gamma_{m\ell}^{(t)}) \right\}^{\tau_{j\ell}}, \qquad (5.8)$$

subject to $\sum_{m} \tau_{im} = 1$ for each i. Given $\tau_{-i} = (\tau_j)_{j \neq i}$, Equation (5.8) defines a mapping to update $\tau_i = f(\tau_{-i})$. This leads to an iterative algorithm that cycles through the updates $\tau_i = f(\tau_{-i})$, for all $i \in [n]$, and converges to a fixed point $\tau^{(t)}$.

The update (5.8) is reminiscent of a conditional sampling step in a Gibbs sampler. Given $\theta^{(t)}$, suppose we wish to develop a Gibbs sampler to draw $[Z \mid Y, \theta^{(t)}]$ by iteratively sampling from the conditional distribution $[Z_i \mid Y, Z_{-i}, \theta^{(t)}]$, for $i = 1, \ldots, n$. Dropping $\theta^{(t)}$ for notational brevity,

$$\mathbb{P}(Z_{im} = 1 \mid Y, Z_{-i}) \propto \mathbb{P}(Z_{im} = 1 \mid Z_{-i})\mathbb{P}(Y \mid Z_{im} = 1, Z_{-i})$$

$$= \pi_m^{(t)} \prod_{j \neq i} \prod_{\ell=1}^{K} \left\{ f(Y_{ij} \mid \gamma_{m\ell}^{(t)}) \right\}^{Z_{j\ell}},$$

which is identical to (5.8) if we replace $Z_{j\ell}$ with its mean $\tau_{j\ell}$. In this sense, we may regard (5.8) as a mean-field approximation to the Gibbs sampler.

In the M-step, given $\tau^{(t)}$, we maximize $L(\theta, \tau^{(t)})$ over θ:

$$\max_{\theta} L(\theta, \tau^{(t)}), \qquad \text{subject to} \ \sum_{m} \pi_m = 1.$$

Assuming the Bernoulli model for binary edges (5.1), the solution $\theta^{(t+1)}$ to the above maximization problem is given by

$$\pi_m^{(t+1)} = \frac{1}{n} \sum_{i=1}^{n} \tau_{im}^{(t)}, \qquad m \in [K], \tag{5.9}$$

$$\gamma_{m\ell}^{(t+1)} = \frac{\sum_{i \neq j} \tau_{im}^{(t)} \tau_{j\ell}^{(t)} Y_{ij}}{\sum_{i \neq j} \tau_{im}^{(t)} \tau_{j\ell}^{(t)}}, \qquad (m, \ell) \in [K] \times [K]. \tag{5.10}$$

Recall that $\tau_{im}^{(t)}$ approximates $\mathbb{E}(Z_{im} \mid Y, \theta^{(t)})$ and can be regarded as the weight of node i in cluster m. Thus, $\sum_i \tau_{im}^{(t)}$ may be viewed as the expected size of community m. This provides an intuitive interpretation for the update of $\pi_m^{(t+1)}$. Similarly, $\tau_{im}^{(t)} \tau_{j\ell}^{(t)}$ approximates

$$\mathbb{E}(Z_{im} Z_{j\ell} \mid Y, \theta^{(t)}) = \mathbb{P}(Z_{im} = 1, Z_{j\ell} = 1 \mid Y, \theta^{(t)}),$$

the posterior probability of node i in cluster m and j in cluster ℓ given Y. The edge variable Y_{ij} indicates whether there is an edge between the two nodes. If we replace τ_{im} with Z_{im} in (5.9) and (5.10), then $(\pi_m^{(t+1)}, \gamma_{m\ell}^{(t+1)})$ is the MLE $\widehat{\theta}$ given the complete data (Y, Z):

$$\widehat{\pi}_m = \frac{1}{n} \sum_{i=1}^{n} Z_{im}, \qquad m \in [K],$$

$$\widehat{\gamma}_{m\ell} = \frac{\sum_{i \neq j} Z_{im} Z_{j\ell} Y_{ij}}{\sum_{i \neq j} Z_{im} Z_{j\ell}}, \qquad (m, \ell) \in [K] \times [K].$$

Remark 5.2. Even for a moderate value of K, there are a large number of connection probabilities $\{\gamma_{m\ell}\}$. To improve the accuracy of parameter estimation, Peng and Zhou (2022) proposed a hierarchical model and an empirical Bayes estimate for $\{\gamma_{m\ell}\}$, which facilitate data sharing among different blocks through shrinkage. It is shown with simulated data that the empirical Bayes estimate can reduce by more than 50% the mean squared error of the variational estimate.

We discuss briefly the consistency of the variational estimator (Bickel *et al.*, 2013). We consider two estimators: the MLE

$$\widehat{\theta}^{\mathrm{ML}} = \operatorname*{argmax}_{\theta} \ell(\theta \mid Y),$$

and the variational estimator

$$\widehat{\theta}^{\mathrm{VR}} = \operatorname*{argmax}_{\theta} \left[\max_{\tau} L(\theta, \tau) \right]$$

defined by maximizing the ELBO (5.6). A key step in the analysis of $\widehat{\theta}^{\mathrm{VR}}$ is to bound the profile log-likelihood $\max_{\tau} L(\theta, \tau)$ by two log-likelihood functions,

$$\log \mathbb{P}(Y, Z = z \mid \theta) \leq \max_{\tau} L(\theta, \tau) \leq \ell(\theta \mid Y), \qquad (5.11)$$

for any z. Note that the left- and right-hand sides of (5.11) are, respectively, the log-likelihood functions of the complete and observed data. We leave it as an exercise to prove the above inequality.

Define the logit transformation of the parameters:

$$\omega_m = \log\{\pi_m / \pi_K\}, \quad m = 1, \dots, K-1,$$

$$\nu_{m\ell} = \log\{\gamma_{m\ell}/(1 - \gamma_{m\ell})\}, \quad m, \ell = 1, \dots, K.$$

Bickel *et al.* (2013) established consistency and asymptotic normality for both estimators as $n \to \infty$, as given by the following theorem.

Theorem 5.1. *Assume the SBM under* (5.1) *with the true parameter* $\theta^* = (\pi^*, \gamma^*)$, *where* γ^* *has no identical columns. Let* $\lambda_n = n\mathbb{P}(Y_{ij} = 1 \mid \theta^*)$, *i.e., the expected degree of a node. If* $\lambda_n / \log n \to \infty$, *then*

$$\sqrt{n}(\widehat{\omega} - \omega^*) \xrightarrow{d} \mathcal{N}(0, \Sigma_1),$$

$$\sqrt{n\lambda_n}(\widehat{\nu} - \nu^*) \xrightarrow{d} \mathcal{N}(0, \Sigma_2),$$

for both $\widehat{\theta}^{\mathrm{VR}}$ *and* $\widehat{\theta}^{\mathrm{ML}}$, *where* Σ_1 *and* Σ_2 *are functions of* θ^*.

This result shows that the variational estimator $\widehat{\theta}^{\mathrm{VR}}$ and the MEL $\widehat{\theta}^{\mathrm{ML}}$ follow the same asymptotic distribution, justifying the use of the variational EM algorithm from a theoretical point of view. One key step in the proof of this theorem is to show that

$$\frac{\mathbb{P}(Y, Z = z^* \mid \theta^*)}{\mathbb{P}(Y \mid \theta^*)} = 1 + o_p(1),$$

where z^* is the true cluster labels of the n nodes. By (5.11), the profile log-likelihood $\max_\tau L(\theta, \tau)$ will be close to $\ell(\theta \mid Y)$ in a neighborhood of θ^* that contains the MLE $\widehat{\theta}^{\mathrm{ML}}$.

5.3.4 *Community detection*

The task of community detection is to predict the cluster labels Z. As for mixture models, cluster labels are only identifiable up to label switching. For any vector $z \in [K]^n$ of the cluster labels, we denote by $[z]$ its equivalence class, i.e., $z' \in [z]$ if and only if z' and z give equivalent groupings of n nodes into K clusters.

Let $\widehat{\theta}$ be an estimator of θ satisfying a certain convergence rate. Celisse *et al.* (2012) showed that the posterior distribution $\mathbb{P}(Z \mid Y, \widehat{\theta})$ concentrates on the true cluster labels $[z^*]$ in the following sense:

$$\frac{\sum_{z \notin [z^*]} \mathbb{P}(Z = z \mid Y, \widehat{\theta})}{\mathbb{P}(Z \in [z^*] \mid Y, \widehat{\theta})} \xrightarrow{p} 0.$$

Therefore, after estimating $\widehat{\theta}$, one may further maximize the posterior $\max_z \mathbb{P}(Z = z \mid Y, \widehat{\theta})$ to predict the cluster labels. However, this maximization problem is difficult to solve in general.

A computationally more tractable method is the spectral clustering method (von Luxburg, 2007), which can achieve a vanishing clustering error rate under certain assumptions (Rohe *et al.*, 2011):

$$\frac{\# \text{ of misclustered nodes}}{n} \xrightarrow{\text{a.s.}} 0.$$

Let $d_i = \sum_j Y_{ij}$ be the degree of node i. Define the normalized graph Laplacian $L = D^{-1/2} A D^{-1/2}$, where $D = \mathrm{diag}(d_1, \ldots, d_n)$. The spectral clustering of $A = (Y_{ij})_{n \times n}$ consists of the following main steps:

(i) Find $X = [X_1 \mid \cdots \mid X_K] \in \mathbb{R}^{n \times K}$, where X_j's are the orthogonal eigenvectors corresponding to the largest K eigenvalues (in absolute value) of L.

(ii) Treating each row of X as a data point in $\mathbb{R}^K$, apply k-means to cluster the n rows into K clusters, $C_1, \ldots, C_K$, which forms a partition of $[n]$.

(iii) Output the predicted labels $\widehat{Z}_{im} = 1$ if $i \in C_m$, for $i \in [n]$.

As can be seen, spectral clustering is quite simple and does not involve any complicated numerical approach. It embeds the n nodes of the graph into a K-dimensional Euclidean space and then solves the node clustering problem through clustering in the Euclidean space.

Why does spectral clustering work? For a rigorous analysis, see Rohe *et al.* (2011). We now demonstrate the key intuition with its population version. Define the population version of the adjacency matrix A by $\mathcal{A} = (\mathcal{A}_{ij})_{n \times n}$, where

$$\mathcal{A}_{ij} = \mathbb{E}(Y_{ij} \mid Z) = \mathbb{P}(Y_{ij} = 1 \mid Z).$$

Here, both the expectation and the probability are w.r.t. the true model, and Z denotes the true cluster labels represented as indicator vectors. Let $B = (\gamma_{m\ell})_{K \times K}$ and $Z = (Z_{im})_{n \times K}$. Then, $\mathcal{A} = ZBZ^\mathsf{T}$. Define the graph Laplacian of $\mathcal{A}$ in a similar way, i.e., $\mathcal{L} = \mathcal{D}^{-1/2} \mathcal{A} \mathcal{D}^{-1/2}$, where $\mathcal{D}_{ii} = \sum_j \mathcal{A}_{ij}$. Since $\mathrm{rank}(\mathcal{A}) = K$, the matrix $\mathcal{L}$ has only K nonzero eigenvalues and the associated eigenvectors $\mathcal{U} = [U_1 \mid \cdots \mid U_K] = (u_{ij}) \in \mathbb{R}^{n \times K}$ satisfy

$$u_i = u_j \Leftrightarrow Z_i = Z_j,$$

where u_i is the ith row of $\mathcal{U}$. That is, nodes in the same cluster all have identical rows in the matrix $\mathcal{U}$. For a finite graph of n nodes, the eigenvectors of L converge to the eigenvectors of $\mathcal{L}$ as $n \to \infty$. In other words, the matrix X calculated in step (i) of the spectral clustering algorithm is a noisy version of the $\mathcal{U}$ matrix in the population version.

Let us consider an example graph over $n = 4$ nodes, for which the connection probabilities and community label indicators are given, respectively, by

$$B = \begin{bmatrix} 0.8 & 0.1 \\ 0.1 & 0.7 \end{bmatrix} \quad \text{and} \quad Z = \begin{bmatrix} 1 & 0 \\ 1 & 0 \\ 0 & 1 \\ 0 & 1 \end{bmatrix}.$$

Given these two matrices, we find the population adjacency matrix and its graph Laplacian:

$$
\mathcal{A} = \begin{bmatrix} 0.8 & 0.8 & 0.1 & 0.1 \\ 0.8 & 0.8 & 0.1 & 0.1 \\ 0.1 & 0.1 & 0.7 & 0.7 \\ 0.1 & 0.1 & 0.7 & 0.7 \end{bmatrix}, \quad \mathcal{L} = \begin{bmatrix} 0.444 & 0.444 & 0.059 & 0.059 \\ 0.444 & 0.444 & 0.059 & 0.059 \\ 0.059 & 0.059 & 0.438 & 0.438 \\ 0.059 & 0.059 & 0.438 & 0.438 \end{bmatrix}.
$$

We see that $\mathcal{L}$ has a rank of 2 and only two nonzero eigenvalues: $\lambda_1 = 1$ and $\lambda_2 = 0.764$. The first two columns of its eigenvectors form the $\mathcal{U}$ matrix:

$$
\mathcal{U} = \begin{bmatrix} -0.514 & 0.485 \\ -0.514 & 0.485 \\ -0.485 & -0.514 \\ -0.485 & -0.514 \end{bmatrix},
$$

which has identical rows for nodes in the same cluster, such as the first two rows ($i = 1, 2$). Applying any clustering algorithm to the rows of $\mathcal{U}$ would recover the two communities perfectly.

5.3.5 *Extensions and discussion*

There are various extensions to the SBM. First, we may model networks with weighted edges by weighted graphs. For example, if the edge variables take integer values, such as counts, we may assume that

$$
Y_{ij} \mid Z_{im} = 1, Z_{j\ell} = 1 \sim \text{Poiss}(\gamma_{m\ell}),
$$

where $\gamma_{m\ell}$ is the rate of the Poisson distribution. One property of the SMB is that every node has the same degree distribution. In some applications, the vertices of a network could possess very different degrees, say a hub node connecting to many other nodes. Then, we may consider a degree-corrected block model:

$$
Y_{ij} \mid Z_{im} = 1, Z_{j\ell} = 1 \sim \text{Poiss}(\gamma_{m\ell}\kappa_i\kappa_j),
$$

where the node-specific parameter κ_i controls the expected degree of node i.

When covariates are observed, we may also generalize the SBM to account for them. There are at least two types of covariates that may arise depending on the application. The first type is node-wise covariates, x_i, $i = 1, \ldots, n$, where each x_i is a vector of covariates for node i. One may incorporate x_i into the distribution for the cluster label Z_i, assuming that

$$
Z_i \sim \text{M}(1, \pi(x_i)),
$$

where the cell probability $\pi_m(x_i) = \mathbb{P}(Z_{im} = 1)$ is a function of x_i for all m. Therefore, two nodes (i.e., individuals in the network) that have similar covariates will have a higher probability to come from the same community. The second type is edge-wise covariates, x_{ij}, $i \neq j$, which are attributes between a pair of nodes (i, j). This type of covariate can be easily included into the conditional mode for $[Y_{ij} \mid Z_{im}, Z_{j\ell}]$. For binary edges, we may use a logistic regression in the form of

$$\text{logit}\,\{\mathbb{P}(Y_{ij} = 1 \mid Z_{im} = 1, Z_{j\ell} = 1)\} = x_{ij}^{\mathsf{T}}\beta + \gamma_{m\ell},$$

and for integer-valued edges, we may use a Poisson regression:

$$Y_{ij} \mid Z_{im} = 1, Z_{j\ell} = 1 \sim \text{Poiss}(\exp(x_{ij}^{\mathsf{T}}\beta + \gamma_{m\ell})).$$

For both models, $\gamma_{m\ell}$ determines the baseline connection probability, while the regression coefficient vector β quantifies the effect of each covariate x_{ij} on the connection probability. For a more detailed review of these further extensions, see Matias and Robin (2014) and the references therein.

5.4   Graphon

In this section, we restrict our attention to simple graphs, i.e., graphs with binary edges and without self-loops. Formally, the edges $Y_{ij} = Y_{ji} \in \{0, 1\}$, for $i \neq j$.

Definition 5.1 (Graphon). Let $g : [0, 1]^2 \to [0, 1]$ be a symmetric function. Define a random simple graph $(Y_{ij}) \in \{0, 1\}^{n \times n}$ given g as follows:

(i) Draw $U_i \overset{\text{iid}}{\sim} \text{Unif}(0, 1)$, for $i = 1, \ldots, n$.
(ii) Draw $Y_{ij} = Y_{ji} \sim \text{Bern}(g(U_i, U_j))$, for all $i \neq j$.

The function g is a *graphon* (Lovasz and Szegedy, 2006).

The SBM can be regarded as a special case of a graphon model. Partition the unit interval $(0, 1)$ into K intervals, J_m for $m = 1, \ldots, K$, so that $|J_m| = \pi_m$. Let $g(u, v) = \gamma_{m\ell}$ if $u \in J_m$ and $v \in J_\ell$, which defines a block-wise constant graphon function. Then, the graphon model defined with g is equivalent to the SBM with parameter $\theta = (\pi, \gamma)$. To see this, let $Z_{im} = I(U_i \in J_m)$, and then

$$\mathbb{P}(Z_{im} = 1) = |J_m| = \pi_m.$$

If $Z_{im} = 1$ and $Z_{j\ell} = 1$, we have

$$g(U_i, U_j) = \gamma_{m\ell},$$

$$Y_{ij} \sim \text{Bern}(g(U_i, U_j)) = \text{Bern}(\gamma_{m\ell}),$$

which coincides with the SBM (5.1).

The graphon naturally arises in infinitely exchangeable graphs.

Definition 5.2 (Exchangeable graphs). A random graph G is said to be exchangeable if its distribution is invariant to any relabeling (or permutation) of its vertex set.

An equivalent definition of an exchangeable graph is that its adjacency matrix $(Y_{ij})_{n \times n}$ is a jointly exchangeable random array, i.e.,

$$\mathbb{P}(Y_{ij} \in A_{ij}, \forall i, j \in [n]) = \mathbb{P}(Y_{\pi(i)\pi(j)} \in A_{ij}, \forall i, j \in [n]), \tag{5.12}$$

for every permutation π of $\{1, \ldots, n\}$ and every collection of measurable sets $\{A_{ij}\}$. We say that $(Y_{ij}) \stackrel{d}{=} (Y_{\pi(i)\pi(j)})$ when (5.12) holds.

We may generalize (5.12) to an infinite array with indices $i, j \in \mathbb{N}$, where $\mathbb{N}$ is the set of all positive integers. The following result provides a nice representation of infinitely exchangeable arrays (Aldous, 1981).

Theorem 5.2 (Aldous–Hoover). *A random array $(X_{ij}), X_{ij} \in \Omega, i, j \in \mathbb{N}$, is jointly exchangeable if and only if there is a random function $F : [0,1]^3 \to \Omega$ such that*

$$(X_{ij}) \stackrel{d}{=} (F(U_i, U_j, U_{ij})), \tag{5.13}$$

where $(U_i)_{i \in \mathbb{N}}$ and $(U_{ij})_{i,j \in \mathbb{N}}$ are, respectively, an infinite sequence and an infinite array of i.i.d. Unif$[0,1]$ random variables independent of F.

Remark 5.3. Note that $(X_{ij})_{i,j \in \mathbb{N}}$ is an infinite two-way array. The exchangeability of X is understood as an assumption on the data source. An exchangeable graph G over n nodes is regarded as a sample of finite size from this data source.

Applying Theorem 5.2 to the adjacency matrix of an infinite simple graph $(Y_{ij})_{\mathbb{N} \times \mathbb{N}}$, we arrive at the following representation for infinitely exchangeable graphs, where the random function g is the graphon.

Corollary 5.1. *A random simple graph G with vertex set $\mathbb{N}$ is exchangeable if and only if there is a random symmetric function $g : [0,1]^2 \to [0,1]$ such that its adjacency matrix*

$$(Y_{ij}) \overset{d}{=} (I(U_{ij} < g(U_i, U_j))), \qquad (5.14)$$

where $(U_i)_{i \in \mathbb{N}}$ and $(U_{ij})_{i,j \in \mathbb{N}}$ are i.i.d. Unif[0,1] and independent of g.

Proof. The sample space for Y_{ij} is $\Omega = \{0,1\}$. Applying (5.13),

$$(Y_{ij}) \overset{d}{=} (F(U_i, U_j, U_{ij})),$$

where the random function $F(x,y,u) \in \{0,1\}$ for all $x,y,u \in [0,1]$ and F is symmetric in (x,y) because $Y_{ij} = Y_{ji}$. Given F, define a function $g : [0,1]^2 \to [0,1]$ using $g(x,x) = 0$ and

$$g(x,y) := \mathbb{P}(F(x,y,U) = 1 \mid F), \qquad \text{for } x \neq y \in [0,1],$$

where $U \sim \text{Unif}[0,1]$ and is independent of F. Then, g is a random symmetric function.

By conditioning, we find that

$$\mathbb{P}(Y_{ij} = 1 \mid U_i, U_j, F) = g(U_i, U_j) \quad \text{(by the definition of } g\text{)}$$

$$= \mathbb{P}(U_{ij} < g(U_i, U_j) \mid U_i, U_j, F).$$

This implies that

$$(Y_{ij}) \overset{d}{=} (I(U_{ij} < g(U_i, U_j)))$$

because $(Y_{ij})_{i,j \in \mathbb{N}}$ are independent given $(U_i)_{i \in \mathbb{N}}$ and F. $\square$

The important implication of this result is that every exchangeable random simple graph G over $\mathbb{N}$ is represented by a random graphon g. This result leads to the following data generation procedure for G:

  (i) Draw g from a distribution ν (over functions $[0,1]^2 \to [0,1]$).
  (ii) Draw $U_i, i \in \mathbb{N}$ independently from Unif[0,1].
  (iii) For every pair $i < j \in \mathbb{N}$, draw

$$Y_{ij} = Y_{ji} \mid g, U_i, U_j \sim \text{Bern}(g(U_i, U_j)).$$

The distribution of G, or (Y_{ij}), is determined by ν. A statistical model of exchangeable simple graphs is parameterized by the graphon g. From a Bayesian view, the distribution ν is the prior for g, and the posterior distribution $[g \mid Y]$ provides the basis for inference of g. See Orbanz and Roy (2015) for a review of Bayesian models for exchangeable random arrays, graphs, and other structures.

5.5 Problems

(1) Let $Y = (Y_{ij})_{n \times n}$ be the adjacency matrix of a random graph generated from a stochastic block model with K communities and (π, γ) as parameters. Let $D_i = \sum_j Y_{ij}$ be the degree of node i. Find $\mathbb{E}(D_i)$.

(2) Prove inequality (5.11).

(3) Suppose that G_n is an n-node random graph drawn from a graphon model, in which the graphon is $g : [0, 1]^2 \to [0, 1]$.

   (a) Show that G_n is exchangeable.
   (b) Find the degree distribution $\mathbb{P}(D_i = m)$, for $m = 0, \ldots, n - 1$, where D_i is the degree of node i.
   (c) Let F be a fixed graph of n nodes. Find $\mathbb{P}(F \subseteq G_n)$, i.e., the probability that the edge set of F is a subset of the edge set of G_n.

(4) Consider a stochastic block model with K communities for a Poisson-valued weighted graph (Y_{ij}) of n nodes:

$$Y_{ij} \mid Z_{im} = 1, Z_{j\ell} = 1 \sim \mathrm{Poiss}(\gamma_{m\ell}).$$

   (a) Derive a variational EM algorithm to estimate the parameters (π, γ) of this model.
   (b) Simulate a network of size $n = 500$ and $K = 3$ communities with the following parameters:

$$\pi = (0.3, 0.4, 0.3), \qquad \gamma = \begin{bmatrix} 5 & 0.5 & 0.2 \\ 0.5 & 4 & 0.3 \\ 0.2 & 0.3 & 6 \end{bmatrix}.$$

   Implement and apply the variational EM algorithm in (a) to the simulated network. Compare the estimated parameters with the true values.
   (c) Let $A_{ij} = I(Y_{ij} > 0)$ for $i, j \in [n]$. Apply spectral clustering to the adjacency matrix $A = (A_{ij})_{n \times n}$ of the network simulated in (b) and calculate the clustering error rate for the detected communities.

PART 2
Causal Graphical Models

Chapter 6

Undirected Graphical Models

We have reviewed conditional independence (CI) and its use in statistical inference in Section 1.3. Starting from this chapter, we will see that CI also plays a central role in graphical models and causal inference. We first show the similarities between graph separation and CI, both viewed as ternary relations, through the concept of the graphoid (Pearl, 1988, Section 3.1.2). Then, we introduce undirected graphs and other representations for CI.

6.1 Graphoid

Let X, Y, and Z be three sets of random variables. A CI statement in the form of $X \perp\!\!\!\perp Y \mid Z$ defines a ternary relation among X, Y, and Z. Let us write this relation as $\langle X, Y \mid Z \rangle$.

Suppose X, Y, Z, and W are disjoint subsets of random variables from a joint distribution $\mathbb{P}$. Let $YW := Y \cup W$. The CI relation satisfies the following inference rules (Pearl, 1988, Section 3.1.2):

(C1) symmetry: $\langle X, Y \mid Z \rangle \Rightarrow \langle Y, X \mid Z \rangle$;
(C2) decomposition: $\langle X, YW \mid Z \rangle \Rightarrow \langle X, Y \mid Z \rangle$;
(C3) weak union: $\langle X, YW \mid Z \rangle \Rightarrow \langle X, Y \mid ZW \rangle$;
(C4) contraction: $\langle X, Y \mid Z \rangle \,\&\, \langle X, W \mid ZY \rangle \Rightarrow \langle X, YW \mid Z \rangle$.

If the joint density of $\mathbb{P}$ with respect to (w.r.t.) a product measure is positive and continuous, then we also have:

(C5) intersection: $\langle X, Y \mid ZW \rangle \,\&\, \langle X, W \mid ZY \rangle \Rightarrow \langle X, YW \mid Z \rangle$.

For example, (C2) means that if $X \perp\!\!\!\perp \{Y, W\} \mid Z$, then $X \perp\!\!\!\perp Y \mid Z$.

Definition 6.1 (Graphoid). A ternary relation $\langle A, B \mid C \rangle$ that satisfies (C1)–(C4) is called a *semi-graphoid*. If (C5) also holds, then it is called a *graphoid*. The inference rules (C1)–(C5) are called the graphoid axioms.

We provide three concrete examples of the graphoid. As stated above, the CI of a joint distribution with positive and continuous density is a graphoid. The second example is graph separation in an undirected graph. Let $\mathcal{G}$ be a graph over a vertex set V and $X, Y, Z \subseteq V$ be disjoint subsets of vertices. We say that Z separates X and Y if all paths from X to Y intersect Z (see Section 6.3 for a detailed discussion). Define a ternary relation $\langle X, Y \mid Z \rangle$ to represent that nodes Z separate X and Y. Then, one may show that this relation is also a graphoid (Pearl, 1988, Section 3.1.2). The third example is partial orthogonality. Let X, Y, and Z be disjoint sets of linearly independent vectors in $\mathbb{R}^n$. We say that they satisfy the relation $\langle X, Y \mid Z \rangle$ if $P_Z^\perp X$ is orthogonal to $P_Z^\perp Y$, where $P_Z^\perp X = (I_n - P_Z)X$ is the residual vector after projecting X onto $\mathrm{span}(Z)$. This relation is also a graphoid (Amini *et al.*, 2022b), and it is closely related to the orthogonal meet of linear subspaces (Lauritzen, 1996, Section 3.1).

The graph separation graphoid is particularly relevant to graphical models, as it provides an intuitive graphical interpretation for the CI axioms. A key idea in graphical models is to use various graph separations to represent conditional independences in a joint distribution. On the other hand, given data from the joint distribution, we may infer CI relations and then use them to learn the structure of the graphs.

Conditional independence is also useful for causal inference. We demonstrate this with a simple example (Pearl, 2000).

Consider an observational study on n individuals. Let $I = 1, \ldots, n$ indicate the individuals. Let X be a treatment received by the n individuals and Y be the outcome. We want to examine the potential causal effect of X on Y for each individual, which can be carried out by testing if $Y \perp\!\!\!\perp X \mid I$. We have observed data $\{(x_i, y_i) : i \in [n]\}$. Obviously, we cannot use the data to do the test since, conditioning on $I = i$ for any i, there is only one data point (x_i, y_i).

However, suppose $Z = Z(I)$ is a set of sufficient covariates such that $Y \perp\!\!\!\perp I \mid (X, Z)$. Then, we can show that

$$Y \perp\!\!\!\perp X \mid I \Leftrightarrow Y \perp\!\!\!\perp X \mid Z. \tag{6.1}$$

The second CI relation is testable based on the data $\{(x_i, y_i, z_i) : i \in [n]\}$. The following proof uses the CI axioms.

Proof. First, note that $Y \perp\!\!\!\perp X \mid I \Leftrightarrow Y \perp\!\!\!\perp X \mid (I, Z)$ because $Z = Z(I)$. Therefore, it suffices to show that

$$Y \perp\!\!\!\perp X \mid (I, Z) \Leftrightarrow Y \perp\!\!\!\perp X \mid Z. \tag{6.2}$$

(i) To show $\Leftarrow$: a sufficient set, i.e., $Y \perp\!\!\!\perp I \mid (X, Z)$, and the right-hand side of (6.2) imply $Y \perp\!\!\!\perp (I, X) \mid Z$ by (C4), which then implies $Y \perp\!\!\!\perp X \mid (I, Z)$ by (C3).
(ii) To show $\Rightarrow$: a sufficient set and the left-hand side of (6.2) imply $Y \perp\!\!\!\perp (X, I) \mid Z$ by (C5) and thus $Y \perp\!\!\!\perp X \mid Z$ by (C2).

$\square$

A graphoid is called a *compositional graphoid* if it additionally satisfies:

(C6) composition: $\langle X, Y \mid Z \rangle \,\&\, \langle X, W \mid Z \rangle \Rightarrow \langle X, YW \mid Z \rangle$.

Examples of compositional graphoids include graph separation in an undirected graph, the CI of a regular Gaussian distribution, and partial orthogonality among linearly independent vectors. For a more detailed discussion of these examples, see Amini *et al.* (2022b) and the references therein.

6.2 Conditional Independence Tests

To learn the structure of a graphical model, a common approach is to infer conditional independence relations from observed data, which is done through conditional independence tests. In general, it is sufficient to test whether two random variables X and Y are conditionally independent given a set of other random variables S. In this test, the null hypothesis is

$$H_0 : X \perp\!\!\!\perp Y \mid S.$$

We focus on CI tests on Gaussian data and discrete data.

If (X, Y, S) follows a joint Gaussian distribution, then H_0 is equivalent to the partial correlation $\mathrm{cor}(X, Y \mid S) = 0$. This test can be performed as follows:

(a) Calculate the sample covariance matrix $\widehat{\Sigma}$ from the observed data of (X, Y, S). If $|S| = d$, $\widehat{\Sigma}$ will be a $(d + 2) \times (d + 2)$ matrix.

(b) Let $\widehat{\Omega} = (\omega_{ij}) = \widehat{\Sigma}^{-1}$ and $\widehat{\rho}_{XY|S} = -\omega_{12}/\sqrt{\omega_{11}\omega_{22}}$, which is the estimated partial correlation coefficient.

(c) Apply the Fisher z-transformation:

$$z(X,Y|S) = \frac{1}{2}\log\left(\frac{1+\widehat{\rho}_{XY|S}}{1-\widehat{\rho}_{XY|S}}\right). \tag{6.3}$$

Then, $\sqrt{n-|S|-3}\cdot z(X,Y|S)\mid H_0 \sim \mathcal{N}(0,1)$ approximately when n is large. We perform the CI test according to this normal distribution.

If X, Y, and S are all finite discrete random variables, let $|X|, |Y|$, and $|S|$ be the corresponding number of possible values for these random variables. Note that if S consists of multiple random variables, $|S|$ is the number of all possible values for the combination of these discrete variables. We first calculate the contingency table for (X, Y) given $S = s$, for each value s. Denote the counts in this table by $(O_{xys})_{|X|\times|Y|}$, which is the number of data points with $X = x$, $Y = y$, and $S = s$. Under H_0, $X \perp\!\!\!\perp Y \mid S = s$, from which we can find the expected counts E_{xys} for $X = x$ and $Y = y$ given $S = s$ from (O_{xys}). Then, we calculate the G^2 statistic for $S = s$:

$$G^2(X,Y;S=s) = 2\sum_{x,y} O_{xys}\log(O_{xys}/E_{xys}),$$

which follows approximately $\chi^2_{(|X|-1)(|Y|-1)}$, with the degree of freedom $(|X|-1)(|Y|-1)$, when the data size n is large. We combine the G^2 statistics for all s to calculate the overall G^2 statistic:

$$G^2(X,Y;S) = \sum_s G^2(X,Y;S=s), \tag{6.4}$$

which follows $\chi^2_{(|X|-1)(|Y|-1)|S|}$ under the null hypothesis H_0.

For some applications, we may assume that

$$X = f(Z) + \varepsilon_X \quad \text{and} \quad Y = g(Z) + \varepsilon_Y,$$

where f and g are possibly nonlinear functions and the error variables $\varepsilon_X, \varepsilon_Y$ are independent of Z. Then, the CI of X and Y given Z is equivalent to $\varepsilon_X \perp\!\!\!\perp \varepsilon_Y$. Thus, the CI test under this model is translated to an independence test between the residuals after nonlinear regression of X and Y onto Z. General and flexible independence test methods (Gretton *et al.*,

2007; Pfister *et al.*, 2018; Shah and Peters, 2020) may be employed to carry out the CI test for this model.

6.3 Undirected Graphs

We review the definition and some terminology of undirected graphs, which are the focus of this chapter. As mentioned in Section 6.1, graphical models use graph separations to represent conditional independences among a set of random variables, in which the vertices of a graph are associated with the random variables. Implications of CI from graph separations are precisely defined by various Markov properties, which will be discussed in the following section.

Definition 6.2 (Undirected graph). A graph $\mathcal{G} = (V, E)$ is defined by a set of vertices (or nodes) $V = \{1, \ldots, p\}$ and a set of edges $E \subseteq V \times V$. If the edges are undirected, then $\mathcal{G}$ is an undirected graph, for which we write an edge $(i, j) \in E$ as $i - j$.

Throughout this chapter, we consider undirected graphs unless otherwise noted, and we regard E as a set of *unordered pairs*. Given a graph $\mathcal{G} = (V, E)$, two vertices i and j are *neighbors* if $(i, j) \in E$. Let ne(i) denote the set of neighbors of i, i.e.,

$$\text{ne}(i) := \{j \in V : (i, j) \in E\}.$$

A *path* of length n from i to j is a sequence $a_0 = i, \ldots, a_n = j$ of distinct vertices so that $(a_{k-1}, a_k) \in E$ for all $k = 1, \ldots, n$. A subset $C \subseteq V$ *separates* two vertices a and b in $\mathcal{G}$ if all paths from a to b intersect C. The subset C separates two subsets of vertices A and B if C separates a and b for every $a \in A$ and $b \in B$, in which case we write $A - C - B$.

Let $N(i) = \text{ne}(i) \cup \{i\}$. By definition, ne$(i)$ separates i and $V \setminus N(i)$ because every path from i to any $j \in V \setminus N(i)$ must pass through a neighbor of i and thus intersect ne(i).

Example 6.1. The graph in Figure 6.1 over $V = \{1, \ldots, 5\}$ has six edges. There are several paths from node 1 to node 5, such as $(1, 2, 4, 5)$ and $(1, 3, 2, 4, 5)$. The neighbor set of node 1 is ne$(1) = \{2, 3\}$, and $\{2, 3\}$ indeed separates node 1 from $\{4, 5\}$. The following list contains all subsets that separate nodes 1 and 5: $\{2, 3\}, \{4\}, \{2, 4\}, \{3, 4\}, \{2, 3, 4\}$.

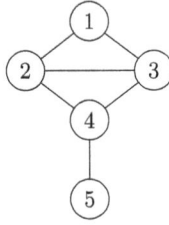

Fig. 6.1. An example undirected graph.

6.4 Markov Properties

To construct a graphical model, we associate the vertex set V of a graph $\mathcal{G}$ to random variables X_i, $i = 1, \ldots, p$, with joint distribution $\mathbb{P}$, and use graph separation to represent conditional independence among $X_1, \ldots, X_p$ through different Markov properties. Then, $(\mathcal{G}, \mathbb{P})$ is called a graphical model. The probability spaces for X_i's are either real finite-dimensional vector spaces or discrete sets. We often use node i and X_i interchangeably.

6.4.1 *Definitions*

We follow the definitions of Markov properties in Lauritzen (1996, Section 3.2).

Definition 6.3 (Markov properties on undirected graphs). Let $\mathcal{G} = (V, E)$ be an undirected graph and $\mathbb{P}$ be a joint distribution over V. We say that $\mathbb{P}$ satisfies:

(P) the pairwise Markov property w.r.t. $\mathcal{G}$ if, for any two vertices $i \neq j \in V$,

$$(i, j) \notin E \Rightarrow i \perp\!\!\!\perp j \mid V \setminus \{i, j\};$$

(L) the local Markov property w.r.t. $\mathcal{G}$ if, for any $i \neq j$,

$$(i, j) \notin E \Rightarrow i \perp\!\!\!\perp j \mid \operatorname{ne}(i);$$

(G) the global Markov property w.r.t. $\mathcal{G}$ if, for any disjoint $A, B, C \subseteq V$,

$$A - C - B \Rightarrow A \perp\!\!\!\perp B \mid C.$$

All the Markov properties in Definition 6.3 are one-way implications, from a graph separation statement to a CI relation in the joint distribution.

Since CI is closely related to how a joint density factorizes into conditional and marginal densities, there is another Markov property on factorization of the joint distribution according to cliques in the graph. A subset $C \subseteq V$ is a *clique* if the subgraph on C is complete, i.e., there is an edge between every pair of nodes in C. A clique that is maximal w.r.t. $\subset$ is called a *maximal clique*. Thus, if C is a maximal clique, then there is no $C' \supset C$ that is also a clique. In the graph in Figure 6.1, $\{1, 2\}$ is a clique but not a maximal clique, $\{1, 2, 3\}$ is a maximal clique, and $\{1, 2, 3, 4\}$ is not a clique.

Definition 6.4 (Factorization via cliques). We say $\mathbb{P}$ factorizes according to $\mathcal{G}$, denoted by (F), if, for every maximal clique $A \subseteq V$, there exists a function $\psi_A(x_A) \geq 0$ such that the joint density of $\mathbb{P}$ has the form

$$f(x) = \prod_{A \in \mathcal{C}} \psi_A(x_A),$$

where $\mathcal{C}$ is the set of maximal cliques of $\mathcal{G}$.

The relations among the four Markov properties are summarized in the following theorem (Lauritzen, 1996, Proposition 3.8).

Theorem 6.1. *For any undirected graph $\mathcal{G}$ and any probability distribution P over V, we have*

$$(F) \Rightarrow (G) \Rightarrow (L) \Rightarrow (P).$$

A complete proof was given by Lauritzen (1996). We leave it as an exercise to show that $(G) \Rightarrow (L) \Rightarrow (P)$.

6.4.2 *Examples*

We demonstrate how to use Markov properties to infer CI relations with a few examples.

Example 6.2 (Markov chain). Suppose X_t, $t = 1, \ldots, n$, is a Markov chain. Then, the joint distribution is

$$\mathbb{P}(X_1, \ldots, X_n) = \mathbb{P}(X_1)\mathbb{P}(X_2 \mid X_1) \cdots \mathbb{P}(X_n \mid X_{n-1}).$$

Let $\mathcal{G}$ be the graph in Figure 6.2, of which the set of maximal cliques is

$$\mathcal{C} = \{\{i, i+1\} : i = 1, \ldots, n-1\}.$$

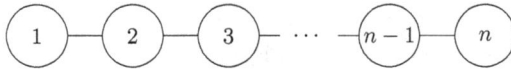

Fig. 6.2. A graphical model for Markov chains.

We see that (F) holds since

$$\mathbb{P}(X_1,\ldots,X_n) = \psi_1(X_1,X_2)\cdots\psi_{n-1}(X_{n-1},X_n),$$

where $\psi_1(X_1,X_2) = \mathbb{P}(X_1)\mathbb{P}(X_2 \mid X_1)$ and $\psi_{i-1}(X_{i-1},X_i) = \mathbb{P}(X_i \mid X_{i-1})$ for $i \geq 3$. By Theorem 6.1, (F) implies (G). Therefore, using the global Markov property (G), we conclude that $X_i \perp\!\!\!\perp X_k \mid X_j$ for any $i < j < k$, since $i - j - k$ (j separates i and k). This gives more CI relations than those in the definition of a Markov chain, $X_t \perp\!\!\!\perp X_i \mid X_{t-1}$, for $i < t - 1$, demonstrating the usefulness of a graphical model.

Example 6.3 (Hidden Markov model). Let us consider a hidden Markov model (HMM), $\{(Z_t,Y_t) : t = 1,\ldots,n\}$, where Z_t is the hidden state and Y_t is the observation (Section 4.1). The joint probability $\mathbb{P}(Y,Z)$ factorizes according to (4.3), i.e.,

$$\mathbb{P}(Y,Z) = \mathbb{P}(Z_1)\mathbb{P}(Y_1 \mid Z_1)\mathbb{P}(Z_2 \mid Z_1)\mathbb{P}(Y_2 \mid Z_2)$$
$$\times \cdots \times \mathbb{P}(Z_n \mid Z_{n-1})\mathbb{P}(Y_n \mid Z_n)$$
$$= \prod_{t=1}^{n-1} f_t(Z_t, Z_{t+1}) \prod_{t=1}^{n} g_t(Z_t, Y_t).$$

Based on this factorization, we can represent the model by the graph in Figure 6.3, where we create a clique of the variables in each function, f_t and g_t. Accordingly, the maximal cliques are $\{Z_t, Z_{t+1}\}, t = 1,\ldots,n-1$ and $\{Z_t, Y_t\}, t = 1,\ldots,n$. By such a construction, the factorization property (F) holds, and thus (G) also holds. Now, by graph separation, we see that V_{t-i}, Y_t, and V_{t+j} are mutually independent conditional on Z_t, for $i, j \geq 1$, where $V_k \in \{Y_k, Z_k\}$.

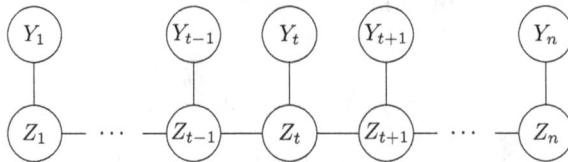

Fig. 6.3. A graphical model for HMMs.

6.4.3 *Equivalence*

We have introduced four Markov properties. A natural question is, under what conditions these Markov properties are equivalent. The answer is given by the following theorem (Lauritzen, 1996, Theorem 3.9).

Theorem 6.2. *If* $\mathbb{P}$ *has a positive and continuous density* f *w.r.t. a product measure, then* (F) $\Leftrightarrow$ (P).

The conclusion implies (F) $\Leftrightarrow$ (G) $\Leftrightarrow$ (L) $\Leftrightarrow$ (P), in light of Theorem 6.1. If $X_j \in \mathbb{R}$ is real, its density is defined w.r.t. the Lebesgue measure. For discrete random variables, the density is defined w.r.t. the counting measure. The more essential condition is that f is positive, while the continuity of f may be relaxed considerably.

To better understand the condition on f, let us discuss a few examples. Suppose that the domain of (X_1, X_2) is $[0, 1] \times [0, 1]$. If the density $f(x_1, x_2)$ is positive and continuous w.r.t. the Lebesgue measure over $[0, 1] \times [0, 1]$, which is a product measure, then the condition in Theorem 6.2 holds. Consider that both X_1 and X_2 are discrete, taking values in $\mathcal{X}_1$ and $\mathcal{X}_2$, respectively. With respect to the counting measure over $\mathcal{X}_1 \times \mathcal{X}_2$, the density f is the probability mass function. Then, the condition in Theorem 6.2 is satisfied as long as $f(x_1, x_2) > 0$ for all $(x_1, x_2) \in \mathcal{X}_1 \times \mathcal{X}_2$. That is, the probability for every possible combination of x_1 and x_2 is positive. It is helpful to consider a counterexample. Suppose X_1 and X_5 follow Bern(0.5) independently, $X_2 = X_1$, $X_4 = X_5$, and $X_3 = X_2 X_4$. This defines a joint distribution $\mathbb{P}$ over five binary random variables $(X_1, \ldots, X_5) \in \{0, 1\}^5$. Let $\mathcal{G}$ be a chain with edge set $E = \{(i, i+1) : i = 1, \ldots, 4\}$. Then, one can verify that (L) holds but (G) does not. For example, $X_2 \perp\!\!\!\perp X_4 \mid (X_1, X_3)$ [by (L)] is true, while $X_2 \perp\!\!\!\perp X_4 \mid X_3$ [by (G)] is false! This is because the probability mass function f is not positive on all possible values of X_i's due to the deterministic relations, e.g., $\mathbb{P}(X_1 = 1, X_2 = 0) = 0$.

6.5 Conditional Independence Graphs

A commonly used undirected graphical model is the conditional independence graph (CIG), which satisfies the pairwise Markov property (P). For a set V and $i, j \in V$, let $V_{-ij} := V \setminus \{i, j\}$.

Definition 6.5 (CIG). A CIG is a graphical model $(\mathcal{G}, \mathbb{P})$ such that the pairwise Markov property (P) holds. That is, for any $i \neq j \in V$,

$$(i, j) \notin E \Rightarrow i \perp\!\!\!\perp j \mid V_{-ij}.$$

A sparser graph $\mathcal{G}$ with fewer edges implies more CI relations in $\mathbb{P}$. Thus, the convention is to define the CIG as the sparsest $\mathcal{G}$ such that (P) holds, which effectively replaces $\Rightarrow$ with $\Leftrightarrow$ in the above definition. If the joint distribution $\mathbb{P}$ satisfies the condition in Theorem 6.2, then we immediately know that the other three stronger Markov properties, (F), (G), and (L), also hold for $(\mathcal{G}, \mathbb{P})$.

Assuming that we have observed data from $\mathbb{P}$, then one may estimate the structure, i.e., the edge set, of $\mathcal{G}$. The estimated graph encodes a set of CI relations learned from the data, which is a main goal of structure learning of graphical models. In this section, we discuss the CIGs for two models: the joint Gaussian distribution and the discrete Ising model.

6.5.1   *Gaussian graphical models*

The Gaussian graphical model (GGM) refers to a CIG for a joint Gaussian distribution, $\mathbb{P} = \mathcal{N}_p(0, \Sigma)$, where the covariance $\Sigma \succ 0$ is positive definite. The pairwise CI relations of $\mathcal{N}_p(0, \Sigma)$ are given by the sparsity pattern of the inverse covariance matrix $\Theta = (\theta_{jk})_{p \times p} = \Sigma^{-1}$. The matrix Θ is also called the precision matrix.

Lemma 6.1. *Suppose* $(X_1, \ldots, X_p) \sim \mathcal{N}_p(0, \Sigma)$, *with* $\Sigma \succ 0$, *and let* $\Theta = (\theta_{jk})_{p \times p} = \Sigma^{-1}$. *Then,*

$$X_j \perp\!\!\!\perp X_k \mid X_{-jk} \Leftrightarrow \theta_{jk} = 0, \tag{6.5}$$

for all $j \neq k \in [p]$.

Proof. The element θ_{jk} of Θ is connected to the partial correlation ρ_{jk} between X_j and X_k given X_{-jk}:

$$\rho_{jk} = -\theta_{jk} / \sqrt{\theta_{jj} \theta_{kk}}.$$

Recall that ρ_{kj} is the correlation calculated from the conditional covariance matrix $\Sigma_{(j,k)|V_{-jk}} = \mathrm{Cov}(X_j, X_k \mid X_{-jk})$. Since $[X_j, X_k \mid X_{-jk}]$ is a Gaussian distribution,

$$X_j \perp\!\!\!\perp X_k \mid X_{-jk} \Leftrightarrow \rho_{jk} = 0 \Leftrightarrow \theta_{jk} = 0.$$

This completes the proof of this lemma. $\qquad\square$

According to (6.5), let us construct a graph $\mathcal{G}$ with the edge set

$$E = \{(j,k) : \theta_{jk} \neq 0\}. \tag{6.6}$$

Then, the pairwise Markov property (P) holds, and thus $\mathcal{G}$ is a CIG. Since $\mathcal{N}_p(0, \Sigma)$ has a continuous and positive density, (L), (G), and (F) also hold. One can verify (F) directly as well, showing that the joint Gaussian density factorizes according to the cliques in $\mathcal{G}$.

Example 6.4. Given the zero pattern of Θ, the graph $\mathcal{G}$ constructed using (6.6) is shown in Figure 6.4.

Suppose we wish to find all S such that $X_1 \perp\!\!\!\perp X_5 \mid S$. By the global Markov property (G), we find all S that separates nodes 1 and 5 in the graph $\mathcal{G}$. The collection of such S is $\{\{2,3\}, \{4\}, \{2,4\}, \{3,4\}, \{2,3,4\}\}$.

Learning GGMs from data is an active research area. Given data $x_i \overset{iid}{\sim} \mathcal{N}_p(0, \Sigma)$, $i = 1, \ldots, n$, our goal is to estimate the structure of $\mathcal{G}$, which is equivalent to estimating the support of Θ:

$$\mathrm{supp}(\Theta) := \{(j,k) : \theta_{jk} \neq 0\}.$$

Dempster (1972) developed a simple method, called covariance selection, which thresholds the MLE of Θ to estimate its support. Setting $\mu = 0$ in (2.11), the log-likelihood of Σ is

$$\ell(\Sigma) = -\frac{n}{2} \log \det(\Sigma) - \frac{1}{2} \mathrm{tr}(S\Sigma^{-1}), \tag{6.7}$$

where $S = \sum_i x_i x_i^{\mathsf{T}}$ is a $p \times p$ matrix and a sufficient statistic for Σ. The MLE $\widehat{\Sigma}^{\mathrm{MLE}} = S/n$ always exists. If $n > p$, S is positive definite almost

$$\Theta = \begin{bmatrix} * & * & * & 0 & 0 \\ * & * & * & * & 0 \\ * & * & * & * & 0 \\ 0 & * & * & * & * \\ 0 & 0 & 0 & * & * \end{bmatrix}$$

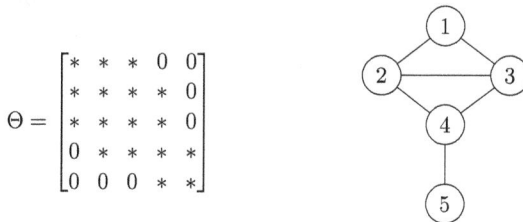

Fig. 6.4. Graph (right) constructed according to the zero pattern of Θ (left).

surely. We invert $\widehat{\Sigma}^{\mathrm{MLE}}$ to obtain the MLE of Θ, $\widehat{\Theta}^{\mathrm{MLE}} = (\widehat{\Sigma}^{\mathrm{MLE}})^{-1}$. Then, construct $\widehat{\mathcal{G}}$ by thresholding:

$$\widehat{E} = \{(j,k) : |\widehat{\theta}_{jk}^{\mathrm{MLE}}| > \tau\},$$

where $\widehat{\theta}_{jk}^{\mathrm{MLE}}$ is the (j,k)th element of $\widehat{\Theta}^{\mathrm{MLE}}$. Let $\Theta_0 = (\theta_{jk}^*)$ denote the true precision matrix and E_0 be the true edge set of the GGM. Under the classical asymptotic setting, assuming Θ_0 is fixed, one may choose a proper τ such that $\widehat{E}$ perfectly recovers E_0 with probability approaching one as $n \to \infty$. Let $\gamma = \min\{|\theta_{jk}^*| : (j,k) \in E_0\}$, which is a positive constant. Since

$$\widehat{\theta}_{jk}^{\mathrm{MLE}} - \theta_{jk}^* = O_p(n^{-1/2}) \qquad \text{for any } (j,k),$$

as long as $n^{-1/2} \ll \tau < \gamma/2$, we have $\mathbb{P}(\widehat{E} = E_0) \to 1$ as $n \to \infty$.

If $p > n$, $\widehat{\Sigma}^{\mathrm{MLE}}$ is not invertible and the MLE $\widehat{\Theta}^{\mathrm{MLE}}$ does not exist. However, we may find a sparse estimate of Θ using ℓ_1 regularization (Yuan and Lin, 2007; Friedman *et al.*, 2008; Banerjee *et al.*, 2008). Define the element-wise ℓ_1 norm, $\|\Theta\|_1 := \sum_{j<k} |\theta_{jk}|$, and the ℓ_1-regularized loss,

$$
\begin{aligned}
f(\Theta) &= -\frac{2}{n}\ell(\Theta^{-1}) + \lambda\|\Theta\|_1 \\
&= -\log\det(\Theta) + \mathrm{tr}(\widehat{\Sigma}^{\mathrm{MLE}}\Theta) + \lambda\|\Theta\|_1,
\end{aligned}
\tag{6.8}
$$

where $\ell(\cdot)$ is the log-likelihood function (6.7) and $\lambda > 0$ is a tuning parameter. The ℓ_1-regularized estimate is

$$\widehat{\Theta} = \operatorname*{argmin}_{\Theta \succ 0} f(\Theta),$$

which is also called the graphical lasso (Friedman *et al.*, 2008). The loss f is convex, and there are efficient algorithms to minimize (6.8). The minimizer $\widehat{\Theta} = (\widehat{\theta}_{jk})$ is well defined even for $p > n$. In general, the solution is sparse, i.e., $\widehat{\theta}_{jk} = 0$ for some (j,k). This naturally leads to an estimated graph with $\widehat{E} = \{(j,k) : \widehat{\theta}_{jk} \neq 0\}$. Under quite strong assumptions, including an irrepresentability condition, it has been shown that $\mathbb{P}(\widehat{E} = E_0) \to 1$. Instead, we may consider thresholding $\widehat{\Theta}$ to construct $\widehat{E} = \{(j,k) : |\widehat{\theta}_{jk}| > \tau\}$. Under somewhat weaker conditions, with high probability, the operator norm error

$$\|\widehat{\Theta} - \Theta_0\|_2 < C\sqrt{d^2\log p/n},\tag{6.9}$$

where d is the maximum degree of $\mathcal{G}$ and $C > 0$ is a constant. See Ravikumar *et al.* (2011) for more details. Based on this result, assuming a certain beta-min condition on γ, one can show that $\mathbb{P}(\widehat{E} = E_0) \to 1$ if $d^2\log p \ll n$, under a high-dimensional asymptotic scaling $p \gg n \to \infty$. In practice, the tuning parameter λ is usually chosen via cross-validation or a model selection criterion.

In addition to the above likelihood-based, or score-based, learning methods, we may also estimate $\mathcal{G}$ through neighborhood regression, i.e., regressing X_j on X_{-j}. Since $X \sim \mathcal{N}_p(0, \Sigma)$, the conditional distribution $[X_j \mid X_{-j}]$ can be represented by a linear regression model:

$$X_j = \sum_{k \neq j} \beta_{kj} X_k + \varepsilon_j, \tag{6.10}$$

where the regression coefficient $\beta_{kj} = -\theta_{jk}/\theta_{jj}$ and ε_j follows a normal distribution. By symmetry, $\beta_{jk} = -\theta_{kj}/\theta_{kk}$. Thus, we have

$$(j, k) \notin E \Leftrightarrow \theta_{jk} = 0 \Leftrightarrow \beta_{kj} = \beta_{jk} = 0. \tag{6.11}$$

This suggests that we may learn the edge set by estimating the support of $\beta_j = (\beta_{kj})$, for $j \in [p]$. This approach is called the neighborhood regression method, which can be regarded as using CI constraints to infer the graph. Let $S_j = \text{supp}(\beta_j)$. Under the linear model (6.10),

$$X_j \perp\!\!\!\perp \{X_k, k \notin S_j\} \mid X_{S_j}.$$

In this sense, finding the support of β_j can be regarded as a way to learn CI relations, which are then used to construct the graph such that $\text{ne}(j) = S_j$. We summarize the neighborhood regression method for undirected graphs:

(i) Apply a model selection method to find the support for each neighborhood regression (6.10). Let $\widehat{S}_j$, $j \in [p]$, be the estimated supports.
(ii) Combine $\widehat{S}_j$'s to define $\widehat{\mathcal{G}}$ via either the AND or the OR rule:

$$\widehat{E} = \{(j, k) : k \in \widehat{S}_j \text{ and } j \in \widehat{S}_k\} \text{ or } \widehat{E} = \{(j, k) : k \in \widehat{S}_j \text{ or } j \in \widehat{S}_k\}.$$

For GGMs, one may use the lasso (Tibshirani, 1996) to estimate the support in (6.10). Such a constructed $\widehat{E}$ recovers the true edge set E_0 with high probability under a high-dimensional scaling $p \gg n \to \infty$ (Meinshausen and Bühlmann, 2006), although the assumptions are restrictive in nature.

6.5.2 *Discrete graphical models*

A class of undirected graphical models for discrete variables is the Ising model. Let $X_i \in \{-1, +1\}$, for all $i \in V = [p]$. Given a graph $\mathcal{G} = (V, E)$, we define a joint distribution:

$$\mathbb{P}(x_1, \ldots, x_p \mid \theta) = \frac{1}{Z(\theta)} \exp \left\{ \sum_{i \in V} \theta_i x_i + \sum_{(j,k) \in E} \theta_{jk} x_j x_k \right\}, \tag{6.12}$$

where the parameter vector $\theta = [(\theta_i), (\theta_{jk})] \in \mathbb{R}^{p+|E|}$ and $Z(\theta) > 0$ is the normalizing constant. The parameter θ_i models the marginal probability $\mathbb{P}(X_i = x)$, $x \in \{-1, 1\}$, while θ_{jk} models the joint probability for X_i and X_k. It is easy to verify that (F) holds, which implies (G), (L), and (P). As an example application, we may model social networks using (6.12), in which X_i is a binary outcome for an individual in the network. If $\theta_{jk} > 0$, then the two individuals j and k, connected by an edge $(j, k) \in E$, tend to show identical outcomes, i.e., $X_j = X_k$.

Example 6.5. Given the $\mathcal{G}$ in Figure 6.5, define $\mathbb{P}(x_1, \ldots, x_6)$ as in (6.12). Denote the maximal cliques $\{1, 2, 3\}, \{1, 2, 6\}, \{1, 4\}, \{1, 5\}$ by $\mathcal{C}_i$, $i = 1, 2, 3, 4$. We see that

$$\mathbb{P}(x_1, \ldots, x_6) = \prod_{i=1}^{4} \psi_i(x_{\mathcal{C}_i}),$$

which verifies (F). We may read off CI statements by (G), such as

$$X_4 \perp\!\!\!\perp X_5 \mid X_1, \quad X_3 \perp\!\!\!\perp X_6 \mid \{X_1, X_2\},$$
$$\{X_2, X_3, X_6\} \perp\!\!\!\perp \{X_4, X_5\} \mid X_1.$$

The Ising model may be generalized to model the finite discrete variables $X_i \in \{1, \ldots, m\}, i \in V = [p]$. Given an undirected graph $\mathcal{G} = (V, E)$, we define a joint distribution,

$$\mathbb{P}(x_1, \ldots, x_p \mid \gamma, \theta) =$$

$$\frac{1}{Z(\gamma, \theta)} \exp \left\{ \sum_{i \in V} \sum_{z=1}^{m} \gamma_{iz} I(x_i = z) + \sum_{(j,k) \in E} \theta_{jk} I(x_j = x_k) \right\},$$

parameterized by γ and θ. Similar to the Ising model (6.12), γ_{iz} models the marginal probabilities of X_i, while θ_{jk} models the joint probability between X_j and X_k.

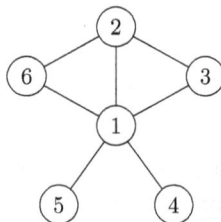

Fig. 6.5. An example graph for Ising model.

To simplify the description, we focus on structure learning of graphs under the Ising model. In principle, one may estimate θ via a likelihood-based approach. However, one challenge with this approach is that the normalizing constant $Z(\theta)$ does not have a closed-form expression, which makes it difficult to find the MLE. A more practical approach is to perform neighborhood regression. From (6.12), we can work out the conditional distribution $[X_i \mid X_{-i}]$, which turns into a logistic regression model for X_i:

$$\log\left[\frac{\mathbb{P}(X_i = 1 \mid X_{-i})}{\mathbb{P}(X_i = -1 \mid X_{-i})}\right] = 2\theta_i + \sum_{j \in \text{ne}(i)} 2\theta_{ij} X_j, \qquad (6.13)$$

where $\text{ne}(i) = \{j \in V : (i,j) \in E\}$ is the set of neighbors of node i in $\mathcal{G}$. This motivates the following neighborhood regression learning method, similar to that for the GGM:

(i) For each $i \in V$, apply logistic regression X_i on X_{-i} with variable selection to learn $\widehat{N}(i)$, the estimated neighbor set of X_i.
(ii) Combine $\{\widehat{N}(i) : i \in V\}$ to construct $\widehat{\mathcal{G}}$ with an estimated edge set,

$$\widehat{E} = \left\{(i,j) : j \in \widehat{N}(i) \text{ and } i \in \widehat{N}(j)\right\},$$

using the AND rule. One may also replace it with the OR rule.

Step (i) can be done with ℓ_1-regularized logistic regression or BIC stepwise selection. Using ℓ_1-regularized logistic regression, the neighborhood regression approach is consistent in structure learning, i.e., $\widehat{\mathcal{G}} = \mathcal{G}$ with high probability (Ravikumar *et al.*, 2010). The sample size requirement is $n = \Omega(d^2 \log p)$, where d is the maximum degree of the true graph (Bento and Montanari, 2009). Here, $a_n = \Omega(b_n)$ if and only if $b_n = O(a_n)$.

6.6 Faithfulness

For a graphical model $(\mathcal{G}, \mathbb{P})$, the distribution $\mathbb{P}$ satisfies certain Markov properties, say (G), w.r.t. the graph $\mathcal{G}$. Then, graph separation implies condition independence but not the other way around. If $\mathbb{P}$ is faithful to $\mathcal{G}$, then CI implies graph separation as well. In this case, we have a one-to-one correspondence between graph separations and CI relations.

Definition 6.6 (Faithfulness). For a graphical model $(\mathcal{G}, \mathbb{P})$, we say that the distribution $\mathbb{P}$ is faithful to the graph $\mathcal{G}$ if, for every triple of disjoint sets $A, B, S \subseteq V$,

$$A \perp\!\!\!\perp B \mid S \text{ in } \mathbb{P} \Leftrightarrow S \text{ separates } A \text{ and } B \text{ in } \mathcal{G}. \tag{6.14}$$

Remark 6.1. When (6.14) holds, $\mathbb{P}$ is sometimes said to be *perfect* w.r.t. $\mathcal{G}$. In this book, we use the two terms faithfulness and perfectness interchangeably.

Faithfulness is a common assumption for structure learning of graphical models, which makes use of CI relations learned from data to infer edges of the underlying graph. A natural question is how likely it is that $\mathbb{P}$ is faithful. We address this question in its most general form in the following chapter after covering directed acyclic graphs and chain graphs. Here, we focus on GGMs, in which $\mathbb{P}$ is $\mathcal{N}_p(0, \Sigma)$. Given an undirected graph $\mathcal{G} = (V, E)$ over $V = [p]$, we say that Σ is *Markov* w.r.t. $\mathcal{G}$ if

$$\text{supp}(\Sigma^{-1}) = E \cup \{(i, i) : i \in [p]\}.$$

For almost all such Σ, the distribution $\mathcal{N}_p(0, \Sigma)$ is faithful to $\mathcal{G}$, in which case we simply say that Σ is faithful to $\mathcal{G}$. Theorem 1 in the work by Amini *et al.* (2022a) offers a precise statement, which is presented here:

Theorem 6.3. *For any undirected graph $\mathcal{G}$ on $[p]$, the set of positive definite matrices $\Theta \in \mathbb{R}^{p \times p}$ for which $\Sigma = \Theta^{-1}$ is Markov but not faithful w.r.t. $\mathcal{G}$ has a Lebesgue measure of zero.*

In this theorem, the Lebesgue measure is of dimension $d = |E| + p$.

Unfaithful Σ arises when Θ satisfies additional equality constraints, which may lead to CI relations in $\mathcal{N}_p(0, \Sigma)$ that are not implied by any graph separation in $\mathcal{G}$. Since equality constraints in general define a manifold of dimension $\leq d - 1$ in the space of Θ, the measure of such a manifold is zero.

6.7 Markov Blanket

The neighborhood regression approach for undirected graphical models is tightly connected to the estimation of a Markov blanket. Let $V = \{X_i\}$ be a set of random variables.

Definition 6.7 (Markov blanket). A *Markov blanket* of $X_i \in V$ is any subset $S \subseteq V_{-i} := V \setminus \{X_i\}$ such that

$$X_i \perp\!\!\!\perp (V_{-i} \setminus S) \mid S. \tag{6.15}$$

A *Markov boundary* is a minimal Markov blanket, i.e., none of its proper subset satisfies (6.15).

A classical general algorithm to find a Markov boundary (MB) is the grow-shrink (G-S) algorithm (Margaritis and Thrun, 1999).

Algorithm 6.1 (The G-S algorithm to find MB of $i \in V$).

1: $S \leftarrow \varnothing$.
2: **while** there is $j \in V_{-i}$ such that $j \not\perp i \mid S$ **do**
3: $\quad S \leftarrow S \cup \{j\}$. ▷ Growing phase
4: **end while**
5: **while** there is $j \in S$ such that $j \perp\!\!\!\perp i \mid S \setminus \{j\}$ **do**
6: $\quad S \leftarrow S \setminus \{j\}$. ▷ Shrinking phase
7: **end while**
8: $\mathrm{MB}(i) \leftarrow S$.

Remark 6.2. Algorithm 6.1 is presented as a population version which assumes perfect knowledge about the CI relations (CI oracle) among the variables. In practice, one would replace the CI statements with the corresponding CI tests performed on observed data.

Theorem 6.4. *Assume the CI of V is a compositional graphoid. Then, the* $\mathrm{MB}(i)$ *output by Algorithm 6.1 is a Markov boundary of i.*

Proof. When the first while loop is broken, we have $j \perp\!\!\!\perp i \mid S$ for any $j \notin S$. By the composition axiom (C6), this shows that S is a Markov blanket of i after the growing phase.

Consider the first time Line 6 is executed, which removes j from S. Since $S \cup \{j\}$ is a Markov blanket, for any $k \notin S \cup \{j\}$, we have

$$i \perp\!\!\!\perp k \mid \{S, j\}.$$

By the condition for the while loop (Line 5), $i \perp\!\!\!\perp j \mid S$. Now, apply the contraction axiom (C4) of CI:

$$i \perp\!\!\!\perp k \mid \{S, j\} \quad \& \quad i \perp\!\!\!\perp j \mid S \quad \Rightarrow \quad i \perp\!\!\!\perp \{k, j\} \mid S.$$

This means that S is still a Markov blanket of i after one iteration of the loop. By induction, S remains a Markov blanket when the shrinking phase is done.

Next, we show that no proper subset of S is a Markov blanket of i, which implies that $\mathrm{MB}(i)$ is a Markov boundary. Suppose that there is a proper subset $T \subset S$ such that T is a Markov blanket of i. Let $S \backslash T = \{k_1, \ldots, k_m\}$, for some $m \geq 1$. Then, $i \perp\!\!\!\perp \{k_1, \ldots, k_m\} \mid T$. The weak union (C3) implies that $i \perp\!\!\!\perp k_m \mid S \backslash \{k_m\}$, which is contradictory to the termination of the while loop on Line 5. □

If the conditional distribution of $[X_i \mid V_{-i}]$ corresponds to a regression model, such as for a joint Gaussian distribution or the Ising model, then the growing phase can be replaced with the lasso or ℓ_1-regularized logistic regression. The support of the regression coefficients is an estimated Markov blanket of X_i. The proof of Theorem 6.4 shows that the output of Algorithm 6.1 is a Markov boundary as long as the S produced by the growing phase is a Markov blanket of X_i.

For an undirected graph model $(\mathcal{G}, \mathbb{P})$, by the local Markov property, the neighbor set $\mathrm{ne}(i)$ is a Markov blanket of i, and it is the unique Markov boundary of i if $\mathbb{P}$ is faithful w.r.t. $\mathcal{G}$. Thus, neighborhood regression can be regarded as a method to find the Markov boundary of i.

6.8 Neighborhood Lattice

Let $\mathcal{I}(\mathbb{P})$ denote the set of CI relations of a distribution $\mathbb{P}$. For an undirected graph $\mathcal{G} = (V, E)$, denote the collection of all separation statements by $\mathcal{I}(\mathcal{G})$, which is also called the independence model induced by $\mathcal{G}$. It is easy to verify that $\mathcal{I}(\mathcal{G})$ is a compositional graphoid (Section 6.1), where $\langle A, B \mid C \rangle$ is defined by the separation statement $A - C - B$. Therefore, the set of CI relations of a distribution $\mathbb{P}$ that is faithful to $\mathcal{G}$ is also a compositional graphoid because $\mathcal{I}(\mathbb{P}) = \mathcal{I}(\mathcal{G})$.

Theorem 6.5. *If a distribution $\mathbb{P}$ is faithful to an undirected graph $\mathcal{G}$, then $\mathcal{I}(\mathbb{P})$ is a compositional graphoid.*

In fact, this result holds for all graphical models in the literature, including directed acyclic graphs, chain graphs, and acyclic directed mixed graphs, which will be discussed in the following few chapters. If $\mathbb{P}$ is faithful to any of these graphs, then $\mathcal{I}(\mathbb{P})$ is a compositional graphoid. See Lauritzen and Sadeghi (2018) for precise statements and a more detailed discussion.

In the absence of a faithful graph, Amini *et al.* (2022b) introduced the neighborhood lattice decomposition to represent $\mathcal{I}(\mathbb{P})$, under the assumption that $\mathcal{I}(\mathbb{P})$ is a compositional graphoid. There exists a distribution $\mathbb{P}$ for which $\mathcal{I}(\mathbb{P})$ is a compositional graphoid, but $\mathbb{P}$ is not faithful to a graph. For example, a Gaussian distribution $\mathcal{N}(0, \Sigma)$ may not be faithful to any graph, but its CI relations form a compositional graphoid as long as Σ is positive definite. Together with Theorem 6.5, this shows that the class of distributions that can be represented by neighborhood lattice decomposition is *strictly larger* than the class of distributions that are faithful to a graph.

Definition 6.8 (Complete lattice). A complete lattice is a partially ordered set, or *poset*, in which every subset has a supremum (join) and an infimum (meet).

Given a set V, the set of all subsets of V ordered by inclusion is a poset, denoted $\mathscr{B}(V)$. It is easy to see that $\mathscr{B}(V)$ is a complete lattice with the supremum and the infimum given by set union and intersection, respectively: for any $S, T \subseteq V$,

$$\sup\{S, T\} := S \cup T, \qquad \inf\{S, T\} := S \cap T.$$

An interval $[A, B] := \{S \subseteq V : A \subseteq S \subseteq B\}$ is also a complete lattice and thus a sublattice of $\mathscr{B}(V)$.

Let $V = [p]$ correspond to random variables $X_1, \ldots, X_p$ following a joint distribution $\mathbb{P}$. If $\mathcal{I}(\mathbb{P})$ is a compositional graphoid, then $A \perp\!\!\!\perp B \mid S$ if and only if $i \perp\!\!\!\perp j \mid S$ for all $i \in A$ and $j \in B$, because of axioms (C2) and (C6). This allows us to focus on CI of the form $i \perp\!\!\!\perp j \mid S$, for $i, j \in V$ and $S \subset V$. For any subset $S \subseteq V_{-j}$, define

$$\mathcal{T}_j(S) := \{U \subseteq S : j \perp\!\!\!\perp (S \setminus U) \mid U\}. \tag{6.16}$$

It follows from (C5) that the infimum of $\mathcal{T}_j(S)$ is well defined, denoted by $m(j; S) = \inf \mathcal{T}_j(S)$, and $m(j; S) \in \mathcal{T}_j(S)$, i.e.,

$$j \perp\!\!\!\perp (S \setminus m(j; S)) \mid m(j; S).$$

Note that $m(j; S)$ is the unique Markov boundary of j relative to S. By convention, $j \perp\!\!\!\perp \varnothing \mid S$ and thus $S \in \mathcal{T}_j(S)$. It is then not difficult to verify that $\mathcal{T}_j(S) = [m(j; S), S]$ is an interval.

Now, we are ready to introduce the neighborhood lattice.

Definition 6.9 (Neighborhood lattice). Let $j \in V$ and $S \subseteq V_{-j}$. Given a compositional graphoid $\mathcal{I}$ over V, the *neighborhood lattice* of j relative to S is

$$\mathscr{L}_j(S) := \{U \subseteq V_{-j} : m(j; U) = m(j; S)\}. \tag{6.17}$$

Remark 6.3. The neighborhood lattice is defined for a general compositional graphoid $\mathcal{I}$, not limited to conditional independence of a distribution. In this context, a CI statement $A \perp\!\!\!\perp B \mid C$ is understood as the ternary relation $\langle A, B \mid C \rangle$ that defines the compositional graphoid $\mathcal{I}$.

The intuition behind the neighborhood lattice can be understood using the linear regression of $X_j \sim X_S$ as an example. The support of the regression coefficient vector is $m(j; S)$. The lattice $\mathscr{L}_j(S)$ consists of all sets U for which the support of $X_j \sim X_U$ is $m(j; S)$.

The main properties of the neighborhood lattice, established by Amini *et al.* (2022b), are summarized in the following two theorems.

Theorem 6.6. *The neighborhood lattice $\mathscr{L}_j(S)$ on a composition graphoid $\mathcal{I}$ over V is an interval and thus a complete sublattice of $\mathscr{B}(V_j)$, with infimum and supremum given by*

$$\inf \mathscr{L}_j(S) = m(j; S),$$
$$\sup \mathscr{L}_j(S) = m(j; S) \cup \{ k \in V_{-j} : j \perp\!\!\!\perp k \mid m(j; S) \}.$$

Furthermore, the neighborhood lattices $\{\mathscr{L}_j(S)\}$ partition the subset lattice $\mathscr{B}(V_j)$:

$$\mathscr{B}(V_j) = \biguplus \{ \mathscr{L}_j(m) : m = m(j; T) \text{ for some } T \subseteq V_{-j} \}. \qquad (6.18)$$

Let $M(j; S) = \sup \mathscr{L}_j(S)$. Then, the neighborhood lattice is

$$\mathscr{L}_j(S) = [m(j; S), M(j; S)] = \{ U : m(j; S) \subseteq U \subseteq M(j; S) \}.$$

Since the infimum is unique, we may index the neighborhood lattice $\mathscr{L}_j(S)$ by its infimum $m = m(j; S)$. The collection $\{\mathscr{L}_j(m)\}$ over all distinct infima m forms a partition of all subsets of V_{-j}, as stated in (6.18). This is the *neighborhood lattice decomposition* for j. Label the distinct lattices $\{\mathscr{L}_j(m)\}$ as $\mathscr{L}_j^1, \ldots, \mathscr{L}_j^K$, where $\mathscr{L}_j^k = [m^k, M^k]$. A CI relation $j \perp\!\!\!\perp i \mid C$ is related to inclusion in a lattice, as given in the following.

Theorem 6.7. *Let $\mathcal{I}$ be a compositional graphoid over V, $i, j \in V$ and $C \subseteq V \setminus \{i, j\}$. Then, $j \perp\!\!\!\perp i \mid C$ if and only if there is a unique lattice $\mathscr{L}_j^k$, for some $k \in [K]$, such that $C \cup \{i\} \in \mathscr{L}_j^k$, $i \notin m^k$, and $m^k \subseteq C$.*

This shows that the lattice decomposition provides an alternative representation of all CI relations of the form $j \perp\!\!\!\perp i \mid C$ in the absence of a faithful graph. To facilitate practical applications, Amini *et al.* (2022b) designed algorithms to estimate lattice decomposition efficiently and consistently via nonparametric CI tests on high-dimensional data.

For easy understanding, we illustrate the neighborhood lattice with a distribution that is faithful to the graph in Figure 6.4. Let $j = 5$ and $S = \{2, 4\}$. Then, the infimum and supremum of $\mathscr{L}_5(S)$ are, respectively,

$$m(5; S) = \{4\}, \quad M(5; S) = \{1, 2, 3, 4\}.$$

The lattice $\mathscr{L}_5(S)$ contains exactly all 2^3 subsets between $m(5; S)$ and $M(5; S)$, such as $\{1, 4\}$ and $\{2, 3, 4\}$. Let $i = 3$ and $C = \{2, 4\}$. As C and i satisfy all the conditions w.r.t. $\mathscr{L}_5(S)$, Theorem 6.7 implies that $X_5 \perp\!\!\!\perp X_3 \mid \{X_2, X_4\}$.

6.9 Problems

(1) Show that (G) $\Rightarrow$ (L) $\Rightarrow$ (P).
(2) Assume that $(X_1, \ldots, X_p) \sim \mathcal{N}_p(0, \Theta^{-1})$ and $\Theta = (\theta_{jk})$. Show that

$$[X_j \mid X_{-j}] = [X_j \mid X_{-\{j,k\}}]$$

if $\theta_{jk} = 0$, and thus $X_j \perp\!\!\!\perp X_k \mid X_{-\{j,k\}}$.
(3) Let $\mathcal{G}$ be the graph in Figure 6.4 and $\mathbb{P} = \mathcal{N}(0, \Theta^{-1})$, assuming $\Theta \succ 0$.

  (a) Find all maximal cliques in $\mathcal{G}$.
  (b) Show that (F) holds for $\mathbb{P}$ w.r.t. $\mathcal{G}$.
  (c) Find the neighborhood lattice $\mathscr{L}_1(\{2, 3, 4\})$ on $\mathcal{G}$.

(4) Show that (6.13) is true for the Ising model.
(5) Implement a modified G-S algorithm for data from a joint Gaussian distribution as follows: (i) Perform a lasso regression of X_i onto X_{-i} with the tuning parameter selected via cross-validation. Let S be the support of the lasso estimate. (ii) Implement the shrinkage phase of Algorithm 6.1 with the partial correlation test (Section 6.2).

  (a) Simulate datasets from a zero-mean Gaussian distribution with a sparse precision matrix $\Theta \succ 0$. Apply this algorithm to the simulated data and compare the estimated MB with the ground truth.
  (b) Fix $\Theta \succ 0$, and let the sample size $n \to \infty$. Let α_n be the significance level of the partial correlation test in (ii). Assume that the lasso satisfies a screening property, i.e., $\mathbb{P}(S \supseteq S^*) \to 1$ as $n \to \infty$, where S^* is the true support of the linear regression of X_i onto X_{-i}. Show that, for some $\alpha_n \to 0$, the modified G-S algorithm is consistent in estimating the MB of X_i.

Chapter 7

Directed Acyclic Graphs

As a class of graphical models, directed acyclic graphs (DAGs) are commonly used for modeling causal relations among random variables, in which a directed edge from X to Y, denoted $X \to Y$, encodes that X is a direct cause of Y. In this chapter, we cover the basic terminology, graph separations, and Markov properties of DAGs and their parameterizations for joint probability distributions. As a generalization, we also briefly introduce chain graphs, which include both undirected graphs and DAGs as special cases. We conclude this chapter with an overview of the topics related to DAGs that will be discussed in the following chapters.

7.1 Terminology for DAGs

Let $\mathcal{G} = (V, E)$ be a directed graph, i.e., all the edges are directed $(i \to j)$ and $E = \{(i, j) : i \to j\}$. We assume that there is at most one edge between any two vertices. Let $i, j \in V$ be two vertices. A *path* of length n from i to j is a sequence $a_0 = i, \dots, a_n = j$ of distinct vertices so that $(a_{k-1}, a_k) \in E$ for all $k = 1, \dots, n$, i.e., $i \to a_1 \to \cdots \to a_{n-1} \to j$. An *n-cycle* is a path of length n with the modification that $i = j$. A cycle is directed if it contains a directed edge. Note that the definitions of path and cycle apply to both undirected and directed graphs.

Definition 7.1 (DAGs). A DAG is a directed graph that does not contain any directed cycles.

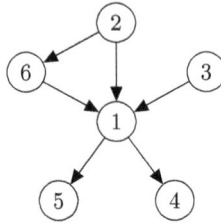

Fig. 7.1.   An example DAG.

Suppose $\mathcal{G}$ is a DAG over V and $i, j \in V$ are two vertices. If $i \to j$, then we say that i is a parent of j and j is a child of i. Denote the set of parents of j by

$$\mathrm{pa}(j) := \{i \in V : i \to j\},$$

and similarly, let $\mathrm{ch}(i)$ be the set of children of i. If there is a path from i to j, we say that i leads to j and write $i \longmapsto j$. The ancestors, descendants, and non-descendants of a vertex are defined, respectively, as

$$\mathrm{an}(i) := \{j \in V : j \longmapsto i\},$$

$$\mathrm{de}(i) := \{j \in V : i \longmapsto j\},$$

$$\mathrm{nd}(i) := V \setminus (\mathrm{de}(i) \cup \{i\}).$$

A *topological sort or ordering* of $\mathcal{G}$ over p vertices is an ordering σ, i.e., a permutation of $\{1, \ldots, p\}$, such that $j \in \mathrm{an}(i)$ implies $j \prec i$ in σ. Due to acyclicity, every DAG has at least one sort.

Example 7.1. Consider the DAG in Figure 7.1. It is easy to see that $\mathrm{pa}(1) = \{2, 3, 6\}$ and $\mathrm{ch}(1) = \{4, 5\}$. Two example paths are $2 \to 6 \to 1 \to 4$ and $3 \to 1 \to 5$, while $2 \to 6 \to 1 \leftarrow 3$ is *not* a path because the edge $1 \leftarrow 3$ is in the opposite direction. Furthermore, $\mathrm{an}(4) = \{2, 6, 3, 1\}$, $\mathrm{de}(6) = \{1, 4, 5\}$, and $\mathrm{nd}(6) = \{2, 3\}$. There are multiple topological sorts for this DAG, such as $(2, 6, 3, 1, 4, 5)$ and $(3, 2, 6, 1, 5, 4)$.

7.2   *d*-Separation

Graph separation in a DAG is defined through the concept of d-separation (Pearl, 1988, Section 3.3), which considers the directionality of edges.

A *chain* of length n from i to j is a sequence $a_0 = i, \ldots, a_n = j$ of distinct vertices so that $a_{k-1} \to a_k$ or $a_k \to a_{k-1}$ for all $k = 1, \ldots, n$, such as $i \leftarrow a_1 \rightarrow a_2 \rightarrow \cdots \rightarrow a_{n-1} \leftarrow j$. The vertices a_0 and a_n are called the endpoints of the chain. A non-endpoint vertex γ is a *collider* on a chain π if the arrows of π meet at γ (i.e., $i \rightarrow \gamma \leftarrow j$). It is a *non-collider* if the arrows of π do *not* meet at γ (i.e., $i \rightarrow \gamma \rightarrow j$ or $i \leftarrow \gamma \rightarrow j$). In Figure 7.1, $2 \rightarrow 6 \rightarrow 1 \leftarrow 3$ is a chain but not a path. On this chain, vertex 1 is a collider and vertex 6 is a non-collider.

Definition 7.2 (d-separation). A chain π from a to b is said to be *blocked* by $S \subset V$ if the chain contains a vertex γ such that either (i) or (ii) holds:

(i) $\gamma \in S$ and γ is a non-collider;
(ii) $\{\gamma \cup \mathrm{de}(\gamma)\} \cap S = \varnothing$ and γ is a collider.

Two subsets A and B are d-separated by S if all chains from any $a \in A$ to any $b \in B$ are blocked by S. Otherwise, we say that A and B are d-connected by S.

Example 7.2. Let us use the DAG in Figure 7.1 as an example. The chain $2 \rightarrow 6 \rightarrow 1 \rightarrow 4$ has no colliders and is blocked by $\{1\}, \{6\}$, or $\{1,6\}$, as each of these subsets satisfies condition (i). The chain $2 \rightarrow 6 \rightarrow 1 \leftarrow 3$ has a collider (vertex 1) and is thus blocked by $\varnothing$ according to condition (ii). But this chain is *not* blocked by $\{1\}$ or any node in $\mathrm{de}(1)= \{4,5\}$, i.e., the chain is d-connected by $\{1\}, \{4\}$, or $\{5\}$.

To d-separate vertices 2 and 4, we find S to block $\pi_1 = \langle 2 \rightarrow 1 \rightarrow 4 \rangle$ and $\pi_2 = \langle 2 \rightarrow 6 \rightarrow 1 \rightarrow 4 \rangle$. Since there are no colliders on either chain, S must include at least one non-collider from each chain. Thus, the valid choices are $S = \{1\}$ and $S = \{1,6\}$.

Let us find a subset S that d-separates vertices 3 and 6. There are two chains between the two vertices: $6 \rightarrow 1 \leftarrow 3$ and $6 \leftarrow 2 \rightarrow 1 \leftarrow 3$. Since vertex 1 is a collider on both chains and $\mathrm{de}(1) = \{4,5\}$, vertices 3 and 6 will be d-separated by S if it does not contain any nodes in $\{1,4,5\}$. This gives us $S = \varnothing$ or $S = \{2\}$.

Example 7.3. Suppose the edge between vertices 1 and 6 in Figure 7.1 is flipped, which leads to the DAG in Figure 7.2. Now, let us find S to d-separate vertices 3 and 6.

There are two chains, $\pi_1 = \langle 3 \rightarrow 1 \rightarrow 6 \rangle$ and $\pi_2 = \langle 3 \rightarrow 1 \leftarrow 2 \rightarrow 6 \rangle$, both of which must be blocked. To block π_1, we must have $1 \in S$. However, vertex 1 is a collider on π_2; therefore, this chain is d-connected by vertex 1. To block π_2, we must include $2 \in S$ as well. Thus, $S = \{1,2\}$ d-separates

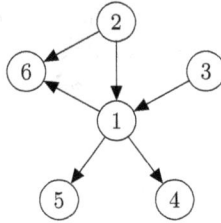

Fig. 7.2. An example to demonstrate d-separation.

vertices 3 and 6. In fact, any S such that $\{1, 2\} \subseteq S \subseteq \{1, 2, 4, 5\}$ d-separates the two vertices.

Remark 7.1. In some literature on DAGs, what we call a chain and a path here are called a path and a directed path, respectively. We adopt the former in this chapter as they are also applicable to undirected graphs and chain graphs, so that we have unified terminology for all three classes of graphs.

7.3   Markov Properties

7.3.1   *Definitions and relations*

Let $\mathcal{G} = (V, E)$ be a DAG and $\mathbb{P}$ be a joint distribution over V. We assume the same setting as in Section 6.4 for $\mathbb{P}$.

Definition 7.3 (Markov properties on DAGs). We say that the joint distribution $\mathbb{P}$:

(DF) admits a recursive factorization according to $\mathcal{G}$ if $\mathbb{P}$ has a density f such that

$$f(x) = \prod_{j \in V} f_j(x_j \mid \mathrm{pa}(j)), \qquad (7.1)$$

where f_j is the density for the conditional distribution $[j \mid \mathrm{pa}(j)]$.

(DG) satisfies the directed global Markov property w.r.t. $\mathcal{G}$ if, for any disjoint subsets $A, B, S \subseteq V$,

$$S \ d\text{-separates } A \text{ and } B \Rightarrow A \perp\!\!\!\perp B \mid S.$$

(DL) satisfies the directed local Markov property w.r.t. $\mathcal{G}$ if

$$i \perp\!\!\!\perp \mathrm{nd}(i) \mid \mathrm{pa}(i), \qquad \text{for all } i \in V.$$

(DP) satisfies the directed pairwise Markov property w.r.t. $\mathcal{G}$ if for any $(i,j) \notin E$ with $j \in \text{nd}(i)$,

$$i \perp\!\!\!\perp j \mid \text{nd}(i) \setminus \{j\}.$$

All of the Markov properties are one-way implications, from graph separations or graphical structures to conditional independence statements in the joint distribution.

The relation among the four Markov properties are summarized as follows.

Theorem 7.1. *For any DAG $\mathcal{G}$ and joint distribution $\mathbb{P}$, we have*

$$(\text{DF}) \Rightarrow (\text{DG}) \Rightarrow (\text{DL}) \Rightarrow (\text{DP}). \tag{7.2}$$

Note that $\text{pa}(i)$ d-separates i and $\text{nd}(i) \setminus \text{pa}(i)$. This shows that (DG) implies (DL). Since $\text{pa}(i) \subseteq \text{nd}(i)$, by CI axiom (C3) in Section 6.1, one sees that (DL) $\Rightarrow$ (DP). The proof of (DF) $\Rightarrow$ (DG) is more technical. We refer the reader to Lauritzen (1996, Section 3.2.2) for details, which also provides a proof of the following equivalence among three Markov properties.

Theorem 7.2. *If $\mathbb{P}$ has a density with respect to a product measure, then* (DF), (DG), *and* (DL) *are equivalent.*

For discrete random variables, the probability mass function, which always exists, is a density w.r.t. the counting measure, and thus the condition for Theorem 7.2 is satisfied. For continuous random variables, the condition may be violated when there are deterministic functional relationships among them. For example, suppose $X_1 \sim \mathcal{N}(0,1)$ and $X_2 = X_1$. Then, the joint distribution of (X_1, X_2) does not have a density w.r.t. the Lebesgue measure over $\mathbb{R}^2$.

Example 7.4 (Markov chain). A Markov chain $\{X_1, \ldots, X_n\}$ is represented by a DAG in Figure 7.3, which is a directed chain.

It is easy to verify (DF) by noting that

$$\mathbb{P}(X_1, \ldots, X_n) = \mathbb{P}(X_1)\mathbb{P}(X_2 \mid X_1) \cdots \mathbb{P}(X_n \mid X_{n-1})$$

and $\text{pa}(i) = i - 1$ for $i = 2, \ldots, n$. Consequently, (DG) holds, and thus we may use d-separation to infer CI relations. In particular, for any $i < j < k$,

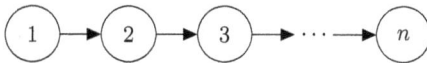

Fig. 7.3. A Markov chain represented as a DAG.

because j d-separates i and k, we have

$$X_i \perp\!\!\!\perp X_k \mid X_j.$$

Example 7.5. Suppose the density $f(x_1, \ldots, x_6)$ factorizes according to the DAG in Figure 7.2. Let us demonstrate how to use the global and local Markov properties to identify the CI relations.

(DG): Since $\{1, 2\}$ d-separates vertices 3 and 6, we have

$$X_3 \perp\!\!\!\perp X_6 \mid \{X_1, X_2\}.$$

As another example, vertices 2 and 3 are d-separated by $\varnothing$, and thus $X_2 \perp\!\!\!\perp X_3$. Is it true that $X_2 \perp\!\!\!\perp X_3 \mid X_5$? The answer is no. This is because vertex 5 is a descendant of a collider (vertex 1) on the chain $2 \to 1 \leftarrow 3$, and thus vertices 2 and 3 are d-connected by vertex 5.

(DL): Since pa(6)= $\{1, 2\}$ and $3 \in$ nd(6), we have $X_3 \perp\!\!\!\perp X_6 \mid \{X_1, X_2\}$. Vertex 4 has no descendant and pa(4)= $\{1\}$. Therefore,

$$X_4 \perp\!\!\!\perp \{X_2, X_3, X_5, X_6\} \mid X_1.$$

7.3.2   *Moral graphs*

Markov properties on DAGs are connected to Markov properties on undirected graphs (Section 6.4) through the concept of moral graphs.

Definition 7.4. Given a DAG $\mathcal{G}$, construct a graph by (i) adding an edge between every pair of parents of each node if they are not connected already in $\mathcal{G}$ and then (ii) ignoring all edge orientations. The resulting undirected graph is the moral graph $\mathcal{G}^m$ of $\mathcal{G}$.

Figure 7.4 shows a DAG $\mathcal{G}$ and its moral graph $\mathcal{G}^m$. To construct $\mathcal{G}^m$, undirected edges are added between all parents (nodes $2, 3$, and 6) of node 1, which are shown in red in (b). Then, the orientations of the original edges in $\mathcal{G}$ are removed, which are the black edges in $\mathcal{G}^m$. By construction, every node and its parent set, i.e., $\{i\} \cup$ pa(i), will form a clique in the moral graph, such as $\{1, 2, 3, 6\}$ in Figure 7.4(b).

Lemma 7.1. *If* $\mathbb{P}$ *admits a recursive factorization according to a DAG* $\mathcal{G}$, *then it also factorizes according to the moral graph* $\mathcal{G}^m$. *That is,* (DF) *w.r.t.* $\mathcal{G} \Rightarrow$ (F) *w.r.t.* $\mathcal{G}^m$.

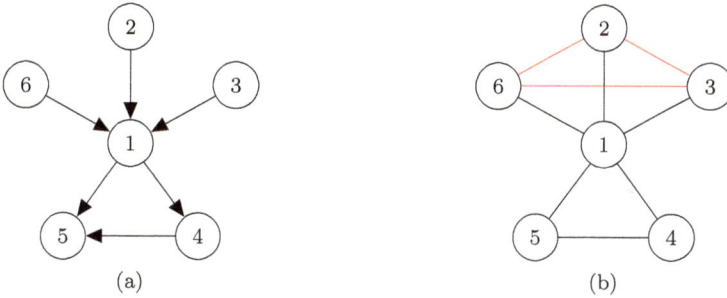

Fig. 7.4. (a) DAG $\mathcal{G}$ and (b) its moral graph $\mathcal{G}^m$.

Proof. Let $N_j = \{j\} \cup \text{pa}(j)$, $j \in V$. By construction, each N_j is a clique and thus a subset of a maximal clique in the moral graph $\mathcal{G}^m$. Assign each N_j to a unique maximal clique $M_j \supseteq N_j$. Let $\mathcal{C}$ be the set of maximal cliques of $\mathcal{G}^m$. For each $A \in \mathcal{C}$, let

$$\psi_A(x_A) = \prod_{j: M_j = A} f_j(x_j \mid \text{pa}(j)).$$

By (7.1), we have

$$f(x) = \prod_{j \in V} f_j(x_j \mid \text{pa}(j)) = \prod_{A \in \mathcal{C}} \psi_A(x_A),$$

which shows that f factorizes according to $\mathcal{G}^m$. □

Remark 7.2. As a consequence of this lemma, $\mathbb{P}$ also satisfies the global (G), local (L), and pairwise (P) Markov properties w.r.t. $\mathcal{G}^m$ if it admits a recursive factorization w.r.t. $\mathcal{G}$.

The proof gives a reason why parents of the same node are connected in the moral graph: the conditional density $f_j(x_j \mid \text{pa}(j))$ depends on all variables in N_j, which would induce a clique in an undirected graph that represents the factorization of $f(x)$.

Let $A \subseteq V$. For a DAG, if $\text{pa}(i) \subseteq A$ for all $i \in A$, then the subset A is an ancestral set. For a subset S of nodes, $\text{An}(S)$ is the smallest ancestral set containing S. That is, $\text{An}(S)$ consists of S and all its ancestors. Let $\mathcal{G}_S = (S, E_S)$ denote the subgraph of $\mathcal{G} = (V, E)$ over the subset S, where $E_S = E \cap S \times S$. Now, we present a connection between separation criteria for a DAG and moral graphs.

$$\boxed{2} \qquad \boxed{3}$$

Fig. 7.5.   Moral graph of the subgraph $\mathcal{G}_{\{2,3\}}$.

Theorem 7.3. *Let A, B, and S be disjoint subsets of nodes in a DAG $\mathcal{G}$. Then, S d-separates A and B in $\mathcal{G}$ if and only if S separates A and B in $(\mathcal{G}_{An(A \cup B \cup S)})^m$.*

See Lauritzen (1996, Section 3.2.2) for a proof.

We use the DAG $\mathcal{G}$ in Figure 7.4(a) to illustrate how to check d-separation from moral graphs. Let us first examine whether nodes 2 and 3 are d-separated by $\varnothing$. Since $An(\{2,3\}) = \{2,3\}$, we consider the graph $(\mathcal{G}_{\{2,3\}})^m$, which contains two disconnected notes, as shown in Figure 7.5. Therefore, the two nodes are separated by $\varnothing$. Indeed, nodes 2 and 3 are d-separated by $\varnothing$ in the DAG in Figure 7.4(a). To check if nodes 2 and 3 are d-separated by node 5, we find $An(\{2,3,5\}) = \{1,2,3,4,5,6\}$, and thus we need to examine the moral graph $\mathcal{G}^m$. As shown in Figure 7.4(b), nodes 2 and 3 are not separated by node 5 in $\mathcal{G}^m$. In fact, nodes 2 and 3 are not separated by any subset since they are adjacent in $\mathcal{G}^m$. By Theorem 7.3, they are not d-separated by node 5 in $\mathcal{G}$. This is consistent with the definition of d-separation, as node 5 is a descendant of a collider (node 1) on the chain $2 \to 1 \leftarrow 3$.

7.3.3   *Markov equivalence*

Different DAGs may encode the same set of d-separations. For example, consider two DAGs $\mathcal{G}_1 : 1 \to 2 \to 3$ and $\mathcal{G}_2 : 1 \leftarrow 2 \leftarrow 3$. Both DAGs imply only one d-separation, i.e., node 2 d-separates nodes 1 and 3. Such DAGs are called Markov equivalent.

Definition 7.5 (Markov equivalence). Two DAGs over the same set of vertices are Markov equivalent if they imply the same set of d-separations.

For a DAG $\mathcal{G} = (V, E)$, its skeleton, denoted $sk(\mathcal{G})$, is the undirected graph over V obtained after removing the orientations of all edges in E. A v-structure is a triplet $\{i, j, k\} \subseteq V$ of the form $i \to k \leftarrow j$. That is, i and j are non-adjacent, and k is an *uncovered collider*. The general condition for Markov equivalence (Verma and Pearl, 1990) involves the skeleton and the v-structures of a DAG:

Theorem 7.4. *Two DAGs over the same set of vertices are Markov equivalent if and only if they have the same skeleton and the same v-structures.*

This theorem can be proved using three key results (Verma, 1991). First, two nodes i and j are adjacent in a DAG if and only if they cannot be separated by any set. Second, a v-structure $i \rightarrow k \leftarrow j$ exists if and only if i and j are separated by some S but not any set containing k. These two results imply that adjacency and v-structures are solely determined by d-separation and hence remain invariant among equivalent DAGs. Third, if two DAGs have the same skeleton and v-structures, then every S-active chain in one DAG corresponds to an S-active chain in the other. We say a chain π is S-active if it is not blocked by S. The last result establishes the sufficiency for equivalence.

Let us verify that the three DAGs $\mathcal{G}_1, \mathcal{G}_2$, and $\mathcal{G}_3$ in Figure 7.6 are Markov equivalent. All of these DAGs have the same skeleton, and they all have one v-structure, $2 \rightarrow 4 \leftarrow 3$. Thus, the three DAGs are indeed Markov equivalent. There are two types of edges, compelled versus reversible edges, among DAGs that are Markov equivalent (or in the same equivalence class). If the orientation of an edge is the same across all DAGs in the equivalence class, then it is a compelled edge. If either direction occurs in at least one equivalent DAG, then we say that the edge is reversible. Reversing either or both of the edges $2 \rightarrow 4$ and $3 \rightarrow 4$ would destroy the v-structure $2 \rightarrow 4 \leftarrow 3$ and lead to a DAG that is not equivalent. Thus, these two edges are compelled edges. Moreover, reversing the edge $4 \rightarrow 5$ would introduce new v-structures formed with edges pointing to node 4; therefore, it is also a compelled edge. Both directions, $1 \rightarrow 2$ and $2 \rightarrow 1$, occur among the three DAGs, and thus this edge is reversible. Similarly, the edge $1 \rightarrow 3$ is also reversible. Although both are reversible, the orientations $2 \rightarrow 1$ and $3 \rightarrow 1$

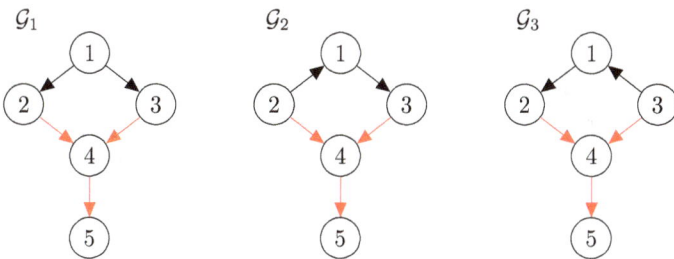

Fig. 7.6. Three Markov-equivalent DAGs. The orientation of the red edges is compelled, while the black edges are reversible.

cannot happen simultaneously in the same DAG because this would create a new v-structure not possessed by the equivalence class.

7.4 Parameterizations

To model a joint distribution with a DAG, the starting point is to construct a DAG according to which the joint distribution admits a recursive factorization. This defines a Bayesian network for the joint distribution.

Definition 7.6 (Bayesian networks). Given a joint distribution $\mathbb{P}$ with density f over $\{X_1, \ldots, X_p\}$ and a permutation $\sigma = (\sigma(1), \ldots, \sigma(p))$ of $[p]$, we factorize

$$f(x) = \prod_{j=1}^{p} f(x_{\sigma(j)} \mid x_{\sigma(1)}, \ldots, x_{\sigma(j-1)})$$

$$= \prod_{j=1}^{p} f(x_{\sigma(j)} \mid x_{A_j}), \tag{7.3}$$

where $A_j \subseteq \{\sigma(1), \ldots, \sigma(j-1)\}$ is the minimum subset such that (7.3) holds. Then, the DAG $\mathcal{G}$ with $\mathrm{pa}(\sigma(j)) = A_j$, for all $j \in [p]$, is a Bayesian network of $\mathbb{P}$.

If $\mathcal{G}$ is a Bayesian network of $\mathbb{P}$, then (DF) holds by construction, and consequently, (DG), (DL), and (DP) also hold. Note that σ is a topological sort of the DAG $\mathcal{G}$.

To parameterize a Bayesian network $\mathcal{G}$, we just need to parameterize the conditional distribution $[X_j \mid \mathrm{pa}(j)]$, as in (7.1). The conditional distribution is sometimes called a local distribution. In this section, we focus primarily on Gaussian and discrete Bayesian networks.

7.4.1 *Gaussian Bayesian networks*

For a Gaussian Bayesian network, $[X_j \mid \mathrm{pa}(j)]$ is parameterized by the following structural equations:

$$X_j = \sum_{i \in \mathrm{pa}(j)} \beta_{ij} X_i + \varepsilon_j, \quad j = 1, \ldots, p. \tag{7.4}$$

Here, we assume that the error $\varepsilon_j \sim \mathcal{N}(0, \omega_j^2)$ is normally distributed with a mean of zero and variance of ω_j^2 and $\varepsilon_j \perp\!\!\!\perp \mathrm{pa}(j)$, for all j. This is also called the Gaussian DAG model.

Put $X = (X_1, \ldots, X_p)$ as a column vector in $\mathbb{R}^p$. We can represent the above structural equations in matrix form by defining $B = (\beta_{ij})_{p \times p}$, where $\beta_{ij} = 0$ if $i \notin \text{pa}(j)$, and $\Omega = \text{diag}(\omega_1^2, \ldots, \omega_p^2)$. This leads to the equation

$$X = B^\mathsf{T} X + \varepsilon, \qquad \varepsilon \sim \mathcal{N}_p(0, \Omega).$$

This representation shows that $X \sim \mathcal{N}_p(0, \Theta^{-1})$, where

$$\Theta = (I_p - B)\Omega^{-1}(I_p - B)^\mathsf{T}. \tag{7.5}$$

Note that (7.5) is the Cholesky decomposition of Θ after permuting the rows and columns of Θ, which relates the covariance matrix of X to the edge coefficients (β_{ij}) in its DAG parameterization.

Let us make this point more explicit using permutation matrices. Let $\{e_1, \ldots, e_p\}$ be the standard basis of $\mathbb{R}^p$. To each permutation π on $[p]$, we associate a permutation matrix P_π whose ith row is $e_{\pi(i)}^\mathsf{T}$. Let $B_\pi = P_\pi B P_\pi^\mathsf{T}$ be the matrix obtained by permuting the rows and columns of B simultaneously according to π. Then, B_π will be a strictly lower-triangular matrix if and only if π is the reversal of a topological ordering of $\mathcal{G}$, i.e., $i \prec j$ in π for $j \in \text{pa}(i)$. Similarly, define Θ_π and Ω_π by permuting Θ and Ω, respectively. Then, Equation (7.5) now becomes

$$\Theta_\pi = (I - B_\pi)\Omega_\pi^{-1}(I - B_\pi)^\mathsf{T}, \tag{7.6}$$

where $(I - B_\pi)$ is lower triangular. This is exactly the Cholesky decomposition of Θ_π. Further details on this relation can be found in the works of van de Geer and Bühlmann (2013) and Aragam and Zhou (2015). Ye *et al.* (2021) used this formulation to encode the acyclicity constraint in the regularized-likelihood estimation of topological orderings.

Let us consider an example DAG with its coefficient matrix $B = (\beta_{ij})_{4 \times 4}$ shown in Figure 7.7. A topological ordering of this DAG is $(2, 3, 1, 4)$. Let $\pi = (4, 1, 3, 2)$, which is the reversal of this ordering. Then, we have

$$P_\pi = \begin{bmatrix} 0 & 0 & 0 & 1 \\ 1 & 0 & 0 & 0 \\ 0 & 0 & 1 & 0 \\ 0 & 1 & 0 & 0 \end{bmatrix}, \quad B_\pi = P_\pi B P_\pi^\mathsf{T} = \begin{bmatrix} 0 & 0 & 0 & 0 \\ 0 & 0 & 0 & 0 \\ \beta_{34} & 0 & 0 & 0 \\ 0 & \beta_{21} & \beta_{23} & 0 \end{bmatrix}.$$

It is seen that, after the permutation, B_π is strictly lower triangular.

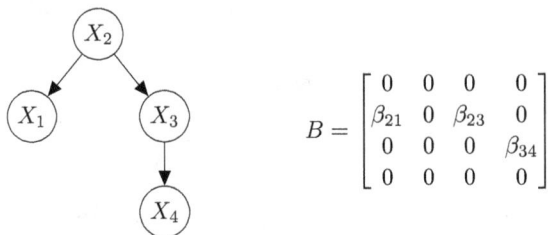

$$B = \begin{bmatrix} 0 & 0 & 0 & 0 \\ \beta_{21} & 0 & \beta_{23} & 0 \\ 0 & 0 & 0 & \beta_{34} \\ 0 & 0 & 0 & 0 \end{bmatrix}$$

Fig. 7.7. A Gaussian DAG with its coefficient matrix $B = (\beta_{ij})$.

7.4.2 Discrete Bayesian networks

A discrete Bayesian network refers to a DAG whose nodes are all discrete random variables. A common choice for the local distribution $[X_j \mid \mathrm{pa}(j)]$ is a set of multinomial distributions parameterized by a conditional probability table with entries

$$\theta_{jkm} = \mathbb{P}(X_j = m \mid X_{\mathrm{pa}(j)} = k), \tag{7.7}$$

where m and k index a possible value for X_j and its parent set, respectively. Assume that each X_j is a finite variable with r_j values and its parent set takes $q_j = \prod_{i \in \mathrm{pa}(j)} r_i$ values. The parameters for $[X_j \mid \mathrm{pa}(j)]$, $j \in [p]$, are arranged into a $q_j \times r_j$ table:

$$\left\{ \theta_{jkm} : \sum_m \theta_{jkm} = 1, k = 1, \ldots, q_j, m = 1, \ldots, r_j \right\}.$$

Consider a simple example in which all variables are binary: $X_j \in \{0,1\}$ and $r_j = 2$, for all j. Suppose $\mathrm{pa}(1) = \{2, 5\}$. There are four possible states for $\mathrm{pa}(1)$, i.e., $q_1 = 4$, corresponding to $(X_2, X_5) = (0,0), (0,1), (1,0), (1,1)$. An example probability table for $[X_1 \mid \mathrm{pa}(1)]$ is shown in Table 7.1, from which we see, for instance, that $\mathbb{P}(X_1 = 0 \mid X_2 = 0, X_5 = 1) = 0.7$.

Table 7.1. A local probability table.

	(X_2, X_5)	$X_1 = 0$	$X_1 = 1$
$X_{\mathrm{pa}(1)} = 1$	$(0,0)$	0.1	0.9
$X_{\mathrm{pa}(1)} = 2$	$(0,1)$	0.7	0.3
$X_{\mathrm{pa}(1)} = 3$	$(1,0)$	0.5	0.5
$X_{\mathrm{pa}(1)} = 4$	$(1,1)$	0.8	0.2

Although commonly used, this traditional multinomial DAG model may end up with too many parameters when r_j and the parent size $|\text{pa}(j)|$ are large for some variables. To reduce the number of parameters, Gu *et al.* (2019) proposed to use a multi-logit regression for $[X_j \mid X_{\text{pa}(j)}]$, in which each variable X_j is encoded by $d_j = r_j - 1$ dummy variables. Given a reference category (which may be arbitrary), the r_j possible values of X_j are encoded by a vector, $\mathbf{z}_j = (z_{jk}, k = 1, \ldots, d_j) \in \{0, 1\}^{d_j}$. If we choose the last value as the reference, then $z_{jk} = I(X_j = k)$ for $k = 1, \ldots, d_j$ so that the reference category is coded as $\mathbf{z}_j = 0$.

Under this parametrization, the conditional distributions take the form

$$\mathbb{P}(X_j = m \mid X_{\text{pa}(j)}) = \frac{\exp\left(\beta_{0mj} + \sum_{i \in \text{pa}(j)} \mathbf{z}_i^\top \boldsymbol{\beta}_{imj}\right)}{\sum_{k=1}^{r_j} \exp\left(\beta_{0kj} + \sum_{i \in \text{pa}(j)} \mathbf{z}_i^\top \boldsymbol{\beta}_{ikj}\right)}, \qquad (7.8)$$

for $m = 1, \ldots, r_j$, where $\boldsymbol{\beta}_{imj} \in \mathbb{R}^{d_i}$ is the coefficient vector for variable X_i to predict the mth level of X_j with intercept β_{0mj}. In this model, there are r_j coefficient vectors $\boldsymbol{\beta}_{imj}$, $m \in [r_j]$ associated with the edge $i \to j$. Let $\boldsymbol{\beta}_{ij} = (\boldsymbol{\beta}_{i1j} \mid \cdots \mid \boldsymbol{\beta}_{ir_j j})$ be a $d_i \times r_j$ coefficient matrix for the node pair (i, j). There does not exist an edge from i to j if $\boldsymbol{\beta}_{ij} = 0$. Thus, the structure of the DAG is given by the sparsity pattern of $\{\boldsymbol{\beta}_{ij}\}$. Based on this formulation, Gu *et al.* (2019) developed a group-norm regularized-likelihood method to learn the structure of a discrete Bayesian network.

Remark 7.3. Both (7.4) and (7.8) are special cases of a general structural equation model (SEM) in the form of

$$X_j = f_j(X_{\text{pa}(j)}, \varepsilon_j), \quad j = 1, \ldots, p, \qquad (7.9)$$

where ε_j are independent error variables. In this general form, the joint distribution of $X_1, \ldots, X_p$ is determined by the function form of f_j and the distribution for ε_j. We may make various parametric and nonparametric assumptions for f_j under this model to fit a DAG model to the data.

7.5 Chain Graphs

7.5.1 *Definition and characterization*

Chain graphs may be regarded as generalizations of DAGs. A chain graph may contain two types of edges: undirected $(i - j)$ and directed $(i \to j)$.

Definition 7.7 (Chain graph). Partition the vertex set V into disjoint subsets $V_t, t = 1, \ldots, T$, i.e., $V = V_1 \cup \cdots \cup V_T$ and $V_s \cap V_t = \varnothing$ for any $s \neq t$. A chain graph over V is a graph such that:

(i) all edges between vertices in the same V_t are undirected;
(ii) all edges between two different subsets V_s and V_t ($s < t$) are directed and pointing from V_s to V_t.

Let $\mathcal{G}$ be a chain graph. If $T = 1$, then $\mathcal{G}$ is an undirected graph. If $|V_t| = 1$ for all t, then $\mathcal{G}$ is a DAG. Therefore, chain graphs include both undirected graphs and DAGs as special cases. As such, chain graphs can represent a larger class of distributions than undirected graphs and DAGs. As a natural generalization of DAGs, chain graphs also have a causal interpretation, which is related to dynamic models with feedback loops (Lauritzen and Richardson, 2002). Moreover, chain graphs are used to represent the Markov equivalence class of a DAG, which will be discussed in Section 9.2.

The definitions of path, cycle, and directed cycle for DAGs (Section 7.1) also apply to chain graphs by noting that the edge set E of a chain graph may include both directed and undirected edges. Recall that a path from i to j is a sequence $a_0 = i, \ldots, a_n = j$ of distinct vertices so that $(a_{k-1}, a_k) \in E$, for all $k = 1, \ldots, n$, such as $i - a_1 \to a_2 \to a_3 - j$. A path is called a *directed path* if it contains at least one directed edge. If there is a path from i to j, we say that i leads to j and write $i \longmapsto j$. If there is no path from i to j, we say that i does not lead to j, written as $i \not\longmapsto j$. If $i \longmapsto j$ and $j \longmapsto i$, then we say that i and j connect, written as $i \leftrightarrow j$. Vertices that connect to a vertex $i \in V$ form an equivalence class $[i] := \{j \in V : i \leftrightarrow j\}$. The equivalence classes defined by connectivity are the *connectivity components* of a graph. For example, if $i - j - k$, then $i \leftrightarrow k$ and $i, j, k \in [i]$. For a DAG, every connectivity component consists of a single node. For a chain graph, the connectivity components are called the *chain components*.

By its definition, a chain graph has no directed cycles, and its chain components induce undirected subgraphs. To find the chain components, we first remove all directed edges and then identify the connectivity components of the resulting graph. Using the chain graph in Figure 7.8 as an example, after removing all directed edges, there are three connectivity components, $V_1 = \{1, 2, 3\}$, $V_2 = \{4\}$, and $V_3 = \{5, 6\}$, which are the chain components. There is a path $2 - 1 - 3$ consisting of only undirected edges, and thus $2 \leftrightarrow 3$. There is a directed path $1 - 3 \to 6 - 5$ that includes a directed edge $3 \to 6$. Thus, $1 \longmapsto 5$ but $5 \not\longmapsto 1$.

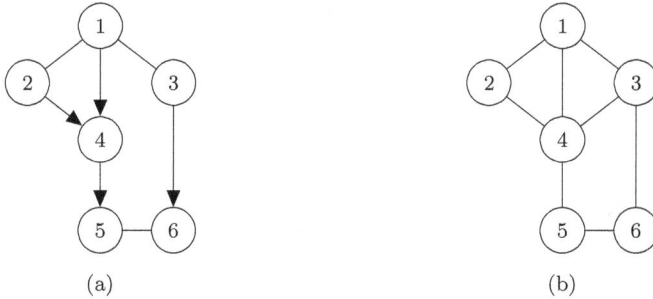

Fig. 7.8. (a) A chain graph over six nodes and (b) its moral graph.

7.5.2 *Markov properties*

To define Markov properties on chain graphs, we generalize a few definitions for DAGs and undirected graphs. The definitions of the parent set pa(i) and the neighbor set ne(i) of a node i apply to chain graphs in an obvious way. The union of the two sets is called the boundary of i:

$$\mathrm{bd}(i) := \mathrm{pa}(i) \cup \mathrm{ne}(i) = \{j \in V : j \to i \text{ or } j - i\}.$$

The set of ancestors of i is

$$\mathrm{an}(i) := \{j \in V : j \longmapsto i, i \not\longmapsto j\},$$

in which the condition $i \not\longmapsto j$ excludes any nodes in the same chain component as i. In other words, j is an ancestor of i if there is a *directed* path from j to i. In the same way, we define the sets of descendants and non-descendants as

$$\mathrm{de}(i) := \{j \in V : i \longmapsto j, j \not\longmapsto i\},$$
$$\mathrm{nd}(i) := V \setminus (\mathrm{de}(i) \cup \{i\}).$$

Let $A \subseteq V$. If bd(i) $\subseteq A$ for all $i \in A$, then A is an ancestral set. For a subset S of nodes, An(S) is the smallest ancestral set containing S. In general, An(S) contains the chain components and the ancestors of every node in S.

Definition 7.8 (Moral graph). The moral graph of a chain graph $\mathcal{G}$ is constructed as follows: (i) For each chain component C, add an undirected edge between every pair of non-adjacent nodes in pa(C) := $\cup_{i \in C}$ pa(i). (ii) Then, ignore all edge directions. The resulting undirected graph, denoted $\mathcal{G}^m$, is the moral graph of $\mathcal{G}$.

We are now ready to define local and global Markov properties on a chain graph $\mathcal{G}$. See Lauritzen (1996, Section 3.2.3) for other Markov properties.

Definition 7.9. Let $\mathcal{G}$ be a chain graph over V. A joint distribution $\mathbb{P}$ over V satisfies:

(CL) the local chain Markov property if $i \perp\!\!\!\perp \mathrm{nd}(i) \mid \mathrm{bd}(i)$ for all $i \in V$.

(CG) the global chain Markov property if, for any disjoint $A, B, S \subseteq V$,

$$S \text{ separates } A \text{ and } B \text{ in } (\mathcal{G}_{\mathrm{An}(A \cup B \cup S)})^m \Rightarrow A \perp\!\!\!\perp B \mid S.$$

Both (CL) and (CG) unify Markov properties for undirected graphs and DAGs. To see this, let us consider (CG) in the special cases where $\mathcal{G}$ is an undirected graph or a DAG. If $\mathcal{G}$ is an undirected graph, then $\mathrm{An}(A \cup B \cup S)$ is the union of the connectivity components of the nodes in $A \cup B \cup S$ and $(\mathcal{G}_{\mathrm{An}(A \cup B \cup S)})^m = \mathcal{G}_{\mathrm{An}(A \cup B \cup S)}$. Consequently, the separation statement in (CG) is equivalent to $A - S - B$ in $\mathcal{G}$. If $\mathcal{G}$ is a DAG, then the separation statement is equivalent to that S d-separates A and B by Theorem 7.3.

Let us continue with the example graph in Figure 7.8. We see that $\mathrm{bd}(1) = \{2, 3\}$, $\mathrm{bd}(4) = \{1, 2\}$, and $\mathrm{bd}(5) = \{4, 6\}$. The descendant set $\mathrm{de}(3) = \{4, 5, 6\}$ due to the existence of directed paths $3 - 1 \to 4$ and $3 \to 6 - 5$, while $\mathrm{de}(5) = \varnothing$. Applying the local Markov property to node 5, we arrive at $5 \perp\!\!\!\perp \{1, 2, 3\} \mid \{4, 6\}$. Since $\mathrm{An}(\{1, 2, 3\}) = \{1, 2, 3\}$ and $(\mathcal{G}_{\{1,2,3\}})^m = \mathcal{G}_{\{1,2,3\}}$, by the global Markov property, we conclude that $2 \perp\!\!\!\perp 3 \mid 1$. If we add 6 to the conditioning set, $S = \{1, 6\}$, we need to exam $\mathcal{G}^m$ because $\mathrm{An}(\{1, 2, 3, 6\}) = V$. The moral graph $\mathcal{G}^m$ is shown in Figure 7.8(b), where an edge $3 - 4$ has been added because both nodes are parents of the chain component $V_3 = \{5, 6\}$. Since $\{1, 6\}$ does not separate nodes 2 and 3 in $\mathcal{G}^m$ due to the path $2 - 4 - 3$, we conclude that $2 \not\perp\!\!\!\perp 3 \mid \{1, 6\}$. Similarly, we see that $1 \perp\!\!\!\perp 6 \mid \{3, 4\}$ from $\mathcal{G}^m$.

7.6   Faithfulness

Faithfulness with respect to (w.r.t.) DAGs is defined in the same way as for undirected graphs (Definition 6.6). It assumes a one-to-one correspondence between d-separations and conditional independence relations.

Definition 7.10 (Faithfulness). For a DAG model $(\mathcal{G}, \mathbb{P})$, we say the distribution $\mathbb{P}$ is faithful to the DAG $\mathcal{G}$ if, for every triple of disjoint sets

$A, B, S \subseteq V$,

$$A \perp\!\!\!\perp B \mid S \text{ in } \mathbb{P} \Leftrightarrow S \text{ } d\text{-separates } A \text{ and } B \text{ in } \mathcal{G}.$$

Remark 7.4. We may define faithfulness for chain graphs in a similar way: replacing the one-way implication "$\Rightarrow$" in (CG) with "$\Leftrightarrow$". This would include, as special cases, the faithfulness assumptions for undirected graphs and DAGs.

It has been shown that almost all linear DAG and Gaussian chain graph models are faithful (Spirtes *et al.*, 2000; Levitz *et al.*, 2001; Peña, 2011). To help understand it, let us elaborate on this general result for the Gaussian DAG model (7.4). Given a DAG $\mathcal{G} = ([p], E)$, choose the coefficient matrix $B = (\beta_{ij})$ such that $\text{supp}(B) = E$ and the error variances $\omega_j^2 > 0$ for all j. Define two parameter spaces:

$$\mathcal{B} = \mathcal{B}(\mathcal{G}) := \{B \in \mathbb{R}^{p \times p} : \text{supp}(B) = E\},$$

$$\mathcal{D} = \mathcal{D}(\mathcal{G}) := \{\text{diag}(\omega_1^2, \ldots, \omega_p^2) : \omega_j^2 > 0 \ \forall j \in [p]\}.$$

The joint distribution defined with $(B, \Omega) \in \mathcal{B} \times \mathcal{D}$ admits a recursive factorization w.r.t. $\mathcal{G}$ and thus satisfies the global Markov property (DG).

Theorem 7.5. *For any DAG $\mathcal{G}$ on $V = [p]$, almost all $(B, \Omega) \in \mathcal{B} \times \mathcal{D}$ define a joint Gaussian distribution that is faithful to $\mathcal{G}$.*

Equivalently, the set of (B, Ω) that defines an unfaithful distribution has a Lebesgue measure of zero. Unfaithfulness occurs when the parameters B and Ω satisfy additional equality constraints that result in CI relations in $\mathbb{P}$ not implied by the d-separations in $\mathcal{G}$. For example, in the DAG $\mathcal{G}_1$ in Figure 7.6, if

$$\beta_{12}\beta_{24} + \beta_{13}\beta_{34} = 0, \tag{7.10}$$

then the correlation between X_1 and X_4 will be zero, and thus $X_1 \perp\!\!\!\perp X_4$, as the joint distribution is Gaussian. However, nodes 1 and 4 are *not d*-separated by $\varnothing$, showing that the joint distribution is not faithful to $\mathcal{G}_1$. The set of B satisfying the equality constraint (7.10) has a measure of zero.

7.7 Overview of Topics

In the following few chapters, we discuss in detail causal inference and causal structure learning using DAGs as a mathematical model for causality. We conclude this chapter with a brief overview of these topics.

Causal inference is a pivotal concept in the realm of statistical analysis and research methodology. At its core, it revolves around modeling causal relationships among variables and identifying causal parent-child connections, $X \to Y$, when the variable X has a direct causal effect on Y, i.e., X is a causal parent of Y. In a graphical model approach, the causality among a set V of variables is distinctly defined by a DAG, together with experimental interventions that manipulate variables and illuminate their causal effects. In short, if the parent set $\mathrm{pa}(i)$ of a variable i is held constant through intervention, then variable i will remain unaffected by any changes in variables in the set $V \setminus \{\mathrm{pa}(i) \cup \{i\}\}$, thus highlighting the causal isolation enabled by interventions. In scenarios where multiple variables $j \in M \subset V$ undergo interventions, the recursive factorization of the joint density $f(x)$ for V will be modified as

$$f(x) = \prod_{j \notin M} f_j(x_j \mid \mathrm{pa}(j)) \prod_{j \in M} g_j(x_j),$$

where g_j is the density of X_j under intervention. This is called the truncated factorization, which illustrates how this causal inference framework accommodates and elucidates the consequences of experimental interventions on a complex system.

Structure learning of DAGs is a fundamental task in probabilistic graphical modeling, aimed at discovering the causal relationships among a set of random variables, the so-called causal discovery problem. The primary objective is to estimate an accurate and informative DAG from data to capture the CI relations among the variables. This process plays a crucial role in various fields, including machine learning, epidemiology, and economics, where understanding causal relationships is essential.

One fundamental principle in structure learning is that the sparser the estimated DAG $\widehat{\mathcal{G}}$, the more CI relations are learned from the observed data. A sparse DAG also provides a more interpretable and parsimonious representation of the underlying causal structure. Several classes of methods have been developed to tackle the problem of DAG structure learning, each with its own strengths and approaches:

- *Score-based methods*: These methods seek to find the DAG $\widehat{\mathcal{G}}$ by optimizing a scoring function over the space of possible DAGs. The scoring function measures how well a given DAG explains the observed data, balancing goodness of fit with model complexity. Regularization techniques are often employed to encourage sparsity in the resulting graph. The idea is to learn simpler DAGs that capture the essential causal relationships while avoiding overfitting.

- *Constraint-based methods*: In contrast, constraint-based methods do not rely on an optimization procedure. Instead, they make use of conditional independence constraints implied by the data. These methods systematically assess whether variables X_i and X_j are conditionally independent given a set of variables X_S, for all possible combinations of i, j, and S. By identifying CI relations, constraint-based methods build a skeleton of the DAG and determine the directions of some edges, which together define an estimated equivalence class of DAGs.
- *Hybrid methods*: These methods combine the strengths of both score-based and constraint-based approaches. They typically start with a constraint-based phase to identify CI relations and eliminate infeasible edges, thus reducing the search space. Subsequently, a score-based method is employed to fine-tune the remaining edges and identify the optimal DAG structure. This two-step process strikes a balance between hypothesis testing and combinatorial optimization.

As we see, structure learning of DAGs is a multifaceted task, involving various methodologies in statistics and machine learning. The choice of method often depends on the specific characteristics of the data and the trade-off between model complexity and computational efficiency.

7.8 Problems

(1) Show that (DG) $\Rightarrow$ (DL) $\Rightarrow$ (DP).
(2) Consider a DAG G on four nodes $\{W, X, Y, Z\}$ with the adjacency matrix

$$
A = \begin{bmatrix} 0 & 1 & 1 & 0 \\ 0 & 0 & 1 & 0 \\ 0 & 0 & 0 & 0 \\ 0 & 0 & 1 & 0 \end{bmatrix}.
$$

(a) Draw all the DAGs that are Markov equivalent to G and identify the compelled edges.
(b) Find all the subsets of nodes that d-separate X and Z.
(c) Use separations in moral graphs to verify the d-separations identified in (b).

(3) The joint distribution of $X_1, \ldots, X_4$ can be factorized as follows:

$$\mathbb{P}(X_1, \ldots, X_4) = \mathbb{P}(X_1) \prod_{i=2}^{4} \mathbb{P}(X_i \mid X_1),$$

where X_i's are binary variables on $\{1, 2\}$. The involved parameters are $\phi = \mathbb{P}(X_1 = 1)$ and $\theta = (\theta_{jk})_{2 \times 2}$, where $\theta_{jk} = \mathbb{P}(X_i = k \mid X_1 = j)$ for $j, k = 1, 2$ and $i = 2, 3, 4$.

   (a) Draw the Bayesian network that represents the given factorization of $\mathbb{P}(X_1, \ldots, X_4)$.
   (b) We have n i.i.d. observations, $(x_{ij})_{n \times 4}$, from this Bayesian network. Find the MLE of the parameters ϕ and θ.

(4) Assume the Gaussian DAG model (7.4) for $X = (X_1, \ldots, X_p)$, where the true DAG is $G_0 = (V, E_0)$. Let $\mathbb{P}$ be the distribution of X.

   (a) Show that, for any permutation π of $[p]$, there exists a DAG $G(\pi)$ such that π is a topological sort of $G(\pi)$ and X follows a Gaussian DAG model with respect to $G(\pi)$.
   (b) Suppose $\mathbb{P}$ is faithful to G_0. Let $E(\pi)$ be the edge set of $G(\pi)$. Show that $|E(\pi)| \geq |E_0|$ for all π.

(5) Since d-separation in a DAG is a compositional graphoid, the neighborhood lattice (Section 6.8) is well defined. Find the lattice $\mathscr{L}_6(S)$ on the DAG in Figure 7.1 for $S = \{1, 3\}$ and list all the d-separation statements implied by this lattice.

Chapter 8

Causal Inference Based on Directed Acyclic Graphs

Causal inference is a fundamental pursuit in understanding the relationships among variables in various domains. Directed acyclic graphs (DAGs) offer a coherent mathematical model for describing, exploring, and learning causal relationships. These graphical models provide a structured way to depict causal pathways and dependencies among variables. In DAG-based causal inference, a primary goal is to identify and calculate causal effects from observed data. By leveraging concepts such as experimental interventions and structural equations, DAGs provide a principled approach to causal inference in complex systems. In this chapter, we discuss the principles and methods of DAG-based causal inference, demonstrating its significance in the identification and estimation of causal effects from observational data.

8.1 Causal DAGs and Intervention

Following Pearl (1995), a causal model among $X_1, \ldots, X_p$ is defined by a DAG $\mathcal{G}$ over $V = [p]$ and a distribution $\mathbb{P}(\varepsilon) = \mathbb{P}(\varepsilon_1, \ldots, \varepsilon_p)$ for background variables $\varepsilon_1, \ldots, \varepsilon_p$. Let $PA_j = \{X_i : i \in \mathrm{pa}(j)\}$, where $\mathrm{pa}(j)$ is the parent set of node j in $\mathcal{G}$. In the context of a causal model, we may call PA_j the causal parents of j or X_j. The causal relation between a variable and its causal parents in $\mathcal{G}$ is specified through a structural equation model (SEM):

$$X_j = f_j(PA_j, \varepsilon_j), \quad j = 1, \ldots, p, \tag{8.1}$$

where f_j is a *deterministic* function and ε_j is a random variable. We assume that the background (error) variables are jointly independent so that

$$\mathbb{P}(\varepsilon_1, \ldots, \varepsilon_p) = \prod_{j=1}^{p} \mathbb{P}(\varepsilon_j). \tag{8.2}$$

Then, the joint distribution $\mathbb{P}(X_1, \ldots, X_p)$ is Markov with respect to (w.r.t.) the DAG $\mathcal{G}$:

$$\mathbb{P}(X_1, \ldots, X_p) = \prod_{j=1}^{p} \mathbb{P}(X_j \mid PA_j), \tag{8.3}$$

i.e., it admits a recursive factorization (DF) w.r.t. $\mathcal{G}$.

The causal effect of one variable on other variables is defined through external interventions. Consider an atomic intervention that forces X_i to some fixed value x_i, which we denote by $\mathrm{do}(X_i = x_i)$, or $\mathrm{do}(x_i)$ for short, using Pearl's do-operator (Pearl, 1995). The effect of $\mathrm{do}(x_i)$ is to replace the SEM for X_i with $X_i = x_i$ and substitute $X_i = x_i$ in the other SEMs (8.1). For two distinct sets of variables X and Y, the causal effect of X on Y is determined using the mapping

$$x \mapsto \mathbb{P}[Y \mid \mathrm{do}(X = x)],$$

i.e., how the distribution $[Y \mid \mathrm{do}(x)]$ changes with the intervention value x. We usually define the causal effect as a summary of the changes in the conditional mean $\mathbb{E}(Y \mid \mathrm{do}(x))$, as illustrated by the following examples. If the SEM is linear, as in (7.4), the causal effect of X on Y is defined by

$$\frac{\partial \mathbb{E}(Y \mid \mathrm{do}(x))}{\partial x} = \mathbb{E}(Y \mid \mathrm{do}(X = x + 1)) - \mathbb{E}(Y \mid \mathrm{do}(X = x)),$$

which is the increase in $\mathbb{E}(Y)$ caused by a one-unit increase in X. If the cause X is a binary variable, say corresponding to a treatment ($X = 1$) and control ($X = 0$), then the causal effect is the average difference in Y between the treatment and control:

$$\mathbb{E}(Y \mid \mathrm{do}(X = 1)) - \mathbb{E}(Y \mid \mathrm{do}(X = 0)).$$

It is sometimes convenient to treat interventions as additional variables (nodes) in a causal DAG. Let F_j indicate whether intervention is applied to X_j. If X_j is fixed to x by intervention, then $F_j = \mathrm{do}(x)$; otherwise, $F_j = \mathrm{idle}$. Accordingly, the SEM for X_j becomes

$$X_j = h_j(PA_j, F_j, \varepsilon_j) = \begin{cases} f_j(PA_j, \varepsilon_j) & \text{if } F_j = \mathrm{idle}, \\ x & \text{if } F_j = \mathrm{do}(x). \end{cases} \tag{8.4}$$

The parent set of X_j is augmented to $PA_j \cup \{F_j\}$. Using discrete variables for illustration,

$$\mathbb{P}(X_j = x_j \mid PA_j, F_j) = \begin{cases} \mathbb{P}(X_j = x_j \mid PA_j) & \text{if } F_j = \mathrm{idle}, \\ I(x_j = x) & \text{if } F_j = \mathrm{do}(x), \end{cases}$$

where $I(x_j = x)$ represents a unit point mass at x.

Let us illustrate the effect of intervention with the DAG in Figure 8.1(a). Suppose the SEMs for $X_1, \ldots, X_4$ are

$$X_1 = \varepsilon_1,$$
$$X_2 = X_1^2 + \varepsilon_2,$$
$$X_3 = -2X_2 + \varepsilon_3,$$
$$X_4 = X_1 + X_3 + \varepsilon_4,$$

where ε_j are independent with a mean of zero. If we fix $X_2 = x$ by intervention, then the SEM for X_2 will be replaced by

$$X_2 = x,$$

while the equations for X_1, X_3, and X_4 will remain the same as before. To determine the effect of $\mathrm{do}(X_2 = x)$ on X_4, we plug $X_1 = \varepsilon_1$ and

$$X_3 = -2X_2 + \varepsilon_3 = -2x + \varepsilon_3$$

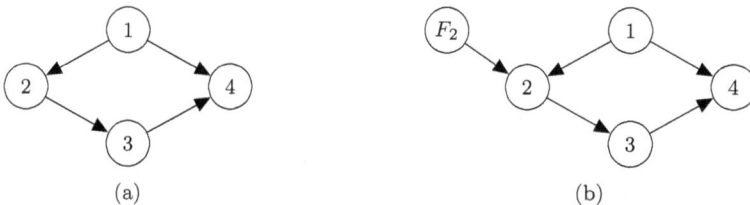

(a)   (b)

Fig. 8.1. A DAG with possible intervention on X_2. (a) A DAG over $\{X_1, X_2, X_3, X_4\}$. (b) Augmented DAG that includes intervention on X_2 as a node F_2.

into the equation for X_4:

$$X_4 = -2x + \varepsilon_1 + \varepsilon_3 + \varepsilon_4.$$

Taking expectation on both sides, we arrive at $\mathbb{E}(X_4 \mid do(X_2 = x)) = -2x$. Thus, the causal effect of X_2 on X_4 is

$$\frac{\partial \mathbb{E}(X_4 \mid do(X_2 = x))}{\partial x} = -2,$$

meaning that a one-unit increase in X_2 *causes* a two-unit decrease in the expectation of X_4.

Figure 8.1(b) shows the augmented DAG with F_2 as an additional parent of X_2. We may use this augmented DAG to infer conditional independence (CI) involving $do(X_2 = x)$. Since X_2 d-separates F_2 and X_3, we have $X_3 \perp\!\!\!\perp F_2 \mid X_2$ and thus

$$\mathbb{P}(X_3 \mid X_2, F_2) = \mathbb{P}(X_3 \mid X_2). \tag{8.5}$$

Now, consider the distribution of X_3 under $do(X_2 = x)$:

$$\mathbb{P}(X_3 \mid do(X_2 = x)) = \mathbb{P}(X_3 \mid X_2 = x, F_2 = do(x))$$
$$= \mathbb{P}(X_3 \mid X_2 = x),$$

where (8.5) is invoked in the second equality to remove F_2 from the conditioning set. We see that the intervention distribution of X_3 given $do(X_2)$ is identical to the ordinary conditional distribution $[X_3 \mid X_2]$. This reasoning leads to the following general result.

Lemma 8.1. *In an augmented DAG with F_i as nodes, if X_i d-separates F_i and X_j, then for any x in the domain of X_i,*

$$[X_j \mid do(X_i = x)] = [X_j \mid X_i = x].$$

This is, however, not the case for X_4 since X_2 does *not* d-separate F_2 and X_4 due to the chain $F_2 \to 2 \leftarrow 1 \to 4$, in which X_2 is a collider. In general, $X_4 \not\perp\!\!\!\perp F_2 \mid X_2$, and thus

$$[X_4 \mid do(X_2 = x)] \neq [X_4 \mid X_2 = x]. \tag{8.6}$$

8.2 Truncated Factorization

If some variables in a DAG are under intervention, the recursive factorization (8.3) must be modified to account for this change in the data generation process. The modified factorization is called the truncated factorization (Pearl, 2000, Section 3.2), also known as the g-formula (Robins, 1986) or the manipulation theorem (Spirtes *et al.*, 2000). It serves as the basis for computing the causal effects of interventions from the joint distribution of observed random variables.

To simplify our notation, let us assume that all X_j are discrete and write $\mathbb{P}(X = x) = P(x)$. Let $pa_j := (x_k : k \in \text{pa}(j))$ be the vector consisting of the values of PA_j.

Definition 8.1 (Truncated factorization). The truncated factorization of $P(x_1, \ldots, x_p)$ given $\text{do}(X_i = x_i^*)$ is

$$P(x_1, \ldots, x_p \mid \text{do}(x_i^*)) = I(x_i = x_i^*) \prod_{j \neq i} P(x_j \mid pa_j). \qquad (8.7)$$

For multiple intervention $\text{do}(X_S = \mathbf{x}^*)$, $S \subseteq \{1, \ldots, p\}$,

$$P(x_1, \ldots, x_p \mid \text{do}(\mathbf{x}^*)) = I(x_S = \mathbf{x}^*) \prod_{j \notin S} P(x_j \mid pa_j). \qquad (8.8)$$

The truncated factorization is an immediate consequence of replacing the SEM for X_i with $X_i = x_i^*$ for $i \in S$.

According to the factorization formula (8.7), X_i is independent of its parents PA_i under $\text{do}(X_i = x_i^*)$. Therefore, $\text{do}(X_i)$ essentially modifies the DAG by deleting the edges $X_j \to X_i$ for all $j \in \text{pa}(i)$. Denote the modified DAG by $\mathcal{G}_{\bar{X}_i}$. To emphasize that the value of X_i is *fixed*, we represent X_i by a square node in $\mathcal{G}_{\bar{X}_i}$. The DAG in Figure 8.1(a) under $\text{do}(X_2)$ is shown in Figure 8.2. As demonstrated in (8.6), $P(x_4 \mid \text{do}(x_2)) \neq P(x_4 \mid x_2)$ in this example. The truncated factorization provides a means to calculate $P(x_j \mid \text{do}(x_i))$ from the joint distribution $P(x_1, \ldots, x_p)$.

Assume that $\mathbb{P}(X_i = x_i^* \mid PA_i = pa_i) > 0$ for any possible pa_i. Putting $x_i = x_i^*$ in (8.7), we have

$$P(x_{-i} \mid \text{do}(x_i^*)) = \prod_{j \neq i} P(x_j \mid pa_j) \cdot \frac{P(x_i^* \mid pa_i)}{P(x_i^* \mid pa_i)}$$

$$= \frac{P(x_1, \ldots, x_p)}{P(x_i^* \mid pa_i)}$$

$$= P(x_j, j \in B \mid x_i^*, pa_i) P(pa_i), \qquad (8.9)$$

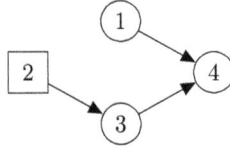

Fig. 8.2.   Modified DAG $\mathcal{G}_{\bar{X}_2}$ under intervention do(X_2).

where $B = [p] \setminus \{i, \mathrm{pa}(i)\}$. Keep in mind that $\mathrm{pa}(i) \subset [p]$ is the parent set of node i and pa_i denotes the value of the random variables PA_i.

Remark 8.1. As the intervention event do(x_i^*) does *not* appear on the right-hand side, Equation (8.9) translates an intervention probability $P(x_{-i} \mid \mathrm{do}(x_i^*))$ into a quantity that involves only the conditional and marginal probabilities (also called pre-intervention probabilities) that are determined by the joint distribution $P(x)$. This provides a means to identify causal effects, defined by intervention probabilities, from pre-intervention probabilities that can be estimated from observational data.

The first method for identifying causal effects is through parent set adjustment (Pearl, 2000, Theorem 3.2.2).

Theorem 8.1 (Parent set adjustment). *Let PA_i be the parents of X_i and Y be any set of other variables in a causal DAG $\mathcal{G}$. If $P(x_i \mid pa_i) > 0$ for every possible value pa_i, then the causal effect of* do($X_i = x_i$) *on Y is given by*

$$P(y \mid \mathrm{do}(x_i)) = \sum_{pa_i} P(y \mid x_i, pa_i) P(pa_i), \qquad (8.10)$$

where $P(y \mid x_i, pa_i)$ and $P(pa_i)$ are pre-intervention probabilities.

Proof. Let $Y = X_A$, for some $A \subseteq B = [p] \setminus \{i, \mathrm{pa}(i)\}$. By definition,

$$[p] \setminus (A \cup \{i\}) = (B \setminus A) \cup \{\mathrm{pa}(i)\}.$$

Marginalize out all $X_j, j \in [p] \setminus (A \cup \{i\})$ on both sides of (8.9) while using the above relationship among the subsets:

$$P(x_A \mid \mathrm{do}(x_i^*)) = \sum_{pa_i} \sum_{x_{B \setminus A}} P(x_j, j \in B \mid x_i^*, pa_i) P(pa_i),$$

$$= \sum_{pa_i} P(x_A \mid x_i^*, pa_i) P(pa_i).$$

Noting that $x_A = y$ completes the proof.   □

Remark 8.2. The significance of Theorem 8.1 is that, once we have knowledge about the causal parents of X_i (the cause), we may use (8.10) to calculate its causal effect on any variable Y. Both $P(y \mid x_i, pa_i)$ and $P(pa_i)$ can
be estimated from observational data on (Y, X_i, PA_i) without any experimental interventions. This shows an example of estimating causal effects
from observational data.

Now, we may use parent set adjustment to calculate $P(x_4 \mid \mathrm{do}(x_2))$ in
Figure 8.1(a). Since pa(2) = {1}, by (8.10), we have

$$P(x_4 \mid \mathrm{do}(x_2)) = \sum_{x_1} P(x_4 \mid x_2, x_1) P(x_1),$$

where the conditional distribution $[X_4 \mid X_2, X_1]$ and the marginal distribution $[X_1]$ can be obtained from the joint distribution of $(X_1, \ldots, X_4)$.
Therefore, given observed data from the joint distribution, we can estimate
both $P(x_4 \mid x_2, x_1)$ and $P(x_1)$, for all possible x_1, and use these estimated
probabilities to calculate $P(x_4 \mid \mathrm{do}(x_2))$.

A simple implication of Theorem 8.1 is that X_i has no causal effect on
its non-descendants. If Y is a set of non-descendants of X_i, then by local
Markov property (DL),

$$Y \perp\!\!\!\perp X_i \mid PA_i,$$

so that $P(y \mid x_i, pa_i) = P(y \mid pa_i)$. Plugging this into (8.10),

$$P(y \mid \mathrm{do}(x_i)) = \sum_{pa_i} P(y \mid x_i, pa_i) P(pa_i)$$

$$= \sum_{pa_i} P(y \mid pa_i) P(pa_i) = P(y),$$

which shows that Y is independent of intervention on X_i. Thus, X_i has no
causal effect on Y.

8.3 Linear Structural Equation Models

Linear SEMs are a class of causal models employed extensively in practice.
By explicitly modeling the child-parent relations (8.1) through a set of
linear equations, linear SEMs enable researchers to investigate the direct
and indirect causal effects through the identification of causal pathways.

More specifically, we assume a linear model for each child-parent relationship in the causal DAG $\mathcal{G}$:

$$X_j = \sum_{i \in \mathrm{pa}(j)} \beta_{ij} X_i + \varepsilon_j, \qquad j = 1, \ldots, p, \qquad (8.11)$$

where ε_j are independent and $\mathbb{E}(\varepsilon_j) = 0$. When it is further assumed that $\varepsilon_j \sim \mathcal{N}(0, \omega_j^2)$ for all $j \in [p]$, the DAG is called a Gaussian DAG, and the graphical model is a Gaussian Bayesian network (Section 7.4.1).

As the causal relation between variables in a linear SEM is linear, the causal effect of X_k on X_j is

$$\gamma_{kj} := \frac{\partial \mathbb{E}(X_j \mid \mathrm{do}(X_k = x))}{\partial x}$$

$$= \mathbb{E}(X_j \mid \mathrm{do}(X_k = c + 1)) - \mathbb{E}(X_j \mid \mathrm{do}(X_k = c)), \qquad (8.12)$$

for any c. In other words, $\mathbb{E}(X_j \mid \mathrm{do}(X_k = x)) = \gamma_{kj} x$ since all variables are centered in (8.11). Let $\mathcal{P}(k, j)$ be the set of all (directed) paths from X_k to X_j in $\mathcal{G}$. If $\mathcal{P}(k, j)$ is not empty, the causal effect γ_{kj} can be represented by the coefficients along these paths (Wright, 1934):

$$\gamma_{kj} = \sum_{\pi \in \mathcal{P}(k,j)} \prod_{(a,b) \in \pi} \beta_{ab}, \qquad (8.13)$$

where we treat a path π as a collection of edges. If $\mathcal{P}(k, j) = \varnothing$, then $\gamma_{kj} = 0$, which happens if and only if j is a non-descendant of k. Using the modified DAG $\mathcal{G}_{\bar{X}_k}$ after the intervention $\mathrm{do}(X_k)$, we have

$$\mathbb{E}(X_j \mid \mathrm{do}(X_k = x)) = \mathbb{E}(X_j \mid X_k = x; \mathcal{G}_{\bar{X}_k}) = \gamma_{kj} x,$$

where $\mathbb{E}(\bullet \, ; \mathcal{G}_{\bar{X}_k})$ takes expectation w.r.t. $\mathcal{G}_{\bar{X}_k}$.

Let us consider the causal effect γ_{25} in Figure 8.3(a), in which

$$\mathcal{P}(2, 5) = \{2 \to 4 \to 5, \ 2 \to 3 \to 4 \to 5\}.$$

Now, by (8.13), we find

$$\gamma_{25} = \beta_{24} \beta_{45} + \beta_{23} \beta_{34} \beta_{45}. \qquad (8.14)$$

To see why this is true, we work with the modified DAG $\mathcal{G}_{\bar{X}_2}$ after performing $\mathrm{do}(X_2 = x_2)$, shown in Figure 8.3(b). We write X_5 as a function of x_2 and ε_j, $j \neq 2$, which turns out to be

$$X_5 = (\beta_{24} \beta_{45} + \beta_{23} \beta_{34} \beta_{45}) x_2 + h(\varepsilon_1, \varepsilon_3, \varepsilon_4, \varepsilon_5),$$

where h is a linear function. This shows that γ_{25} is indeed given by (8.14).

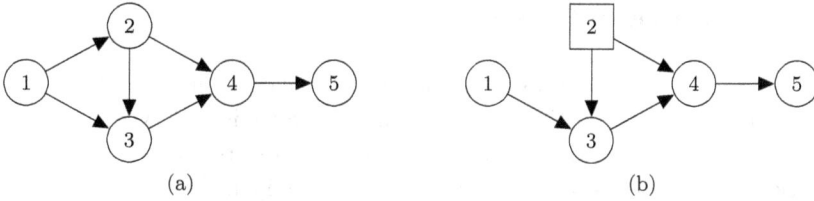

Fig. 8.3. (a) A DAG $\mathcal{G}$ over five variables $\{X_1, \ldots, X_5\}$. (b) Modified DAG $\mathcal{G}_{\bar{X}_2}$ under $do(X_2)$.

When only observational data are available, we apply Theorem 8.1 to find γ_{kj}. Let $Z = PA_k$ and z denote the value of PA_k. Then,

$$p(x_j \mid do(X_k = x_k)) = \int_z p(x_j \mid x_k, z)p(z)dz,$$

where the densities on the right-hand side are given by the pre-intervention distribution $p(x_1, \ldots, x_p)$. This equation implies that

$$\mathbb{E}(X_j \mid do(X_k = x_k)) = \int_z \mathbb{E}(X_j \mid x_k, z)p(z)dz.$$

For linear SEMs,

$$\mathbb{E}(X_j \mid X_k, Z) = \beta X_k + \alpha^\mathsf{T} Z,$$

where β and α are constant coefficients, and thus the causal effect is

$$
\begin{aligned}
\gamma_{kj} &= \frac{\partial}{\partial x_k} \mathbb{E}(X_j \mid do(X_k = x_k)) \\
&= \frac{\partial}{\partial x_k} \int_z \left\{ \beta x_k + \alpha^\mathsf{T} z \right\} p(z)dz = \beta.
\end{aligned}
$$

Consequently, to calculate γ_{kj}, we perform a linear regression of X_j on (X_k, PA_k), where the regression coefficient β of X_k is the causal effect γ_{kj}. For short, we write

$$\gamma_{kj} = \beta_{X_k}(X_j \sim X_k + PA_k). \tag{8.15}$$

We summarize this result in a corollary.

Corollary 8.1. *For the linear SEM* (8.11), *the causal effect of X_k on X_j can be computed using* (8.15).

When observed data are collected, $\widehat{\gamma}_{kj}$ is usually obtained via the corresponding least-squares regression. Applying this method to the DAG in Figure 8.3(a), the causal effect is $\gamma_{25} = \beta_{X_2}(X_5 \sim X_2 + X_1)$.

8.4   Estimation of Causal Effect

Theorem 8.1 shows that we may estimate any causal effect from observational data as long as the parent set of the cause is observed. However, this method is not applicable if some of the parents are unobserved. In this section, we discuss several general methods for identifying causal effects from observational data, allowing the presence of latent variables in the underlying causal DAG.

Definition 8.2 (Identifiable causal effect). Let X and Y be subsets of observed variables in a causal DAG $\mathcal{G}$. Given $\mathcal{G}$, if the distribution $[Y \mid \mathrm{do}(X)]$ can be uniquely computed from the (pre-intervention) distributions of observed variables in $\mathcal{G}$, then we say that the causal effect of X on Y is identifiable.

Note that we allow unobserved nodes in $\mathcal{G}$, and we assume that only observational data are collected.

Figure 8.4 shows a DAG over four variables, an example in Pearl (2000, Section 3.3), where X, Y, and Z are observed and U is hidden. The hidden node U is a common parent of X and Y, usually called a latent confounder. Suppose we have access to an infinite amount of observational data for (X, Y, Z). Can we estimate the causal effects of X on Z, Z on Y, and X on Y from these data? If we can construct consistent estimates, then these causal effects are identifiable.

8.4.1   *Back-door adjustment*

Theorem 8.1 is a special case of back-door adjustment: PA_X blocks all back-door paths from X to Y and satisfies the so-called back-door criterion relative to X and Y.

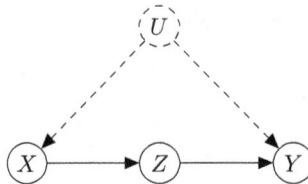

Fig. 8.4.   A DAG over three observed variables (X, Y, Z) and a hidden confounder U.

Definition 8.3 (Back-door criterion). A set of variables Z satisfies the *back-door criterion* relative to an ordered pair of variables (X, Y) in a DAG $\mathcal{G}$ if:

(i) no node in Z is a descendant of X;
(ii) Z blocks every chain between X and Y that contains an arrow into X (back-door path).

If the back-door criterion is satisfied, then we may use back-door adjustment (Pearl, 2000, Theorem 3.3.2) to identify the causal effects.

Theorem 8.2 (Back-door adjustment). *If Z satisfies the back-door criterion relative to an ordered pair of variables (X, Y) in a DAG, then the causal effect of X on Y is given by*

$$P(y \mid \mathrm{do}(x)) = \sum_z P(y \mid x, z) P(z). \tag{8.16}$$

Proof. Let us add the intervention variable F_X and the edge $F_X \to X$ to the DAG. Since Z does not contain any descendant of X, $P(z \mid \mathrm{do}(x)) = P(z)$. Then,

$$P(y \mid \mathrm{do}(x)) = \sum_z P(y \mid \mathrm{do}(x), z) P(z \mid \mathrm{do}(x))$$

$$= \sum_z P(y \mid F_X = \mathrm{do}(x), x, z) P(z). \tag{8.17}$$

There are two possible sets of chains from F_X to Y, in the form of either $\mathcal{P}_1 = \{F_X \to X \to \cdots Y\}$ or $\mathcal{P}_2 = \{F_X \to X \leftarrow \cdots Y\}$. The chains in $\mathcal{P}_1$ are all blocked by X. Every chain in $\mathcal{P}_2$ contains a back-door path $X \leftarrow \cdots Y$ and is thus blocked by Z. These two observations imply that (X, Z) d-separates F_X and Y, and thus

$$P(y \mid \mathrm{do}(x), z) = P(y \mid F_X = \mathrm{do}(x), x, z) = P(y \mid x, z). \tag{8.18}$$

Plugging this into (8.17) completes the proof. $\qquad\square$

For linear SEMs, the back-door adjustment (8.16) is implemented via regression of Y on (X, Z), similar to (8.15). The causal effect is

$$\gamma_{X \to Y} := \frac{\partial}{\partial x} \mathbb{E}(Y \mid \mathrm{do}(x)) = \beta_X (Y \sim X + Z). \tag{8.19}$$

We may further work out $\mathrm{Var}(Y \mid \mathrm{do}(x))$ by generalizing Proposition 2 given by Huang and Zhou (2023).

Corollary 8.2. *Assume that Z satisfies the back-door criterion relative to two variables (X, Y) in a linear SEM (8.11). Then, the causal effect $\gamma_{X \to Y}$*

is identifiable via (8.19) *and*

$$\mathrm{Var}(Y \mid \mathrm{do}(x)) = \mathrm{Var}(Y - \gamma_{X \to Y} X), \qquad (8.20)$$

where the variance on the right-hand side is taken w.r.t. the pre-intervention distribution of (X, Y).

Proof. We calculate $\mathrm{Var}(Y \mid \mathrm{do}(x))$ by conditioning on Z:

$$\mathrm{Var}(Y \mid \mathrm{do}(x)) = \mathbb{E}[\mathrm{Var}(Y \mid Z, \mathrm{do}(x)) \mid \mathrm{do}(x)] + \mathrm{Var}[\mathbb{E}(Y \mid Z, \mathrm{do}(x)) \mid \mathrm{do}(x)]$$

$$= \mathbb{E}[\mathrm{Var}(Y \mid Z, x) \mid \mathrm{do}(x)] + \mathrm{Var}[\mathbb{E}(Y \mid Z, x) \mid \mathrm{do}(x)],$$

where we used (8.18) for the second equality. The conditional distribution $[Y \mid Z, X]$ is given by a linear regression:

$$Y = \gamma_{X \to Y} X + \alpha^{\mathsf{T}} Z + \varepsilon, \qquad (8.21)$$

where $\varepsilon \perp\!\!\!\perp X, Z$ with a variance of σ^2. Therefore,

$$\mathrm{Var}(Y \mid Z, x) = \sigma^2, \qquad \mathbb{E}(Y \mid Z, x) = \gamma_{X \to Y} x + \alpha^{\mathsf{T}} Z.$$

Now, we have

$$\mathrm{Var}(Y \mid \mathrm{do}(x)) = \mathbb{E}[\sigma^2 \mid \mathrm{do}(x)] + \mathrm{Var}[\gamma_{X \to Y} x + \alpha^{\mathsf{T}} Z \mid \mathrm{do}(x)]$$

$$= \sigma^2 + \mathrm{Var}(\alpha^{\mathsf{T}} Z)$$

since $p(z \mid \mathrm{do}(x)) = p(z)$. According to the linear regression model (8.21),

$$\mathrm{Var}(Y - \gamma_{X \to Y} X) = \mathrm{Var}(\alpha^{\mathsf{T}} Z + \varepsilon) = \sigma^2 + \mathrm{Var}(\alpha^{\mathsf{T}} Z),$$

and thus (8.20) follows. $\square$

     Suppose we have observed data for the random variables (X, Y, Z), where Z could be a random vector, and want to estimate the causal effect X on Y from the data.

(i) If the variables are discrete, we estimate $P(y \mid x, z)$ and $P(z)$ for each (x, y, z), say, by the respective maximum likelihood estimate. Then, we plug these estimated probabilities into (8.16) to estimate $P(y \mid \mathrm{do}(x))$.

(ii) For linear SEMs, we perform regression of Y on (X, Z) to obtain

$$\widehat{\gamma}_{X \to Y} = \widehat{\beta}_X (Y \sim X + Z).$$

Consider the causal effect of X_3 on X_5 in the DAG in Figure 8.3(a), assuming a linear SEM. There are two back-door paths, $3 \leftarrow 1 \rightarrow 2 \rightarrow 4 \rightarrow 5$ and $3 \leftarrow 2 \rightarrow 4 \rightarrow 5$, from X_3 to X_5, both blocked by X_2. Thus,

$$\gamma_{35} = \beta_{X_3}(X_5 \sim X_3 + X_2).$$

This shows that we can identify the causal effect γ_{35} even if X_1 or X_4 is hidden. Since $\mathrm{pa}(3) = \{1, 2\}$, applying parent set adjustment (8.15), we arrive at

$$\gamma_{35} = \beta_{X_3}(X_5 \sim X_3 + X_1 + X_2). \tag{8.22}$$

In fact, the above two linear regression models are equivalent at the population level. Note that $\{X_2, X_3\}$ d-separates X_1 and X_5 in the DAG; therefore, $X_5 \perp\!\!\!\perp X_1 \mid \{X_2, X_3\}$. This implies that the regression coefficient of X_1 in (8.22) is zero, and thus X_1 can be removed from the model.

For the DAG in Figure 8.4, we can identify the causal effect of X on Z and that of Z on Y using back-door adjustment without observing U:

$$P(z \mid \mathrm{do}(x)) = P(z \mid x), \tag{8.23}$$

$$P(y \mid \mathrm{do}(z)) = \sum_x P(y \mid x, z) P(x). \tag{8.24}$$

It is easy to see that X satisfies the back-door criterion relative to (Z, Y). We may view (8.23) as a special case of an empty adjustment set, as the only back-door path $X \leftarrow U \rightarrow Y \leftarrow Z$ is blocked by $\varnothing$ due to the collider Y.

8.4.2 *Front-door adjustment*

Now, let us examine if $P(y \mid \mathrm{do}(x))$ is identifiable in Figure 8.4. Since both $P(z \mid \mathrm{do}(x))$ and $P(y \mid \mathrm{do}(z))$ are identifiable, we may use Z as an intermediate variable to help the calculation through averaging over z:

$$P(y \mid \mathrm{do}(x)) = \sum_z P(y, z \mid \mathrm{do}(x))$$

$$= \sum_z P(z \mid \mathrm{do}(x)) P(y \mid z, \mathrm{do}(x)).$$

In the modified graph $\mathcal{G}_{\bar{X}}$ augmented with F_Z, it is easy to see that Z d-separates F_Z and Y, and thus $Y \perp\!\!\!\perp F_Z \mid \{Z, \mathrm{do}(x)\}$. It then follows that

$$P(y \mid z, \mathrm{do}(x)) = P(y \mid \mathrm{do}(z), \mathrm{do}(x)) = P(y \mid \mathrm{do}(z))$$

and, consequently,

$$P(y \mid \mathrm{do}(x)) = \sum_z P(z \mid \mathrm{do}(x)) P(y \mid \mathrm{do}(z)). \qquad (8.25)$$

Now, we plug (8.23) and (8.24) into (8.25) to get

$$P(y \mid \mathrm{do}(x)) = \sum_z P(z \mid x) \sum_{x'} P(y \mid x', z) P(x'), \qquad (8.26)$$

where x' ranges over the domain of X.

Equation (8.26) is an example of *front-door adjustment* (Pearl, 2000, Theorem 3.3.4).

Definition 8.4 (Front-door criterion). A set of variables Z satisfies the front-door criterion relative to (X, Y) if:

(a) Z intercepts all directed paths from X to Y;
(b) there is no back-door path from X to Z;
(c) all back-door paths from Z to Y are blocked by X.

Theorem 8.3 (Front-door adjustment). *If Z satisfies the front-door criterion relative to (X, Y), then*

$$P(y \mid \mathrm{do}(x)) = \sum_z P(z \mid x) \sum_{x'} P(y \mid x', z) P(x'). \qquad (8.27)$$

Proof. By similar reasoning for (8.25), Condition (a) implies

$$P(y \mid \mathrm{do}(x)) = \sum_z P(z \mid \mathrm{do}(x)) P(y \mid \mathrm{do}(z)).$$

Back-door adjustment with (b) shows that $P(z \mid \mathrm{do}(x)) = P(z \mid x)$. Condition (b) also implies that X does not contain any descendant of Z because otherwise there would be a back-door path from some node in X to Z, such as $Z \rightarrow \cdots \rightarrow X_i$ for some $X_i \in X$. Together with (c), we see that X satisfies the back-door criterion relative to (Z, Y). Now, applying back-door adjustment, we have

$$P(y \mid \mathrm{do}(z)) = \sum_{x'} P(y \mid x', z) P(x').$$

This completes the proof. □

For linear SEMs, Theorem 8.3 can be used to find

$$\gamma_{X \to Y} = \gamma_{X \to Z} \times \gamma_{Z \to Y} = \beta_X(Z \sim X) \times \beta_Z(Y \sim Z + X).$$

That is, the causal effect $\gamma_{X \to Y}$ can be identified through two linear regression problems: $Z \sim X$ and $Y \sim Z + X$.

Pearl (2000, Section 3.4) developed the do-calculus, which is a set of inference rules for transforming intervention and pre-intervention probabilities. These rules may be applied to translate causal effects, defined by intervention distributions, to conditional distributions, such as in front-door adjustments (8.27). This allows us to estimate causal effects from observational data by estimating the involved conditional distributions.

8.4.3 *Instrumental variables*

The instrumental variable formula (Bowden and Day, 1984) is a powerful tool in causal inference. It is used to estimate causal effects in observational studies with hidden confounding variables. The instrumental variable method relies on a special type of variable, known as an instrumental variable, that affects the treatment variable of interest but is unrelated to the outcome except through its influence on the treatment. By using this instrument, we can isolate and estimate the causal effect of the treatment, even in the presence of confounding. This method has been applied in various fields, such as economics, epidemiology, and social sciences.

To simplify exposition, we assume linear SEMs throughout this section. The basic idea of an instrumental variable is shown in Figure 8.5. In this DAG, X, Y, and Z are observed, while U is a hidden common parent of X and Y. The coefficients in the linear SEM along the edges $Z \to X$ and $X \to Y$ are, respectively, α_1 and α_2. We want to identify the causal effect of X on Y, i.e., $\gamma_{X \to Y}$. Using the path coefficient representation (8.13), we see that $\gamma_{X \to Y} = \alpha_2$ and $\gamma_{Z \to Y} = \alpha_1 \alpha_2$.

For this DAG, neither back-door adjustment nor front-door adjustment can be applied to identify $\gamma_{X \to Y}$. However, since there are no back-door

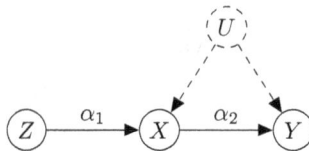

Fig. 8.5. Instrumental variable Z for estimating the causal effect of X on Y.

paths between Z and (X, Y), the causal effects of Z on X and Y are identifiable. Then, we can find $\gamma_{X \to Y}$ using the simple identity

$$\gamma_{X \to Y} = \alpha_2 = \frac{\alpha_1 \alpha_2}{\alpha_1} = \frac{\gamma_{Z \to Y}}{\gamma_{Z \to X}}.$$

This idea motivates a three-step procedure that uses Z as an instrumental variable:

(i) Since Z has no parents, α_1 is identifiable by regressing X on Z:

$$\alpha_1 = \beta_Z(X \sim Z).$$

(ii) Similarly, the causal effect of Z on Y, $\alpha_1 \alpha_2$, is also identifiable:

$$\alpha_1 \alpha_2 = \beta_Z(Y \sim Z).$$

(iii) Combined, we arrive at the *instrumental variable formula*:

$$\alpha_2 = \frac{\beta_Z(Y \sim Z)}{\beta_Z(X \sim Z)} = \frac{\text{Cov}(Y, Z)}{\text{Cov}(X, Z)}. \tag{8.28}$$

As a practical example, suppose we wish to study the causal effect of college education (X) on the job after college (Y). In this problem, X and Y are confounded by the family's social and educational background (U), which is usually latent or difficult to measure. Therefore, the causal effect cannot be identified from observational data of (X, Y). Let Z denote a randomly assigned high-school fellowship exclusively for college preparation and application. Since it is randomly assigned, $Z \perp\!\!\!\perp U$ and any causal effect of Z on Y must happen through its effect on X. The relation among the four variables is consistent with the DAG in Figure 8.5, in which Z serves as an instrumental variable.

In practice, the instrumental variable formula is implemented conveniently via a two-stage least-squares procedure:

(i) Regress X on Z to get $\alpha_1 = \beta_Z(X \sim Z)$, and let $\widehat{X} = \alpha_1 Z$.
(ii) Regress Y on $\widehat{X}$. Then, $\alpha_2 = \beta_{\widehat{X}}(Y \sim \widehat{X})$.

To see why this procedure works, note that

$$\beta_{\widehat{X}}(Y \sim \widehat{X}) = \frac{\text{Cov}(Y, \alpha_1 Z)}{\text{Var}(\alpha_1 Z)} = \frac{\text{Cov}(Y, Z)}{\alpha_1 \text{Var}(Z)}$$
$$= \frac{\beta_Z(Y \sim Z)}{\alpha_1} = \frac{\alpha_1 \alpha_2}{\alpha_1} = \alpha_2.$$

As usual, to estimate α_2 from samples of (X, Y, Z), we replace the involved regression coefficients with their corresponding least-squares estimates.

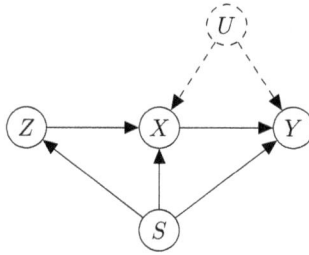

Fig. 8.6. Example DAG to demonstrate conditional instrumental variables.

Brito and Pearl (2002a) generalized the instrumental variable method to conditional instrumental variables. A variable Z is said to be a *conditional instrumental variable* given a subset S of variables relative to (X, Y) if:

(i) S contains no descendants of X or Y;
(ii) S d-separates Z from Y but not from X in the graph obtained after deleting all edges emerging from X.

Then, the causal effect of X on Y is

$$\gamma_{X \to Y} = \frac{\beta_Z(Y \sim Z + S)}{\beta_Z(X \sim Z + S)} = \frac{\mathrm{Cov}(Y, Z \mid S)}{\mathrm{Cov}(X, Z \mid S)}, \tag{8.29}$$

where S is used as a back-door adjustment set for identifying the causal effects of Z on Y and on X. When $S = \varnothing$, the above equation reduces to the instrumental variable formula (8.28).

In the DAG in Figure 8.6, Z is a conditional instrumental variable given S relative to (X, Y). Although the confounder U is unobserved, Equation (8.29) provides a means to identify the causal effect of X on Y.

8.5 Potential Outcome Approach

A different approach to causal inference is the potential outcome framework (Rubin, 1990). The key idea is that each individual or unit in a study has multiple potential outcomes that would happen under different treatment conditions or experimental interventions. These potential outcomes are often denoted as $Y(x)$, a counterfactual entity representing the outcome of Y had X been x. Suppose $X \in \{0, 1\}$ indicates control ($X = 0$) versus treatment ($X = 1$). Then, $Y(1)$ denotes the outcome under treatment, while $Y(0)$ denotes the outcome under control (without treatment).

Table 8.1.   Dataset with potential outcomes.

units	X	Y	$Y(1)$	$Y(0)$	Z (covariates)
1	1	y_1	y_1	?	$\checkmark,\dots,\checkmark$
2	1	y_2	y_2	?	$\checkmark,\dots,\checkmark$
3	1	y_3	y_3	?	$\checkmark,\dots,\checkmark$
4	0	y_4	?	y_4	$\checkmark,\dots,\checkmark$
5	0	y_5	?	y_5	$\checkmark,\dots,\checkmark$

Note: "?" indicates missing data.

By comparing these potential outcomes, we can estimate causal effects and make informed decisions, helping us understand the impact of experimental interventions on the outcomes of interest. Under this setting, the causal effects of interest include the average treatment effect $\mathbb{E}[Y(1) - Y(0)]$, and the average treatment effect on the treated $\mathbb{E}[Y(1) - Y(0) \mid X = 1]$.

An example dataset is shown in Table 8.1, where X is the treatment indicator and Y is the actual observed outcome. Suppose our goal is to estimate the average treatment effect. For each unit under treatment $i \in \{1, 2, 3\}$, we observe its actual outcome $Y_i(1) = y_i$, but its potential outcome without treatment $Y_i(0)$ is missing. For each unit in the control group $i \in \{4, 5\}$, $Y_i(0) = y_i$ is observed while $Y_i(1)$ is missing. Moreover, we have collected a set of covariates Z_i for each unit i in this study. Note that X is not randomly assigned and could depend on the covariates Z. If all outcomes were observed, then the average treatment effect (ATE) would be estimated as

$$\widehat{\text{ATE}} = \frac{1}{n}\sum_{i=1}^{n}[Y_i(1) - Y_i(0)].$$

However, in reality, half of these outcomes are missing. Therefore, one must specify a model to predict or impute the unobserved potential outcomes.

We use $\mathbb{P}$ to denote the distribution of observed outcomes and variables. Let $\mathbb{P}^*[Y(x)]$ be the distribution of the potential outcome $Y(x)$, where $\mathbb{P}^*$ could be different from the distribution $\mathbb{P}$ of the observed outcome, such as $Y(1)$ for the treated. Compared to the definition of do-operators, $\mathbb{P}^*[Y(x)]$ corresponds to $\mathbb{P}(Y \mid \text{do}(x))$, but we are not assuming a causal DAG model under the potential outcome framework. Instead, additional assumptions are made to calculate $\mathbb{P}^*[Y(x)]$. For example, when $X = x$ is actually observed, we must have $\mathbb{P}^*[Y(x) = y] = \mathbb{P}(Y = y \mid x)$, consistent with the observed data distribution. Another common assumption is *conditional ignorability* (Rosenbaum and Rubin, 1983):

$$Y(x) \perp\!\!\!\perp X \mid Z, \qquad x \in \{0, 1\}.$$

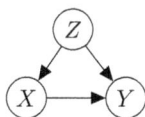

Fig. 8.7. A DAG model for conditional ignorability.

That is, given the covariates Z, the potential outcomes $\{Y(1), Y(0)\}$ are independent of the received treatment X. Under this assumption,

$$\mathbb{P}^*(Y(x) = y) = \sum_z \mathbb{P}^*(Y(x) = y \mid z)P(z)$$

$$= \sum_z \mathbb{P}^*(Y(x) = y \mid x, z)P(z) \quad \text{(by conditional ignorability)}$$

$$= \sum_z \mathbb{P}(Y = y \mid x, z)P(z), \tag{8.30}$$

which coincides with back-door adjustment (Theorem 8.2). Assume that for the observed data, $[Y \mid X, Z]$ is given by a linear model:

$$Y = \gamma X + \alpha^\mathsf{T} Z + \varepsilon. \tag{8.31}$$

Then, by (8.30),

$$\mathbb{E}[Y(x)] = \int (\gamma x + \alpha^\mathsf{T} z)p(z)dz = \gamma x + \alpha^\mathsf{T}\mathbb{E}(Z),$$

and consequently,

$$\text{ATE} = \mathbb{E}[Y(1) - Y(0)] = \gamma.$$

This means that we can estimate ATE through the linear regression (8.31) using the observed data (X, Y, Z), which is identical to the back-door adjustment for linear SEMs.

The conditional ignorability assumption is closely related to the back-door criterion. Suppose that (X, Y, Z) is generated by a causal DAG model shown in Figure 8.7 with the following SEMs:

$$Z = f_Z(\varepsilon_Z), \quad X = f_X(Z, \varepsilon_X), \quad Y = f_Y(X, Z, \varepsilon_Y),$$

where $\varepsilon_X, \varepsilon_Y$, and ε_Z are mutually independent. Note that $Y(x)$ is equivalent to Y under $\text{do}(x)$. Then, the potential outcome

$$Y(x) = f_Y(x, Z, \varepsilon_Y) = h_x(Z, \varepsilon_Y)$$

is a function of Z and ε_Y for any fixed x. Since ε_X, ε_Y, and Z are mutually independent, we have

$$h_x(Z, \varepsilon_Y) \perp\!\!\!\perp f_X(Z, \varepsilon_X) \mid Z,$$

which recovers the conditional ignorability assumption.

8.6   Problems

(1) For each pair of nodes (i, j) in Figure 8.1(a), determine whether

$$[X_j \mid do(X_i = x)] = [X_j \mid X_i = x].$$

(2) Assume a linear SEM for the variables in the DAG in Figure 8.6, where U is a latent confounder. Modify the two-stage least-squares method to identify the causal effect of X on Y.

(3) In the following DAG, W, X, Y, Z are observed while U is hidden. Is the causal effect of X on Y identifiable? If so, express $P(y \mid do(x))$ with pre-intervention probabilities. If not, disprove the statement.

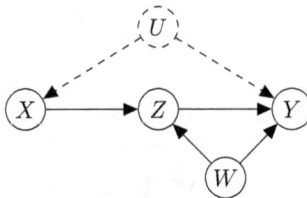

Fig. 8.8.   A DAG with a hidden confounder U.

(4) Suppose both U and W are unobserved in Figure 8.8. Show that the causal effect of X on Y is not identifiable in general.

Hint: Let θ be the parameter of the DAG. Construct an example in which the mapping from θ to the distribution $\mathbb{P}(X, Y, Z)$ is not one-to-one.

Chapter 9

Structure Learning of Directed Acyclic Graphs

Structure learning of directed acyclic graphs (DAGs) is an important problem in causal discovery, aimed at identification of causal relationships among variables in complex systems from data. By encoding causality with a DAG, causal discovery is achieved through the learning of the graphical structure of the causal DAG that generates the data. This approach often relies on statistical and computational methods to analyze observational or experimental data and to infer the most plausible graphs. Structure learning of DAGs plays a crucial role in many fields, such as biology, epidemiology, economics, and artificial intelligence, offering insights into causal relationships that help decision-making and predictive modeling.

From observational data alone, we may *not* be able to identify the orientation of all edges in the underlying causal DAG. What we can learn is typically an equivalence class of DAGs, in which a subset of the edges has invariant orientations. These directed edges encode the causality learned from observational data. The orientation of the other edges cannot be identified using purely observational data but may be estimated using experimental data.

9.1 Overview and Assumptions

We first give a brief overview of classical structure learning methods, following the discussion by Aragam *et al.* (2019b). The various algorithms in the literature for structure learning of DAGs or Bayesian networks fall into three main categories: constraint-based methods, score-based methods, and hybrid methods.

Constraint-based methods rely on repeated conditional independence (CI) tests in order to learn the structure of a DAG. The main idea is to determine which edges cannot exist in a DAG using statistical tests of independence, a procedure which is justified under the faithfulness assumption. These algorithms first use CI tests to learn the skeleton of the DAG and then orient v-structures along with the rest of the edges. Because of the existence of Markov-equivalent DAGs, the direction of some edges may not be decided. The PC algorithm proposed by Spirtes and Glymour (1991) is a well-known example of this kind of method. Another example is the fast causal inference (FCI) algorithm (Spirtes *et al.*, 2000; Zhang, 2008b), which allows for latent variables in the DAG. The output of these algorithms is a partially directed graph, which means that there may be some undirected edges in the estimated graph.

Score-based methods rely on scoring functions such as the log-likelihood or some other loss function. The goal of these algorithms is to find a DAG that optimizes a given scoring function. Some popular scoring functions include several Bayesian Dirichlet metrics (Buntine, 1991; Cooper and Herskovits, 1992; Heckerman *et al.*, 1995), Bayesian information criterion (Chickering and Heckerman, 1997), minimum description length (Bouckaert, 1993; Lam and Bacchus, 1994; Suzuki, 1993), and entropy (Herskovits and Cooper, 1990). One of the classical score-based methods is the *greedy hill-climbing* (HC) algorithm (Gámez *et al.*, 2011). This algorithm is fast but tends to predict too many edges in high-dimensional settings. For discrete networks, the K2 algorithm (Cooper and Herskovits, 1992) is another popular method; however, this method requires prior knowledge about the ordering of the network, which is often unavailable in applications. The greedy equivalence search algorithm (Chickering, 2002) identifies the optimal graph in the space of equivalence classes of DAGs through a two-stage greedy search. There are also Monte Carlo methods (Ellis and Wong, 2008; Niinimäki *et al.*, 2016; Zhou, 2011), which are quite accurate but also computationally demanding.

Finally, there are hybrid methods which combine constraint-based and score-based methods. Hybrid methods first prune the search space by using a constraint-based search and then learn an optimal DAG structure via a score-based search (Gámez *et al.*, 2011; Perrier *et al.*, 2008; Tsamardinos *et al.*, 2006). The max-min hill-climbing (MMHC) algorithm proposed by Tsamardinos *et al.* (2006) is a powerful method of this kind.

Most of the above methods were developed for structure learning on observational data. We introduce typical examples of these classical algorithms in greater detail, followed by recent developments that target learning high-dimensional and large graphs on a mix of observational and experimental data.

Throughout this chapter, we assume that the joint distribution $\mathbb{P}$ of a set of random variables $V = \{X_1, \ldots, X_p\}$ is defined by a causal model with a DAG, $\mathcal{G}$, and a set of SEMs,

$$X_i = f_i(PA_i, \varepsilon_i), \qquad i \in V, \tag{9.1}$$

where the background variables ε_i are independent. Given data from this causal model, the goal of structure learning is to estimate the DAG $\mathcal{G}$.

The two main assumptions for structure learning are causal sufficiency and faithfulness.

Definition 9.1 (Causal sufficiency). A set of variables V is causally sufficient if every common cause of any two or more variables in V is also in V.

For a causal DAG model over V, if every common ancestor of two or more nodes is observed, then causal sufficiency holds. Causal inference and structure learning without causal sufficiency will be discussed in Chapter 11.

Under the faithfulness assumption (Section 7.6), there is a one-to-one mapping between CI in the distribution $\mathbb{P}$ and d-separation in the DAG $\mathcal{G}$. Therefore, one may use CI relations learned from data to infer edges of $\mathcal{G}$, which is the basic idea behind structure learning.

9.2 Equivalence Class and Completed Partially DAGs

In general, there are two types of data for structure learning of causal DAGs: observational data and experimental data. Observational data are collected from naturally occurring circumstances without any experimental intervention or manipulation. Under our causal DAG model, they are regarded as samples from the joint distribution $\mathbb{P}$. In contrast, experimental data are gathered through controlled experiments, where researchers deliberately introduce changes to a subset of variables $S \subset V$. That is, experimental data are generated from the intervention distribution under $\mathrm{do}(S)$.

We first focus on structure learning from observational data, which are more common than experimental data in most applied domains. The first question to answer is: what can be learned from observational data? Due to Markov equivalence among DAGs (Section 7.3.3), in general, we cannot learn the full structure of the underlying causal DAG even with an infinite amount of observational data.

If X and Y are d-separated by Z in a DAG $\mathcal{G}$, we write $\mathcal{D}_{\mathcal{G}}(X, Y \mid Z)$. By Definition 7.5, two DAGs $\mathcal{G}$ and $\mathcal{G}'$ on the same set of nodes V are (Markov) equivalent if $\mathcal{D}_{\mathcal{G}}(X, Y|Z) \Leftrightarrow \mathcal{D}_{\mathcal{G}'}(X, Y|Z)$ for any $X, Y \in V$ and $Z \subseteq V \setminus \{X, Y\}$. Recall that a v-structure is a triplet $\{i, j, k\} \subseteq V$ of the form $i \to k \leftarrow j$, where i and j are non-adjacent and k is called an uncovered collider. Two DAGs are Markov equivalent if and only if they have the same skeletons and the same v-structures. Denote by $\mathcal{G}' \simeq \mathcal{G}$ if the two DAGs are equivalent. Equivalent DAGs form an equivalence class:

$$[\mathcal{G}] := \{\mathcal{G}' : \mathcal{G}' \simeq \mathcal{G}\}.$$

DAGs in the same equivalence class cannot be distinguished from observational data, unless we restrict the SEM (9.1) to a certain family of functions. Thus, in general, we can only learn the equivalence class of $\mathcal{G}$ from observational data.

A partially directed acyclic graph (PDAG) is a graph that contains both directed and undirected edges but does not have directed cycles. For a DAG $\mathcal{G}$, its equivalence class $[\mathcal{G}]$ can be represented by a completed partially DAG (CPDAG), which is also called an essential graph. There are two types of edges in $\mathcal{G}$:

(i) A directed edge $i \to j$ is *compelled* in $\mathcal{G}$ if, for every DAG $\mathcal{G}'$ equivalent to $\mathcal{G}$, the edge $i \to j$ exists in $\mathcal{G}'$.

(ii) If an edge is not compelled in $\mathcal{G}$, then it is *reversible*.

Definition 9.2 (CPDAG or essential graph). The CPDAG of an equivalence class is the PDAG consisting of a directed edge for every compelled edge in the equivalence class and an undirected edge for every reversible edge in the equivalence class.

The three DAGs in Figure 7.6 are Markov equivalent. The equivalence class $[\mathcal{G}_1] = \{\mathcal{G}_1, \mathcal{G}_2, \mathcal{G}_3\}$. The compelled edges and the reversible edges are shown, respectively, in red and black. Accordingly, the CPDAG $\mathcal{C}$ that represents this equivalence class is shown in Figure 9.1, which contains two undirected and three directed edges.

An undirected edge $i - j$ in a CPDAG implies that there are DAGs $\mathcal{G}$ and $\mathcal{G}'$ in the equivalence class such that $i \to j$ in $\mathcal{G}$ and $i \leftarrow j$ in $\mathcal{G}'$. However, an arbitrary orientation of two or more undirected edges can produce a

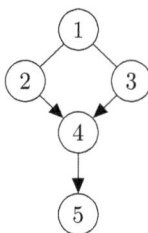

Fig. 9.1. The CPDAG of the equivalence class of DAGs in Figure 7.6.

DAG outside of the equivalence class. Using the CPDAG in Figure 9.1 as an example, if we oriented the two undirected edges as $2 \to 1$ and $3 \to 1$, then the resulting DAG would have an additional v-structure $2 \to 1 \leftarrow 3$, thus not in the equivalence class represented by this CPDAG.

The output of a structure learning algorithm, such as the PC algorithm, on a dataset may be a PDAG. We may want to check whether the PDAG is indeed a CPDAG. Andersson *et al.* (1997) provided a set of necessary and sufficient conditions that characterize a CPDAG or essential graph, summarized in Theorem 9.1. For a subset $S \subseteq V$ of nodes, we denote by $\mathcal{G}_S$ the induced subgraph over S.

Theorem 9.1. *A graph* $\mathcal{G}$ *is a CPDAG for some DAG if and only if* $\mathcal{G}$ *satisfies the following conditions:*

(1) $\mathcal{G}$ *is a chain graph.*
(2) $\mathcal{G}_\tau$ *is chordal for every chain component* τ *of* $\mathcal{G}$.
(3) *The configuration* $a \to b - c$ *does not occur as an induced subgraph of* $\mathcal{G}$.
(4) *Every arrow* $a \to b$ *in* $\mathcal{G}$ *is strongly protected.*

An undirected graph is *chordal* if every cycle of length ≥ 4 possesses a chord, which is an edge between two non-adjacent vertices on the cycle. A chordal graph is also called a triangulated graph since the chords within a cycle effectively "triangulate" the cycle by connecting non-adjacent vertices within it. An arrow $a \to b$ is *strongly protected* in $\mathcal{G}$ if it occurs in at least one of the four configurations in Figure 9.2 as an induced subgraph.

We refer the interested reader to Andersson *et al.* (1997) for a proof of Theorem 9.1. Here, we discuss several counterexamples to provide some intuition behind this characterization of a CPDAG. Consider the four-node graph in Figure 9.3(a). First, this graph is not a chain graph. If it were, the chain components would be $V_1 = \{a, d\}$ and $V_2 = \{b, c\}$. However, there are directed edges from V_1 to V_2 and also from V_2 to V_1, violating

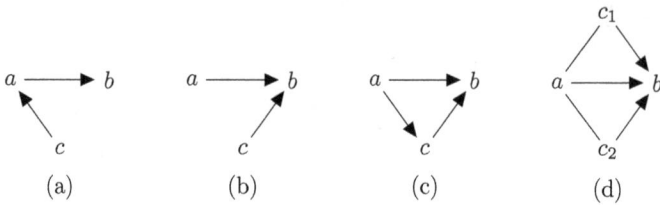

Fig. 9.2.   Four possible configures for a strongly protected arrow $a \to b$ in a CPDAG.

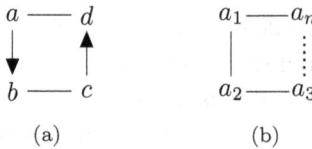

Fig. 9.3.   Example graphs that are not a CPDAG.

the definition of chain graphs (Section 7.5). To see why this graph cannot be a CPDAG, consider possible ways to orient the undirected edges $a - d$ and $b - c$ into a DAG that is in the same equivalence class. This means that we should not create any new v-structures in the orientation. This is only possible by orienting the two edges as $d \to a$ and $b \to c$, which would create a directed cycle. Thus, this graph is not the CPDAG of any DAG. To see why the subgraph $\mathcal{G}_\tau$ over any chain component of a CPDAG must be chordal, let us consider a cycle of $n \geq 4$ nodes, $a_1, \ldots, a_n$, shown in Figure 9.3(b). Suppose this graph is a chain component of a CPDAG. To orient this undirected graph into a DAG, we may start with the orientation $a_1 \to a_2$ without loss of generality. Once this edge is oriented, to avoid any new v-structure, we must orient all the subsequent edges in this cycle into $a_i \to a_{i+1}$, $i = 2, \ldots, n - 1$, and $a_n \to a_1$. Again, this would lead to a directed cycle. Thus, an n-cycle cannot be oriented into a part of a DAG.

9.3   Constraint-Based Learning

We start with a fundamental result, due to Spirtes *et al.* (2000), upon which constraint-based structure learning is built.

Theorem 9.2. *Suppose a distribution $\mathbb{P}$ is faithful to a DAG $\mathcal{G}$ over V. Then, there is no edge between a pair of nodes $X, Y \in V$ in $\mathcal{G}$ if and only if there exists a subset $Z \subseteq V \setminus \{X, Y\}$ such that $X \perp\!\!\!\perp Y \mid Z$ in $\mathbb{P}$.*

This theorem follows from the fact that X and Y are non-adjacent if and only if there exists a subset Z of nodes that d-separates X and Y, which is equivalent to $X \perp\!\!\!\perp Y \mid Z$ under faithfulness. This result translates the problem of learning the edge set of $\mathcal{G}$ into finding CI relations among variables. The latter can be accomplished, at least in principle, with a substantial amount of data from $\mathbb{P}$.

9.3.1 *The PC algorithm*

We discuss the well-known PC algorithm (Spirtes and Glymour, 1991) as a prototype of constraint-based methods. The PC algorithm consists of two main steps:

(i) find the skeleton of the underlying DAG through CI tests;
(ii) identify v-structures and orient other edges.

The output of this algorithm is, in general, a PDAG. However, in the large-sample limit, it will construct the true CPDAG with probability approaching one.

An outline of the PC algorithm is given in Algorithm 9.1. The goal is to learn a CPDAG over a set of random variables $V = \{X_1, \ldots, X_p\}$. This is the so-called population version, assuming that we have knowledge about the CI relations among $X_1, \ldots, X_p$. In practice, the CI statement used in Line 5 will be replaced with a hypothesis test for CI using observed data from the joint distribution $\mathbb{P}$.

Algorithm 9.1 (Outline of the PC algorithm).

1: $E \leftarrow$ edge set of a complete undirected graph on V.
2: $\ell \leftarrow 0$.
3: **for** $\ell \leq \ell_{\max}$ **do**
4:     **for** $(i, j) \in E$ **do**
5:         Search for size-ℓ set $S \subseteq N_i(E) \setminus \{j\}$ such that $X_i \perp\!\!\!\perp X_j \mid S$.
6:         If found, $E \leftarrow E \setminus \{(i,j), (j,i)\}$ and store $S_{ij} = S_{ji} \leftarrow S$.
7:     **end for**
8:     $\ell = \ell + 1$.
9: **end for**
10: Identify v-structures based on E and $\{S_{ij}\}$.
11: Orient as many edges in E as possible by Meek's rules.

The algorithm starts with the edge set E of a complete undirected graph over V, i.e., $E = \{(i,j) \in V \times V : i \neq j\}$. The for loop from Line 3 to Line 9 deletes the edges in E sequentially according to CI relations. In Line 5, $N_i(E) := \{X_k : (i,k) \in E\}$ is the set of neighbors of X_i in the current graph (V,E). For any $(i,j) \in E$, if there is a subset S of $N_i(E) \setminus \{j\}$ such that $X_i \perp\!\!\!\perp X_j \mid S$, we remove the edge $i - j$ from E and store the conditioning set $S_{ij} = S_{ji} = S$. For computational efficiency, the search for conditioning sets S_{ij} is implemented in ascending order of the size $\ell = |S_{ij}|$, for $\ell = 0, \ldots, \ell_{\max}$, where $\ell_{\max}$ is prespecified as an input. When $\ell = k$, the algorithm only searches for a size-k subset S to make X_i and X_j conditionally independent (Line 5). When this is done for all $(i,j) \in E$, the algorithm moves to $\ell = k + 1$. Note that when $\ell = 0$, we are checking whether $X_i \perp\!\!\!\perp X_j$ marginally. At the end of Line 9, the algorithm completes the estimation of the skeleton, denoted $\text{sk}(\widehat{\mathcal{G}})$, of the underlying DAG $\mathcal{G}$.

Given $\text{sk}(\widehat{\mathcal{G}})$, Line 10 identifies v-structures according to the following rule: for each non-adjacent pair (i,j) with a common neighbor k, orient $i - k - j$ as $i \rightarrow k \leftarrow j$ if $k \notin S_{ij}$. The validity of this rule can be proven by contradiction. Suppose the orientation is $i \rightarrow k \rightarrow j$, which creates a path from i to j with k as a non-collider. Since $k \notin S_{ij}$, this implies that i and j are *not* d-separated by S_{ij} and thus, by faithfulness, $X_i \not\perp\!\!\!\perp X_j \mid S_{ij}$, contradictory to Line 5. Using a similar argument, we can show that the orientations $i \leftarrow k \leftarrow j$ and $i \leftarrow k \rightarrow j$ will lead to the same contradiction. Therefore, $i \rightarrow k \leftarrow j$ is the only possible orientation. After this step, the algorithm constructs a PDAG.

In the resulting PDAG, Line 11 orients as many undirected edges as possible by repeated application of four rules (Figure 9.4), called Meek's rules (Meek, 1995), until none of them apply. The basic idea behind these orientation rules is the following. For the population version, the PDAG obtained after Line 10 has the same skeleton and v-structures as the true DAG and thus already defines the equivalence class. Meek's rules are then used to orient other compelled edges in this equivalence class. If orienting an undirected edge $i - j$ as $i \rightarrow j$ would result in additional v-structures or directed cycles, then we orient it as $i \leftarrow j$.

Take R3 as an example. If we oriented $a - c$ as $c \rightarrow a$, to avoid directed cycles, we would have to orient $b \rightarrow a$ and $d \rightarrow a$, which would create a new v-structure $b \rightarrow a \leftarrow d$. Thus, we must orient the edge as $a \rightarrow c$. Note that $a \rightarrow c \leftarrow b$ is *not* a v-structure because a and b are connected by an edge, in which case we call c a *covered collider*. It was shown by Meek (1995) that the four rules are sound and complete in the population version of

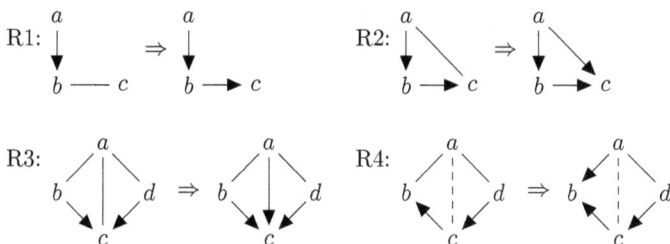

Fig. 9.4. Meek's rules for edge orientation. The dashed line in R4 can be an undirected edge or a directed edge with either orientation.

the PC algorithm, meaning that they will correctly orient all and only the compelled edges in the true equivalence class.

When we apply the PC algorithm to a dataset, the CI checks will be replaced by hypothesis tests for $X_i \perp\!\!\!\perp X_j \mid S_{ij}$. Depending on the assumed joint distribution $\mathbb{P}$, we usually perform the partial correlation test for Gaussian data and the G^2 test for multinomial data. See Section 6.2 for the details and more general CI tests.

Lastly, we present theoretical results regarding the correctness and consistency of the PC algorithm. Suppose that the joint distribution $\mathbb{P}$ of $X_1, \ldots, X_p$ is faithful to a DAG $\mathcal{G}$. Let $\mathcal{C}$ be the CPDAG of $\mathcal{G}$. The first result (Meek, 1995; Spirtes *et al.*, 2000) is on the population version, with the perfect knowledge of the CI relations in $\mathbb{P}$, which we call a CI oracle.

Theorem 9.3. *Assume faithfulness of $\mathbb{P}$ with respect to $\mathcal{G}$. Let $\widehat{\mathcal{G}}$ be the graph constructed using the PC algorithm with a CI oracle. Then, $\widehat{\mathcal{G}} = \mathcal{C}$ and all the found separating sets $|S_{ij}| \leq \max\{|PA_i|, |PA_j|\}$, where $|PA_i|$ is the size of the parent set PA_i.*

Proof. We give a brief proof outline for this result. By Theorem 9.2, any edge of the true DAG $\mathcal{G}$ will not be deleted during the outer for loop in Algorithm 9.1, and thus $PA_i \subseteq N_i(E)$, for all $i \in V$, at every iteration. By the local Markov property, if i and j are not adjacent, they are d-separated by either pa(i) or pa(j). Without loss of generality, assume that they are d-separated by pa(i). Then, there exists at least one subset PA_i such that $X_i \perp\!\!\!\perp X_j \mid PA_i$, which leads to the deletion of the edge (i, j) in E. The above argument shows that by the end of the outer for loop, the set E is the edge set of sk($\mathcal{G}$) and all the stored separating (conditioning) sets are valid. The soundness and completeness of the v-structure identification and Meek's orientation rules then guarantee that $\widehat{\mathcal{G}} = \mathcal{C}$. The conclusion that

$|S_{ij}| \leq \max\{|PA_i|, |PA_j|\}$ follows from the fact that we consider potential separating sets in ascending order of their sizes.   □

Let $\widehat{\mathcal{G}}_n$ be the estimated graph by the PC algorithm applied to a sample of size n from $\mathbb{P}$. Our second result concerns the consistency of $\widehat{\mathcal{G}}_n$ in the large-sample limit as $n \to \infty$, while the distribution $\mathbb{P}$ is fixed. The only difference from the population version is the use of CI tests instead of a CI oracle in Line 5. Let T_n be a test for $H_0 : \theta \in \Theta_0$, against the alternative $H_a : \theta \in \Theta_1$, and denote by $\alpha(T_n \mid \theta_0)$, $\theta_0 \in \Theta_0$, and $\beta(T_n \mid \theta_1)$, $\theta_1 \in \Theta_1$, its type I and type II error probabilities, respectively. We say that T_n with a significance level α is consistent (against a fixed alternative) if

$$\lim_{n \to \infty} \beta(T_n \mid \theta_1) = 0, \qquad \text{for all } \theta_1 \in \Theta_1. \tag{9.2}$$

We say that T_n is point-wise consistent if

$$\lim_{n \to \infty} \alpha(T_n \mid \theta_0) = 0 \quad \text{and} \quad \lim_{n \to \infty} \beta(T_n \mid \theta_1) = 0, \tag{9.3}$$

for all $\theta_0 \in \Theta_0$ and $\theta_1 \in \Theta_1$. Assuming point-wise consistency, all the involved CI tests will be correct, with probability approaching one as $n \to \infty$, which implies that the PC algorithm estimates the CPDAG of $\mathcal{G}$ consistently.

Theorem 9.4. *Assume the faithfulness of $\mathbb{P}$ with respect to $\mathcal{G}$. Let $\widehat{\mathcal{G}}_n$ be the graph estimated by the PC algorithm with point-wise consistent CI tests on a sample of size n from the distribution $\mathbb{P}$. Then,*

$$\lim_{n \to \infty} \mathbb{P}(\widehat{\mathcal{G}}_n = \mathcal{C}) = 1. \tag{9.4}$$

Remark 9.1. As the definition of point-wise consistency (9.3) suggests, the significance level $\alpha = \alpha_n$ of the CI tests should depend on the sample size n such that $\alpha_n \to 0$ as $n \to \infty$.

9.3.2   *p-value adjacency thresholding*

In practical applications of the PC algorithm, the choice of the significance level α of CI tests can substantially impact the sparsity and quality of the resulting estimate. Empirically, the optimal choice of α varies depending on factors such as the sample size and the structure and parameters of the underlying DAG, and no universally well-performing value is known.

To address this difficulty, Huang and Zhou (2022) developed the p-value adjacency thresholding (PATH) algorithm to generate and select from a CPDAG solution path across various values of α from a single execution of the PC algorithm, or indeed from any constraint-based structure learning algorithm that is able to obtain the following. Define Φ_{ij} to be the maximum p-value obtained by the CI tests, with a significance level of α, between X_i and X_j across all the conditioning sets $S \in \mathcal{K}_{ij}$ visited in the PC algorithm. To be concrete, assume that each CI test is performed with the G^2 test (6.4) for discrete data. For all distinct node pairs i, j, we define

$$\Phi_{ij} = \Phi_{ji} := \max_{S \in \mathcal{K}_{ij}} \mathbb{P}\left[\chi_f^2 > G^2(X_i, X_j; S)\right],$$

$$\mathbf{S}(i,j) = \mathbf{S}(j,i) := \operatorname*{argmax}_{S \in \mathcal{K}_{ij}} \mathbb{P}\left[\chi_f^2 > G^2(X_i, X_j; S)\right], \tag{9.5}$$

where f is the degree of freedom, as in (6.4). For a connected node pair $i-j$, $\mathbf{S}(i,j)$ may be considered the conditioning set closest to separating i and j, and Φ_{ij} measures how close they are. Put $\Phi = (\Phi_{ij})$ and $\mathbf{S} = (\mathbf{S}(i,j))$.

The process itself is straightforward. For a decreasing sequence of significance levels $\{\alpha^{(t)}\}$ with $\alpha^{(1)} = \alpha$, we obtain the updated PDAG estimates $\mathcal{G}^{(t)}$ by thresholding the maximum achieved p-values Φ to obtain skeleton estimates with the edge set

$$E^{(t)} = \{(i,j) : \Phi_{ij} \le \alpha^{(t)}\}$$

and then orienting them to CPDAGs according to the last two lines of Algorithm 9.1 with the corresponding separation sets

$$\mathbf{S}^{(t)} = \{\mathbf{S}(i,j) : \Phi_{ij} > \alpha^{(t)}\}.$$

The quality of the estimates is then evaluated based on a scoring criterion, and the highest-scoring network is returned. In what follows, we discuss the choice of the threshold values and present the strategy for estimate generation and selection.

We begin with an estimated graph by executing the PC algorithm with some maximal threshold value α. The goal is to start with the densest graph so that the elements of Φ and $\mathbf{S}$ in (9.5) are maximized over a larger number of visited conditioning sets. We generate a sequence of τ values decreasing from $\alpha^{(1)} = \alpha$ to some minimum threshold value $\alpha^{(\tau)}$. This sequence may be incremented according to some linear or log-linear scale; however, we choose to achieve maximal difference in sparsity among the estimated graphs by utilizing the information in Φ. Given each $\alpha^{(t)}$ corresponding to $\mathcal{G}^{(t)} = (V, E^{(t)})$, we choose $\alpha^{(t+1)}$ such that

$$|E^{(t)}| - |E^{(t+1)}| \approx \frac{|E^{(1)}| - |E^{(\tau)}|}{\tau - 1}. \tag{9.6}$$

Noting that $|E^{(t)}| = \sum_{i<j} I(\Phi_{ij} \le \alpha^{(t)})$, where $I(\cdot)$ is the indicator function, it is easy to see that the sequence $\alpha^{(1)}, \ldots, \alpha^{(\tau)}$ can be straightforwardly obtained using the order statistics of $\{\Phi_{ij}\}$.

Once a solution path of CPDAG estimates $\{\mathcal{G}^{(t)} : t = 1, \ldots, \tau\}$ is obtained, we select the final estimated graph as the one that maximizes a certain score. Let $\mathcal{S}(G \mid \mathbf{D})$ be a score for a graph G given data $\mathbf{D}$, such as the *Bayesian information criterion* (BIC) score; see Section 9.4 for a detailed discussion. The best solution is then obtained according to

$$t^* = \underset{t \in \{1, \ldots, \tau\}}{\operatorname{argmax}} \mathcal{S}(\mathcal{G}^{(t)} \mid \mathbf{D}).$$

Using a decomposable score (Definition 9.3, Section 9.4), one may calculate the scores $\mathcal{S}(\mathcal{G}^{(t)} \mid \mathbf{D})$, for all t, on the solution path very efficiently. See Huang and Zhou (2022) for an elaborate discussion of this point. We summarize the PATH algorithm in Algorithm 9.2, given the maximum p-values Φ, the separation sets $\mathbf{S}$, the number of estimates τ, the significance level α for CI tests, and the minimum threshold $\alpha^{(\tau)}$.

Algorithm 9.2 (PATH(Φ, S, τ, α, $\alpha^{(\tau)}$, D)).

1: Generate decreasing $\{\alpha^{(t)}\}$ from $\alpha^{(1)} = \alpha$ to $\alpha^{(\tau)}$ by (9.6).
2: **for** $t = 1, 2, \ldots, \tau$ **do**
3:     $E^{(t)} \leftarrow \{(i, j) : \Phi_{ij} \le \alpha^{(t)}\}$ and let $\mathcal{G}^{(t)} = (V, E^{(t)})$.
4:     $\mathbf{S}^{(t)} \leftarrow \{\mathbf{S}(i, j) : \Phi_{ij} > \alpha^{(t)}\}$.
5:     Identify v-structures and maximally orient $\mathcal{G}^{(t)}$.
6:     Compute score $\mathcal{S}(\mathcal{G}^{(t)} \mid \mathbf{D})$.
7: **end for**
8: select $t^* = \operatorname{argmax}_{t \in \{1, \ldots, \tau\}} \mathcal{S}(\mathcal{G}^{(t)} \mid \mathbf{D})$.
9: return $\mathcal{G}^{(t^*)}$.

As discussed in Remark 9.1, the p-value threshold α_n has to decrease to zero in a data-dependent way as $n \to \infty$ for the PC algorithm to be structure estimation consistent. Since the choice of α_n is not practically informative due to its implicit dependence on n, Algorithm 9.2 contributes an element of accessibility to this asymptotic result in the form of the following theorem implied by Theorem 3 in the work of Huang and Zhou (2022).

Theorem 9.5. *Suppose the distribution* $\mathbb{P}$ *is faithful to a DAG with CPDAG* $\mathcal{G}^*$ *and* $\mathbf{D}$ *is an i.i.d. sample of size* n *from* $\mathbb{P}$. *Let* $\Phi_n = (\Phi_{n,ij})$ *and* $\mathbf{S}_n$ *be the maximum p-values and corresponding separation sets obtained using the PC algorithm with CI tests of significance level* α. *Let* $\{\mathcal{G}_n^{(t)}\}$ *be the set of estimated graphs generated by Algorithm 9.2 with* $\alpha^{(1)} = \alpha$, $\alpha^{(\tau)} = 0$, *and* $\tau = 1 + \sum_{i<j} I(\Phi_{n,ij} \leq \alpha)$. *If all the CI tests are consistent* (9.2), *then*

$$\lim_{n \to \infty} \mathbb{P}\left[\mathcal{G}^* \in \{\mathcal{G}_n^{(t)} : t = 1, \ldots, \tau\}\right] = 1. \tag{9.7}$$

This result implies that we may apply the PC algorithm with a fixed significance level α and then use the PATH algorithm to generate a solution path which will, with high probability, include the true CPDAG. Then, using a consistent score (Definition 9.5, Section 9.4) one may identify $\mathcal{G}^*$.

In the finite-sample setting, the thresholded estimate $\mathcal{G}^{(t)}$ corresponding to $\alpha^{(t)} < \alpha^{(1)}$ typically does not exactly match the estimate obtained by directly executing the PC algorithm with a significance level of $\alpha^{(t)}$. Adjacencies that would be disconnected in the earlier stages of the learning process with threshold $\alpha^{(t)}$ may survive to the later stages with a more lenient threshold of $\alpha^{(1)}$. The enlarged neighborhoods may persist as additional (potentially false positive) connections in the output structure but will also cause more conditioning sets to be considered and thus lead to the deletion of edges that would not have been deleted in the execution with $\alpha^{(t)}$. Nonetheless, the empirical results reported by Huang and Zhou (2022) demonstrate the potential of the solution path generated by thresholding, showing that PATH applied to the PC algorithm is able to produce estimates of competitive quality to the best of those obtained by multiple executions with various α.

To the best of our knowledge, there is currently no easy way to choose the optimal threshold α_n for the PC algorithm, and thus repeated executions are often needed for parameter tuning. The application of PATH pragmatically allows for a single execution of PC with the fixed threshold α while returning estimates with both theoretical guarantees and empirically good performance.

9.4 Score-Based Learning

A score-based method estimates a DAG by maximizing a scoring function $\mathcal{S}$ over a space of graphs $\mathcal{H}$:

$$\widehat{\mathcal{G}} = \underset{G \in \mathcal{H}}{\operatorname{argmax}}\, \mathcal{S}(G \mid \mathbf{D}), \tag{9.8}$$

where $\mathbf{D} = (x_{ij})_{n \times p} = [x_1 \mid \ldots \mid x_p]$ is an i.i.d. sample of size n from a causal DAG model $(\mathcal{G}, \mathbb{P})$ over random variables $X_1, \ldots, X_p$.

The scoring function $S(G \mid \mathbf{D})$ is usually defined based on the log-likelihood of a DAG G with a penalty term on the model complexity measured by the number of edges or number of free parameters. For example, a commonly used score is the BIC score:

$$\mathcal{S}_{\mathrm{BIC}}(G \mid \mathbf{D}) = \log p(\mathbf{D} \mid \widehat{\theta}, G) - \frac{d}{2} \log n, \tag{9.9}$$

where $\widehat{\theta}$ is the MLE of the model parameters under the DAG G and d denotes the dimension of θ.

In what follows, we discuss in detail the scoring mechanisms and provide a brief introduction to search strategies employed for various graph spaces.

9.4.1 *Scoring functions*

We often use the BIC score for structure learning of Gaussian linear DAGs, parameterized via the linear SEM (7.4). Under this model, the data column $x_j \in \mathbb{R}^n, j \in [p]$ follows a Gaussian linear regression model given the data of its parents $\mathrm{pa}(j)$:

$$x_j = \sum_{i \in \mathrm{pa}(j)} \beta_{ij} x_i + \varepsilon_j, \qquad \varepsilon_j \sim \mathcal{N}_n(0, \omega_j^2 I_n). \tag{9.10}$$

Given a candidate DAG G, let $\mathrm{pa}_G(j)$ be the parent set of node j in G and $\mathrm{pa}_j^G := [x_i : i \in \mathrm{pa}_G(j)]$ be the data matrix of the parent variables. Write $\beta_j = (\beta_{ij} : i \in \mathrm{pa}_G(j))$, which implicitly depends on the DAG G. Using least squares, we find the MLE of (β_j, ω_j^2):

$$(\widehat{\beta}_j, \widehat{\omega}_j^2) = \underset{\beta_j, \omega_j^2}{\operatorname{argmax}} \left[-\frac{n}{2} \log(\omega_j^2) - \frac{1}{2\omega_j^2} \left\| x_j - \mathrm{pa}_j^G \beta_j \right\|^2 \right],$$

for all $j = 1, \ldots, p$. Then, the BIC score is

$$\mathcal{S}_{\text{BIC}}(G \mid \mathbf{D}) = \sum_{j=1}^{p} \log p(x_j \mid pa_j^G, \widehat{\beta}_j, \widehat{\omega}_j^2) - \frac{1}{2}|pa_G(j)| \log n$$

$$= \sum_{j=1}^{p} f_j(x_j, pa_j^G), \qquad (9.11)$$

where f_j is a function of x_j and pa_j^G. This shows that the BIC score for Gaussian linear DAGs is decomposable.

Definition 9.3. A scoring function for DAGs is *decomposable* if it can be expressed as a sum of local scores, each being a function of only the data of a single node and its parents.

The Bayesian Dirichlet (BD) score (Heckerman *et al.*, 1995) is commonly used for discrete DAGs. This scoring function is defined by the posterior probability of a DAG under the multinomial model for $[X_i \mid PA_i]$ introduced in Section 7.4.2. Recall that

$$\theta_{ijk} = \mathbb{P}(X_i = k \mid PA_i = j)$$

and the parameters for $[X_i \mid PA_i]$ are arranged into a $q_i \times r_i$ table,

$$\Theta_i = \left\{ \theta_{ijk} : j \in [q_i], k \in [r_i], \text{such that} \sum_{k=1}^{r_i} \theta_{ijk} = 1 \right\}.$$

For each i and j, put $\theta_{ij} = (\theta_{ij1}, \ldots, \theta_{ijr_i})$, and we assume an independent conjugate prior:

$$\theta_{ij} \mid pa_G(i) \sim \text{Dir}(\alpha_{ij1}, \ldots, \alpha_{ijr_i}). \qquad (9.12)$$

In short, we write this as a product-Dirichlet prior over Θ_i:

$$\Theta_i \mid pa_G(i) \sim \text{Prod-Dir}((\alpha_{ijk})_{q_i \times r_i}), \qquad i = 1, \ldots, p.$$

By default, we choose $\alpha_{ijk} = \alpha/(r_i \cdot q_i)$ for some constant $\alpha > 0$. We may regard α as the total pseudo-count added to each $q_i \times r_i$ contingency table for $[X_i \mid PA_i]$. All the above prior distributions are specified given a candidate DAG G. The last component of our prior is a structural prior over DAGs:

$$\pi(G) \propto \lambda^{d(G)}, \qquad \lambda \in (0, 1),$$

where $d(G) = \sum_{i=1}^{p} r_i q_i$ is the total number of parameters.

Given a candidate DAG G, write $PA_i \equiv PA_i^G$ for notational brevity. From the discrete data matrix $\mathbf{D}$ of size $n \times p$, we first generate p contingency tables, one for each node. Let N_{ijk} be the number of rows in $\mathbf{D}$ such that $X_i = k$ and $PA_i = j$. For node i, we arrange the associated counts into a $q_i \times r_i$ table,

$$N_i = \{N_{ijk} : j \in [q_i], k \in [r_i]\}, \qquad (9.13)$$

which is a sufficient statistic for Θ_i. Let $N_{ij} = (N_{ijk} : k \in [r_i])$ be the jth row of N_i and $N_{ij\bullet} = \sum_k N_{ijk}$ be the row sum.

The marginal probability of N_{ij}, after averaging over θ_{ij}, is

$$P(N_{ij} \mid \mathrm{pa}_G(i)) = \int P(N_{ij} \mid \theta_{ij}) \pi(\theta_{ij} \mid \mathrm{pa}_G(i)) d\theta_{ij}$$

$$= \frac{\Gamma(\alpha/q_i)}{\Gamma(N_{ij\bullet} + \alpha/q_i)} \prod_{k=1}^{r_i} \frac{\Gamma(N_{ijk} + \alpha/(q_i r_i))}{\Gamma(\alpha/(q_i r_i))},$$

where $\pi(\theta_{ij} \mid \mathrm{pa}_G(i))$ is the Dirichlet prior (9.12) and $\Gamma(\cdot)$ is the gamma function. The marginal probability of N_i, the whole table for node i, is

$$P(N_i \mid \mathrm{pa}_G(i)) = \prod_{j=1}^{q_i} P(N_{ij} \mid \mathrm{pa}_G(i)).$$

Combing all p tables, we obtain the marginal probability of the data $\mathbf{D}$ given a DAG G as

$$P(\mathbf{D} \mid G) = \prod_{i=1}^{p} P(N_i \mid \mathrm{pa}_G(i)).$$

Finally, the posterior distribution is

$$P(G \mid \mathbf{D}) \propto \pi(G) P(\mathbf{D} \mid G)$$

$$= \prod_{i=1}^{p} \lambda^{q_i r_i} \prod_{j=1}^{q_i} \frac{\Gamma(\alpha/q_i)}{\Gamma(N_{ij\bullet} + \alpha/q_i)} \prod_{k=1}^{r_i} \frac{\Gamma(N_{ijk} + \alpha/(q_i r_i))}{\Gamma(\alpha/(q_i r_i))}. \qquad (9.14)$$

The BD score is then defined as

$$\mathcal{S}_{\mathrm{BD}}(G \mid \mathbf{D}) := \log \pi(G) + \log P(\mathbf{D} \mid G) = \sum_{i=1}^{p} f_i(N_i), \qquad (9.15)$$

where f_i is a function of N_i only. Thus, the BD score is decomposable as well.

9.4.2 Consistency

The scoring functions (9.11) and (9.15) satisfy a few desirable properties.

Definition 9.4. A scoring function $\mathcal{S}(G \mid \mathbf{D})$ is *score-equivalent* if

$$\mathcal{S}(G_1 \mid \mathbf{D}) = \mathcal{S}(G_2 \mid \mathbf{D})$$

for any two Markov-equivalent DAGs G_1 and G_2.

Chickering (1995) showed that the BIC, the Akaike information criterion (AIC), and BD scores are all score-equivalent.

A DAG G under certain parameterization defines a family of distributions, i.e., the class of distributions that can be parameterized as a DAG model associated with G. We say a joint distribution $\mathbb{P} \in G$ if $\mathbb{P}$ is in the family of distributions defined by G. For example, a Gaussian linear DAG G defines a class of joint Gaussian distributions that can be parameterized by a set of linear regression models (7.4) with pa(j) as the predictors for $j \in [p]$. Let $d(G)$ denote the number of parameters in the DAG model. Following Chickering (2002), we define a consistent scoring function as follows.

Definition 9.5 (Consistent score). A scoring function $\mathcal{S}(G \mid \mathbf{D})$ is *consistent* if the following two properties hold for $\mathbf{D}_n \overset{\text{iid}}{\sim} \mathbb{P}$:

(i) If $\mathbb{P} \in G \setminus H$, then $\mathbb{P}\{\mathcal{S}(G \mid \mathbf{D}_n) > \mathcal{S}(H \mid \mathbf{D}_n)\} \to 1$ as $n \to \infty$.
(ii) If $\mathbb{P} \in G \cap H$ and $d(G) < d(H)$, then $\mathbb{P}\{S(G \mid \mathbf{D}_n) > S(H \mid \mathbf{D}_n)\} \to 1$ as $n \to \infty$.

Property (i) states that if $\mathbb{P}$ can be parameterized by G but not by H, then $\mathcal{S}(G \mid \mathbf{D}_n) > \mathcal{S}(H \mid \mathbf{D}_n)$ in the large-sample limit, which means that $\mathcal{S}$ can distinguish true models for the data $\mathbf{D}_n$ from incorrect models. If both G and H are correct models for the data, then $\mathcal{S}$ prefers the simpler one with fewer parameters, as stated in Property (ii).

Haughton (1988) established that the BIC is consistent for exponential families, and it approximates the posterior probability of a model on a logarithmic scale. Note that both Gaussian linear DAGs and discrete multinomial DAGs are exponential families. Therefore, the results given by Haughton (1988) applied to the scoring functions introduced here imply the following conclusions: First, $\mathcal{S}_{\text{BIC}}(G \mid \mathbf{D})$ (9.9) is consistent. Second, $\mathcal{S}_{\text{BD}}(G \mid \mathbf{D}_n)$ is also consistent since

$$\frac{1}{n}\mathcal{S}_{\text{BD}}(G \mid \mathbf{D}_n) = \frac{1}{n}\mathcal{S}_{\text{BIC}}(G \mid \mathbf{D}_n) + O_p(1/n) = O_p(1) + O_p(1/n).$$

In particular, both (9.11) and (9.15) are consistent scoring functions.

The consistency of score-based learning is implied immediately by the properties of a consistent scoring function.

Theorem 9.6. *Suppose* $\mathbb{P}$ *is faithful to a DAG* $\mathcal{G}$ *and the data* $\mathbf{D}_n \overset{\text{iid}}{\sim} \mathbb{P}$. *If the scoring function* $\mathcal{S}$ *is consistent and score-equivalent, then*

$$\lim_{n \to \infty} \mathbb{P}\left\{ \underset{G}{\operatorname{argmax}}\, \mathcal{S}(G \mid \mathbf{D}_n) = \mathcal{C} \right\} = 1, \qquad (9.16)$$

where $\mathcal{C} = [\mathcal{G}]$ *is the Markov-equivalence class of* $\mathcal{G}$.

In (9.16), argmax returns the set of all maximizers of $\mathcal{S}(G \mid \mathbf{D}_n)$, which is the equivalence class $\mathcal{C}$ with probability approaching one.

9.4.3　*Search strategies*

Three types of search spaces are encountered in score-based learning:

(1) the space of DAGs,
(2) the space of equivalence classes (or CPDAGs),
(3) the space of topological sorts.

Note that for the latter two categories, the scoring function is defined over a set of DAGs. To navigate these diverse graph spaces effectively, various search strategies have been developed.

For the DAG space, a common search strategy is greedy HC (Gámez *et al.*, 2011; Heckerman *et al.*, 1995). Given the current DAG G^t, a set of neighboring DAGs is defined as DAGs obtainable by the addition, deletion, or reversal of a single (directed) edge. One chooses the neighboring DAG that maximally increases the score, which is the updated DAG G^{t+1}. This move is repeated until no improvement can be achieved. While widely applied and regarded as efficient and well-performing, the locality of the HC search in the DAG space unavoidably risks the common problem of accepting locally optimal yet globally suboptimal solutions. As reviewed and discussed by Huang and Zhou (2022), hill-climbing can be augmented to more thoroughly search the DAG space with one or both of *tabu lists* and *random restarts*, respectively governed by the parameters (t_0, t_1) and (r_0, r_1). In what is known as the tabu search, a solution is obtained through HC while a tabu list stores the last t_1 DAG structures visited during the search. Then, the HC procedure is continued for up to t_0 iterations while allowing for minimal score decrements, with a local neighborhood restricted by the tabu list to avoid previously visited structures. In HC with random restarts, the HC procedure is repeated r_0 times after the initial execution by perturbing the current solution with r_1 random local changes. Stochastic search is another effective way to alleviate the issue of local trapping. For example, Zhou (2011) developed a multi-domain sampler to draw random samples of DAGs from a posterior distribution $p(G \mid \mathbf{D})$. Two types of moves were designed: the same set of local moves as used in HC and a global move to draw DAGs from the attraction domains of identified local modes. As expected, the global move greatly reduces the risk of being trapped in a local mode.

The greedy equivalence search (GES) algorithm (Chickering, 2002) is a classical example of score-based learning methods that search the space of CPDAGs. Let $\mathcal{E}$ be an equivalence class of DAGs, i.e., $\mathcal{E} = [\mathcal{G}]$ for some DAG $\mathcal{G}$. If a scoring function $\mathcal{S}(G \mid \mathbf{D})$ is score-equivalent, then we define the score for an equivalence class $\mathcal{E}$ as

$$\mathcal{S}(\mathcal{E} \mid \mathbf{D}) := \mathcal{S}(G \mid \mathbf{D}), \qquad \forall\, G \in \mathcal{E}.$$

Define two sets of neighbors for $\mathcal{E}$, denoted as $\mathcal{N}^+(\mathcal{E})$ and $\mathcal{N}^-(\mathcal{E})$. We say that $\mathcal{E}' \in \mathcal{N}^+(\mathcal{E})$ if and only if there is a DAG $G \in \mathcal{E}$ to which a single edge addition results in a $G' \in \mathcal{E}'$. The set $\mathcal{N}^-(\mathcal{E})$ is defined in a completely analogous way, consisting of equivalence classes that are obtained by deleting a single edge from DAGs in $\mathcal{E}$. GES (Algorithm 9.3) proceeds in two phases from an initial empty graph. In the first phase, it repeatedly adds a single edge to the current CPDAG that maximally increases the score until the score cannot be further improved. Then, in the second phase, GES repeatedly removes a single edge that maximally improves the score until a local maximum is reached.

Algorithm 9.3 (GES).

Phase 1: Let $\mathcal{E}^1$ be an empty graph.

Do $\mathcal{E}^{t+1} \leftarrow \operatorname{argmax}\{\mathcal{S}(\mathcal{E} \mid \mathbf{D}) : \mathcal{E} \in \mathcal{N}^+(\mathcal{E}^t)\}$ repeatedly for $t = 1, 2, \ldots$ until the score does not improve, i.e., $\mathcal{S}(\mathcal{E}^{t+1} \mid \mathbf{D}) \leq \mathcal{S}(\mathcal{E}^t \mid \mathbf{D})$.

$\bar{\mathcal{E}} \leftarrow \mathcal{E}^t$ (graph at the end of phase 1).

Phase 2: $s \leftarrow t$ (initialize phase 2 with $\bar{\mathcal{E}}$).

Do $\mathcal{E}^{t+1} \leftarrow \operatorname{argmax}\{\mathcal{S}(\mathcal{E} \mid \mathbf{D}) : \mathcal{E} \in \mathcal{N}^-(\mathcal{E}^t)\}$ repeatedly for $t = s, s+1, \ldots$ until the score does not improve.

Output: The final equivalence class $\widehat{\mathcal{E}} \leftarrow \mathcal{E}^t$.

As stated by Chickering (2002), in the large-sample limit, $\bar{\mathcal{E}}$ at the end of Phase 1 is the equivalence class of a supergraph of the true DAG, and $\widehat{\mathcal{E}}$ found by GES is the true equivalence class. We summarize the consistency of GES in the following theorem.

Theorem 9.7. *Suppose $\mathbb{P}$ is faithful to a DAG $\mathcal{G}$ and the data $\mathbf{D}_n \overset{\text{iid}}{\sim} \mathbb{P}$. Let $\widehat{\mathcal{E}}$ be the output equivalence class by GES using the BIC or the BD score on the data $\mathbf{D}_n$. Then,*

$$\lim_{n \to \infty} \mathbb{P}\left(\widehat{\mathcal{E}} = [\mathcal{G}]\right) = 1.$$

The last category of methods searches the space of topological sorts or orderings (Larranaga *et al.*, 1996; Teyssier and Koller, 2005). After an optimal ordering is identified, we further learn a DAG structure that is compatible with the ordering. Let $\mathcal{P}$ be the set of permutations of $[p]$, where p is the number of nodes. Define a score for a sort $\pi \in \mathcal{P}$ by

$$\mathcal{S}(\pi \mid \mathbf{D}) := \max_{G \in \mathcal{D}(\pi)} \mathcal{S}(G \mid \mathbf{D}),$$

where $\mathcal{D}(\pi)$ is the set of DAGs that can be sorted by π. Then, a search strategy is applied to find

$$\widehat{\pi} = \operatorname*{argmax}_{\pi \in \mathcal{P}} \mathcal{S}(\pi \mid \mathbf{D})$$

through optimization over the permutation space. The evaluation of the score $\mathcal{S}(\pi \mid \mathbf{D})$ is non-trivial, as it involves optimization over all DAGs in $\mathcal{D}(\pi)$. If we assume $|\operatorname{pa}(i)| \leq d$ for all $i \in [p]$, where d is quite small, then $\mathcal{S}(\pi \mid \mathbf{D})$ can be calculated through dynamic programming. Alternatively, we may employ continuous relaxation of the score $S(G \mid \mathbf{D})$ to ease the optimization (Ye *et al.*, 2021). The search for an optimal sort in the permutation space is also challenging. Stochastic optimization methods, such as genetic algorithms or simulated annealing, have been applied to this problem.

9.5 Experimental Data

Experimental data serve as a gold standard for making causal inferences. Such data are generated when some variables are under experimental intervention, i.e., $\operatorname{do}(X_S)$, for some subset $S \subset V$. The strength of experimental data lies in their ability to provide strong evidence of causation.

To incorporate experimental data in score-based learning, we need to modify the likelihood of a graph since the joint distribution is now an intervention distribution. The starting point is the truncated factorization (8.8) of the joint density $p(x_1, \ldots, x_p)$ under $\text{do}(X_S = \mathbf{x}^*)$:

$$p(x_1, \ldots, x_p \mid \text{do}(\mathbf{x}^*)) = I(x_S = \mathbf{x}^*) \prod_{j \notin S} p(x_j \mid pa_j).$$

This shows that only x_j's that are *not* under intervention depend on their parent pa_j. If X_i is under intervention, say $\text{do}(X_i = x^*)$, this effectively removes the edges $X_k \to X_i$ for all $k \in \text{pa}(i)$. Thus, observing $X_i = x^*$ does not provide any information about its parents; consequently, it does not contribute to the likelihood.

Suppose our data $\mathbf{D} = (x_{ij})_{n \times p}$ contain experimental data under different interventions and also some observational data. Let $\mathcal{O}_j$ be the row indices of the data matrix $\mathbf{D}$ for which node X_j is *not* under intervention (i.e., observational) and put $\mathcal{O} = \{\mathcal{O}_j : j \in [p]\}$. Denote by $p(x_{ij} \mid pa_{ij}, \theta_j)$ the conditional density for $x_{ij}, i \in \mathcal{O}_j$, given the parent data $pa_{ij} := \{x_{ik} : k \in \text{pa}(j)\}$ with the parameter θ_j. Then, the likelihood of the DAG model parameter $\Theta = (\theta_1, \ldots, \theta_p)$ given the data $\mathbf{D}$ (and $\mathcal{O}$) are

$$p(\mathbf{D} \mid \mathcal{O}, \Theta) \propto \prod_{j=1}^{p} \prod_{i \in \mathcal{O}_j} p(x_{ij} \mid pa_{ij}, \theta_j). \tag{9.17}$$

When we work with experimental data, we always assume $\mathcal{O}$ is given, and thus we drop it from conditioning to simplify notation.

Suppose $|\mathcal{O}_j| = m_j$ and $|\text{pa}(j)| = d_j$. Let $x_{\mathcal{O}_j, k} := (x_{ik}, i \in \mathcal{O}_j) \in \mathbb{R}^{m_j}$ for any $k \in [p]$ and $pa_{\mathcal{O}_j, j} := (x_{ik} : i \in \mathcal{O}_j, k \in \text{pa}(j)) \in \mathbb{R}^{m_j \times d_j}$. For a Gaussian linear DAG, the regression (9.10) is replaced with

$$x_{\mathcal{O}_j, j} = \sum_{k \in \text{pa}(j)} \beta_{kj} x_{\mathcal{O}_j, k} + \varepsilon_j, \qquad \varepsilon_j \sim \mathcal{N}_{m_j}(0, \omega_j^2 I_n),$$

which is used to find $(\widehat{\beta}_j, \widehat{\omega}_j^2)$. The BIC (9.11) is modified accordingly to

$$\mathcal{S}_{\text{BIC}}(G \mid \mathbf{D}) = \sum_{j=1}^{p} \log p(x_{\mathcal{O}, j} \mid pa_{\mathcal{O}_j, j}^G, \widehat{\beta}_j, \widehat{\omega}_j^2) - \frac{1}{2} |\text{pa}_G(j)| \log n. \tag{9.18}$$

For discrete data, we simply modify the counts N_{ijk} in (9.13), for each $i \in [p]$, using only data points for which X_i is observational. That is,

$$N_{ijk} = \sum_{r \in \mathcal{O}_i} I(x_{ri} = k, pa_{ri} = j)$$

is now the number of rows with indices in $\mathcal{O}_i$ such that $X_i = k$ and $PA_i = j$. Once these contingency tables are calculated, the subsequent calculations, in particular, the posterior probability (9.14) and the BD score (9.15), all remain the same.

As discussed in Section 9.4.2, commonly used scores, such as (9.11) and (9.15), cannot distinguish equivalent DAGs given purely observational data. In such cases, the DAG parameter Θ is not identifiable (Remark 3.1). Suppose there are M DAGs $\mathcal{G}_1, \ldots, \mathcal{G}_M$ in the equivalence class of the true DAG $\mathcal{G}_1$ and the true parameter associated with $\mathcal{G}_1$ is Θ_1. Let $\mathbb{P}_{\Theta_1}$ be the joint distribution for $X_1, \ldots, X_p$ defined by the DAG $\mathcal{G}_1$ with parameter Θ_1. For the Gaussian linear and the multinomial DAG models, associated with each $\mathcal{G}_m$, there exists a parameter Θ_m such that $\mathbb{P}_{\Theta_m} = \mathbb{P}_{\Theta_1}$, for $m = 2, \ldots, M$. As a result, even with an infinite amount of observational data, we cannot distinguish Θ_1 from other Θ_m based on the likelihood.

However, given sufficient experimental data, the true DAG can be distinguished from its equivalent DAGs and thus becomes identifiable. Under an intervention $do(X_j = x^*)$, experimental data are generated from the intervention distribution $\mathbb{P}(X \mid do(X_j = x^*))$. Suppose that there is a reversible edge $X_j - X_k$ in the CPDAG. If the true edge orientation is $X_j \to X_k$, from the experimental data, we see a dependence of X_k on X_j under $do(X_j)$. If the true edge is $X_j \leftarrow X_k$, the edge is effectively removed by $do(X_j)$, and thus the two variables become independent. This difference will lead to different likelihoods of the two edge orientations given experimental data under $do(X_j)$.

Let us formalize the idea of using experimental data to identify causal DAGs in a simple setting where each single node is under intervention. The data $\mathbf{D}_n$ consist of $(p + 1)$ blocks. The first p blocks, each of size n_j, contain experimental data under $do(X_j = x_j^*)$, with the value x_j^* randomly drawn from the domain of X_j, for $j = 1, \ldots, p$. The last block contains observational data of size n_0. Denote the log-likelihood by

$$\ell(\Theta \mid \mathbf{D}_n) = \log p(\mathbf{D}_n \mid \mathcal{O}, \Theta),$$

as in (9.17). Let $\mathcal{G}^*$ and Θ^* be the true DAG and its parameter, respectively.

Theorem 1 given by Gu *et al.* (2019) provides a set of sufficient conditions to identify Θ^*:

(A1) The distribution $\mathbb{P}_{\Theta^*}$ is faithful to $\mathcal{G}^*$.
(A2) The parameter for $[X_j \mid PA_j^G]$ is identifiable for all j and candidate DAG G.

Theorem 9.8. *Assume* (A1) *and* (A2). *If* $n_j \gg \sqrt{n}$ *for all* $j \in [p]$, *then for any* $\Theta \neq \Theta^*$,

$$\lim_{n \to \infty} \mathbb{P}\{\ell(\Theta^* \mid \mathbf{D}_n) > \ell(\Theta \mid \mathbf{D}_n)\} = 1.$$

Assumption (A2) means that the model for X_j given a candidate parent set PA_j^G is identifiable. For Gaussian linear DAGs, the coefficient vector β_j in the linear regression (9.10) is identifiable, as long as the predictors $x_i, i \in \text{pa}_G(j)$ are linearly independent. The requirement $n_j \gg \sqrt{n}$ guarantees that the statistical variation in $\ell(\Theta \mid \mathbf{D}_n)$ is dominated by the difference in log-likelihood among equivalent DAGs provided by experimental data. In practical applications, we usually have a much larger amount of observational data so that n_0 is much larger than n_j and is close to the total sample size n. This theorem provides theoretical guidance on how much experimental data we need to generate relative to n_0.

In fact, we do not have to generate experimental data under the intervention of each node, which is also quite restrictive in practice. We may first learn a CPDAG $\widehat{\mathcal{C}}$ from observational data of size n_0. Let $T \subseteq V$ be the subset of nodes that are connected to at least one undirected edge in $\widehat{\mathcal{C}}$. Then, we generate experimental data under $\text{do}(X_j)$ of size n_j for each $j \in T$. By a similar reasoning to Theorem 9.8, if the sample size of the experimental data under each intervention satisfies $n_j \gg \sqrt{n_0}$, the causal DAG can be identified as $n_0 \to \infty$.

9.6 Continuous Optimization

The score-based methods discussed so far search in a graph space for an optimal structure according to a scoring criterion. The optimization in a graph space is discrete and combinatorial in nature. As the number of variables increases, this problem becomes notoriously difficult, as it is non-convex and NP-hard (Chickering, 1996). There have been recent efforts to improve the efficiency of structure learning through continuous relaxation of a scoring function. The key idea is to introduce sparse regularization via the ℓ_1 or a concave penalty function, which is a continuous function. We also briefly review some recent work that reformulates the acyclicity constraint, which restricts the search to a highly nonconvex space of DAGs, as a differentiable constraint.

9.6.1 *Maximum penalized likelihood*

For continuous relaxation of scoring functions, we primarily focus on Gaussian DAG learning in the works of Fu and Zhou (2013) and Aragam and Zhou (2015). While the traditional approach to structure learning uses ℓ_0-based penalties such as the BIC, Fu and Zhou (2013) introduced the idea of using continuous penalties via an adaptive ℓ_1 penalty and showed that it can be competitive in practice. They combined a novel method of enforcing acyclicity with a block coordinate descent algorithm in order to compute an ℓ_1-penalized maximum likelihood estimator for structure learning.

Aragam and Zhou (2015) further generalized this idea to the use of concave penalty functions with a reparameterization of the Gaussian log-likelihood. In what follows, we review this method in detail. Let $\beta_j = (\beta_{1j}, \ldots, \beta_{pj}) \in \mathbb{R}^p$ such that $\beta_{ij} \neq 0$ if and only if $i \in \text{pa}(j)$ and write $B = [\beta_1 | \cdots | \beta_p]$ for an arbitrary DAG. The support of the matrix B, defined by $\text{supp}(B) := \{(i, j) : \beta_{ij} \neq 0\}$, corresponds to the edge set of the DAG. We say B is acyclic if its support satisfies the acyclicity constraint. There are two unknown parameters in the Gaussian linear SEM (9.10):

$$B = (\beta_{ij}) \in \mathbb{R}^{p \times p}, \qquad \Omega = \text{diag}(\omega_1^2, \ldots, \omega_p^2) \in \mathbb{R}^{p \times p}.$$

If $X = [x_1 | \cdots | x_p]$ is an $n \times p$ data matrix of i.i.d. observations from this model, the *negative* log-likelihood of (B, Ω), which is our loss function, is easily seen to be

$$L(B, \Omega \mid X) = \sum_{j=1}^{p} \left[\frac{n}{2} \log(\omega_j^2) + \frac{1}{2\omega_j^2} \|x_j - X\beta_j\|^2 \right]. \qquad (9.19)$$

Observe that the loss function in (9.19) is nonconvex. However, we can exploit a reparameterization of the problem, introduced by Städler *et al.* (2010), which leads to a convex loss. The idea is to define new parameters $\eta_j = 1/\omega_j$ and $\phi_j = (\phi_{1j}, \ldots, \phi_{pj}) = \beta_j/\omega_j$, which yield the reparameterized negative log-likelihood

$$L(\Phi, R \mid X) = \sum_{j=1}^{p} \left[-n \log(\eta_j) + \frac{1}{2} \|\eta_j x_j - X\phi_j\|^2 \right], \qquad (9.20)$$

where $\Phi = [\phi_1 | \cdots | \phi_p]$ and $R = \mathrm{diag}(\eta_1, \ldots, \eta_p)$. The loss function in (9.20) is easily seen to be convex. Furthermore, if we interpret Φ as the weighted adjacency matrix of a directed graph, then Φ has exactly the same edges and nonzero entries as B, and thus, in particular, Φ is acyclic if and only if B is acyclic.

In order to promote sparsity and avoid overfitting, we minimize a penalized loss. In what follows, let $\rho_\lambda : [0, \infty) \mapsto \mathbb{R}$ be a nonnegative and nondecreasing penalty function that depends on the tuning parameter λ and possibly one or more additional shape parameters. Given a penalty function, our estimator is

$$(\widehat{\Phi}, \widehat{R}) = \underset{\Phi \in \mathcal{D}, R}{\mathrm{argmin}}\, f(\Phi, R) := \frac{1}{n} L(\Phi, R \mid X) + \sum_{i,j} \rho_\lambda(|\phi_{ij}|), \qquad (9.21)$$

where $\mathcal{D}$ is the set of $p \times p$ weighted adjacency matrices of DAGs and f is called a penalized loss function.

The traditional BIC score (9.11) is equivalent to using the ℓ_0 penalty for ρ_λ, which results in a discrete optimization problem. Examples of continuous penalty functions that can produce sparse solutions include the ℓ_1 penalty (Tibshirani, 1996), the capped-ℓ_1 penalty, and the minimax concave penalty (MCP) (Zhang, 2010), the latter two of which are concave. The capped-ℓ_1 flattens the ℓ_1 penalty at a certain cutoff value, and the MCP penalty represents a smooth interpolation between the ℓ_1 and ℓ_0 penalties. See Figure 9.5 for a visual comparison of these penalties. The key difference between the ℓ_1 penalty and MCP is the flat part of the penalty, which helps reduce bias caused by ℓ_1 regularization.

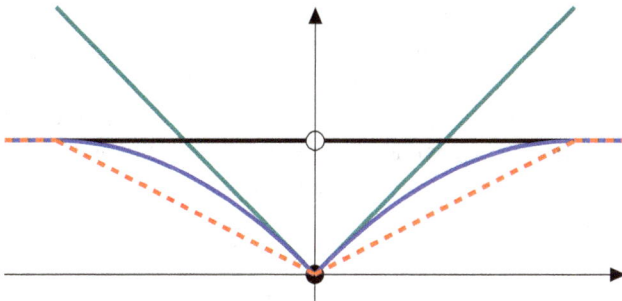

Fig. 9.5. Visualization of four penalty functions: ℓ_0 penalty (black), ℓ_1 penalty (teal), MCP (blue), and capped-ℓ_1 penalty (red dashed).

After choosing a penalty ρ_λ, we apply coordinate descent to minimize the penalized loss (9.21). The idea behind cyclic coordinate descent is quite simple: instead of minimizing the objective function over the entire parameter space simultaneously, we restrict our attention to one variable at a time, perform the minimization in that variable while holding all others constant (hereafter referred to as a *single parameter update*), and cycle through the remaining variables. This procedure is repeated until convergence. Coordinate descent is ideal in situations in which each single parameter update can be performed quickly and efficiently. Moreover, due to acyclicity, we know *a priori* that the parameters ϕ_{kj} and ϕ_{jk} cannot simultaneously be nonzero for $k \neq j$. This suggests performing block-wise coordinate descent, minimizing over $\{\phi_{kj}, \phi_{jk}\}$ simultaneously.

In order to enforce acyclicity, we use a simple heuristic: For each block $\{\phi_{kj}, \phi_{jk}\}$, we check to see if adding an edge from $X_k \rightarrow X_j$ induces a directed cycle in the estimated DAG. If so, we set $\phi_{kj} = 0$ and minimize with respect to ϕ_{jk}. Alternatively, if the edge $X_j \rightarrow X_k$ induces a directed cycle, we set $\phi_{jk} = 0$ and minimize with respect to ϕ_{kj}. If neither edge induces a cycle, we minimize over both parameters simultaneously.

Now, define the single-variable functions

$$Q_1(\phi_{kj}) = \frac{1}{2n} \left\| \eta_j x_j - \sum_{i=1}^{p} \phi_{ij} x_i \right\|^2 + \rho_\lambda(|\phi_{kj}|), \tag{9.22}$$

$$Q_2(\eta_j) = -\log \eta_j + \frac{1}{2n} \left\| \eta_j x_j - \sum_{i=1}^{p} \phi_{ij} x_i \right\|^2. \tag{9.23}$$

The function Q_1 is the objective function in (9.21) considered as a function of the single parameter ϕ_{kj}, while holding all the other parameters fixed and ignoring terms that do not depend on ϕ_{kj}, and Q_2 is the corresponding function for the parameter η_j. We express the dependence of Q_1 and Q_2 on k and/or j implicitly through their respective arguments, ϕ_{kj} or η_j. The minimizer of Q_1 is given by a so-called threshold function associated with the penalty function ρ_λ, while the minimizer of Q_2 has a simple closed-form expression. We refer the reader to Aragam and Zhou (2015) for the technical details.

An overview of the algorithm is given in Algorithm 9.4. We use the notation $\phi_{kj} \Leftarrow 0$ to mean that ϕ_{kj} must be set to zero due to acyclicity, as discussed above.

Algorithm 9.4 (Blockwise coordinate descent for DAG learning).

Input: Initial estimates (Φ^0, R^0).

1. Cycle through ρ_j for $j = 1, \ldots, p$, minimizing Q_2 with respect to ρ_j at each step.

2. Cycle through the $p(p-1)/2$ blocks $\{\phi_{kj}, \phi_{jk}\}$ for $j, k = 1, \ldots, p$, $j \neq k$, minimizing with respect to each block:

   (a) If $\phi_{kj} \Leftarrow 0$, then minimize Q_1 with respect to ϕ_{jk} and set $(\phi_{kj}, \phi_{jk}) = (0, \phi_{jk}^*)$, where $\phi_{jk}^* = \operatorname{argmin} Q_1(\phi_{jk})$;

   (b) If $\phi_{jk} \Leftarrow 0$, then minimize Q_1 with respect to ϕ_{kj} and set $(\phi_{kj}, \phi_{jk}) = (\phi_{kj}^*, 0)$, where $\phi_{kj}^* = \operatorname{argmin} Q_1(\phi_{kj})$;

   (c) If neither 2(a) nor 2(b) applies, then choose the update which leads to a smaller value of the overall objective $f(\Phi, R)$.

3. Repeat steps 1 and 2 until certain convergence or stop criterion is met.

Gu *et al.* (2019) generalized the block-wise coordinate descent method to discrete Bayesian networks based on the multi-logit regression formulation (7.8) for $[X_j \mid PA_j]$. Recall that $\boldsymbol{\beta}_{ij} = (\boldsymbol{\beta}_{i1j} \mid \ldots \mid \boldsymbol{\beta}_{ir_jj})$ is the coefficient matrix of size $d_i \times r_j$ associated with the node pair (i, j). There does not exist an edge from i to j in the DAG if and only if $\boldsymbol{\beta}_{ij} = 0$. Therefore, the DAG structure is given by the sparsity pattern of $\{\boldsymbol{\beta}_{ij}\}$. In order to learn a sparse DAG from data, we estimate $\boldsymbol{\beta} = \{\boldsymbol{\beta}_{ij} : i, j \in [p]\}$ either by maximizing a penalized likelihood or minimizing a penalized loss, as in (9.21). The regular ℓ_1 or MCP is inappropriate for this purpose since it penalizes each component of $\boldsymbol{\beta}$ separately. We instead penalize the matrix $\boldsymbol{\beta}_{ij}$ as a whole via the use of a group penalty, such as the group ℓ_1 penalty (Yuan and Lin, 2006), $\rho_\lambda(\boldsymbol{\beta}_{ij}) = \lambda \|\boldsymbol{\beta}_{ij}\|_F$, where $\|\cdot\|_F$ is the Frobenius norm of a matrix. Define the support of $\boldsymbol{\beta}$ by $\operatorname{supp}(\boldsymbol{\beta}) := \{(i, j) : \boldsymbol{\beta}_{ij} \neq 0\}$. Similarly, we say that $\boldsymbol{\beta} \in \mathcal{D}$ if $\operatorname{supp}(\boldsymbol{\beta})$ is the edge set of a DAG. Let $L(\boldsymbol{\beta} \mid X)$ be the *negative* log-likelihood given the data X. Then, we estimate $\boldsymbol{\beta}$, and thus the DAG structure, by

$$\widehat{\boldsymbol{\beta}} = \operatorname*{argmin}_{\boldsymbol{\beta} \in \mathcal{D}} \left[\frac{1}{n} L(\boldsymbol{\beta} \mid X) + \lambda \sum_{i,j} \|\boldsymbol{\beta}_{ij}\|_F \right].$$

A similar but more complex block-wise coordinate descent algorithm was developed to cycle through $\{\beta_{ij}, \beta_{ji}\}$ to solve the above minimization problem. See Gu *et al.* (2019) for more technical details.

In addition to coordinate descent over pairs of edges, we may also utilize the penalized loss $f(\Phi, R)$ in (9.21) for learning topological orderings of Gaussian DAGs. Let $\mathcal{D}(\pi)$ denote the set of weighted adjacency matrices compatible with π, i.e., π is a topological sort of the DAG with an adjacency matrix $B \in \mathcal{D}(\pi)$. Then, we define a loss for the topological sort:

$$f(\pi) := \min_{\Phi \in \mathcal{D}(\pi), R} f(\Phi, R). \tag{9.24}$$

Once $\pi = (\pi(1), \ldots, \pi(p))$ is fixed, the minimization of $f(\Phi, R)$ can be translated to $(p-1)$ separate penalized linear regression problems, regressing $x_{\pi(j)}$ onto $x_{\pi(1)}, \ldots, x_{\pi(j-1)}$ for $j = 2, \ldots, p$.

Let $\widehat{\Sigma} = \frac{1}{n} X^{\mathsf{T}} X$ be the sample covariance of the data matrix X. The minimization (9.24) is also linked to the sparse Cholesky factorization of the inverse sample covariance $\widehat{\Sigma}^{-1}$ according to the ordering π. To see this, let

$$F = (I - B)\Omega^{-1/2} = R - \Phi \in \mathbb{R}^{p \times p}.$$

The p columns of F are $F_j = \eta_j e_j - \phi_j$, where e_j is the jth standard basis for $\mathbb{R}^p$, $j = 1, \ldots, p$. After some simple algebra similar to the derivation for (2.10), the negative log-likelihood (9.20) can be expressed as

$$L(F \mid \widehat{\Sigma}) = \frac{n}{2} \operatorname{tr}(\widehat{\Sigma} F F^{\mathsf{T}}) - n \log |F|.$$

Suppose that π is an ordering for B. Then, after permuting its columns according to π, F will become an upper-triangular matrix. The minimizer $\widehat{F}$ of $L(F \mid \widehat{\Sigma})$ satisfies that $\widehat{\Sigma}^{-1} = \widehat{F}\widehat{F}^{\mathsf{T}}$, which is equivalent to the Cholesky factor of $\widehat{\Sigma}^{-1}$ after permuting its rows and columns by π. Based on this connection, we developed a proximal gradient algorithm to calculate $f(\pi)$, which is called the regularized Cholesky score, and incorporated it into simulated annealing for optimizing over π (Ye *et al.*, 2021).

9.6.2 *Differentiable acyclicity constraint*

A computational challenge in score-based learning is to enforce the acyclicity constraint, e.g., $\Phi \in \mathcal{D}$ in (9.21). In Algorithm 9.4, we must check if acyclicity is reserved for each potential edge addition. To alleviate such computational bottlenecks, Zheng *et al.* (2018) proposed a reformulation of the acyclicity constraint so that score-based learning can be carried out through continuous optimization subject to a differentiable constraint.

For a directed graph with edge set E, a *directed walk* of length k from node i to node j is a sequence $v_0 = i, \ldots, v_k = j$ of vertices so that $(v_{m-1}, v_m) \in E$ for all $m = 1, \ldots, k$. Note that vertices in a walk may not be distinct.

Lemma 9.1. *Suppose that a binary matrix $A = (a_{ij})_{p \times p}$ is the adjacency matrix of a directed graph. Then, the (i, j)th entry $a_{ij}^{(k)}$ in the matrix power A^k, $k \geq 1$, counts the number of directed walks of length k from node i to node j.*

Proof. We prove this result by induction using

$$a_{ij}^{(k)} = \sum_{m=1}^{p} a_{im}^{(k-1)} a_{mj}. \tag{9.25}$$

The conclusion is obvious for $k = 1$. Suppose $a_{im}^{(k-1)}$ is the number of directed walks of length $k-1$ from i to m. Then, $a_{im}^{(k-1)} a_{mj}$ is the number of length-k walks from i to j that visit node m immediately before reaching j. Since every walk of length k from i to j must visit some node $m \in [p]$ before its last edge $m \to j$, summing over $m \in [p]$ in (9.25) counts all such walks. (Note that m could be i or j when loops are formed along a walk.) □

In particular, a diagonal entry, $a_{ii}^{(k)}$, of A^k counts the number of directed walks of length k from node i to itself. Therefore, $\mathrm{tr}(A^k) = 0$ if and only if there are no directed k-cycles in the graph. Let e^A be the matrix exponential of A defined as

$$e^A := \sum_{k=0}^{\infty} \frac{1}{k!} A^k.$$

Then, it is easy to see that

$$\mathrm{tr}(e^A) = \mathrm{tr}(I) + \sum_{k=1}^{\infty} \frac{1}{k!} \mathrm{tr}(A^k) \geq p$$

for any binary adjacency matrix A, and

$$\mathrm{tr}(e^A) = p \quad \text{if and only if } A \text{ corresponds to a DAG.} \tag{9.26}$$

However, a continuous relaxation of a DAG score is often defined with respect to a weighted adjacency matrix W, such as the matrix B in (9.19) or Φ in (9.20). To apply the above equality constraint to W, we replace A with the Hadamard (element) product $W \circ W$, whose elements are all nonnegative. This leads to the following main result derived by Zheng *et al.* (2018).

Theorem 9.9. *A weighted adjacency matrix $W \in \mathbb{R}^{p \times p}$ is acyclic if and only if*

$$h(W) := \operatorname{tr}(e^{W \circ W}) - p = 0. \tag{9.27}$$

Moreover, the gradient of h is

$$\nabla h(W) = (e^{W \circ W})^{\mathsf{T}} \circ 2W.$$

Consequently, we may optimize a continuous score $\mathcal{S}(W \mid X)$, such as (9.20), subject to the equality constraint (9.27). To give a concrete example, for Gaussian linear SEMs, one may estimate W by an equality-constrained program:

$$\begin{cases} \min_W \mathcal{S}(W \mid X) := \frac{1}{2n}\|X - XW\|_F^2 + \lambda \|W\|_1 \\ \text{subject to } h(W) = 0, \end{cases} \tag{9.28}$$

where $\|A\|_F$ is the Frobenius norm and $\|A\|_1 := \sum_{i,j} |a_{ij}|$ for $A = (a_{ij})$. The loss function consists of a least-squares loss and an ℓ_1 penalty to encourage sparsity in W. The main advantage of program (9.28) is its amenability to large classes of optimization techniques for continuous and differential problems. However, since the feasible set $\{W : h(W) = 0\}$ is a nonconvex set, the program (9.28) is still a nonconvex optimization problem.

In the following few chapters, we introduce some recent developments that make use of similar algebraic acyclicity constraints.

9.7 Problems

(1) Draw the CPDAG that represents the equivalence class of the DAG in Figure 8.8.
(2) Let C be a chain component of a CPDAG $\mathcal{G}$ and pa(i) be the parent set of node i in $\mathcal{G}$. Is it true that pa(i) = pa(j) for any two nodes $i, j \in C$? Prove or disprove it.

(3) Following the instruction in Section 6.1 of Aragam *et al.* (2019b) to install the package **sparsebn** and load the dataset `cytometry Discrete`. The dataset contains 5,400 data points for 11 variables. Some data were generated under experimental interventions, while the remaining were purely observational.

   (a) Use the default setting of the function `estimate.dag` to learn a solution path of DAGs from this dataset. Report the parent set of each variable in the DAG with 15 edges (or the closest one). Next, use the *observational data* (for which none of the 11 variables was under intervention) to answer the following two questions, assuming $[X_j \mid PA_j]$ is a multinomial distribution for each node.

   (b) Test the hypotheses (i) pip3 ⊥⊥ pip2 | plc and (ii) erk ⊥⊥ mek | jnk at a significance level of $\alpha = 0.01$ with the G^2 statistic.

   (c) Suppose we have chosen the DAG in Figure 6(c) in Aragam *et al.* (2019b) as our estimated causal DAG. Given this DAG, estimate $\mathbb{P}[\text{jnk} = j \mid \text{do}(\text{akt} = k)]$ for $k, j \in \{0, 1, 2\}$, i.e., the causal effect of the variable akt on jnk.

(4) Let $\mathcal{G}$ be a DAG on $V = \{X_1, \ldots, X_n\}$ and denote by E^* the edge set of the skeleton of $\mathcal{G}$. Suppose that the joint distribution $\mathbb{P}$ of $X_1, \ldots, X_n$ is faithful to $\mathcal{G}$ and is given so that one can correctly examine all conditional independence relations among any variables. Construct an undirected graph $G = (V, E_0)$ for $\mathbb{P}$ as follows:

$$(i, j) \notin E_0 \text{ if and only if } X_i \perp\!\!\!\perp X_j \mid V \setminus \{X_i, X_j\}, \qquad \text{for all } i \neq j.$$

Consider the following algorithm:

1: $E \leftarrow E_0$.
2: **for** $(i, j) \in E$ **do**
3:     Check if there exists an $S_{ij} \subseteq N_i(E)$ such that $X_i \perp\!\!\!\perp X_j \mid S_{ij}$.
4:     If such an S_{ij} exists, then $E \leftarrow E \setminus \{(i, j), (j, i)\}$.
5: **end for**
6: $\widehat{E} \leftarrow E$.

In the for loop, $N_i(E) = \{X_k : (i, k) \in E\}$ is the current neighbor set of X_i, and an exhaustive search over all subsets of $N_i(E)$ is applied until an S_{ij} is found. Show that $\widehat{E} = E^*$, that is, the above algorithm correctly constructs the skeleton of $\mathcal{G}$.

Chapter 10

Learning Generalized Directed Acyclic Graphical Models

In previous chapters, we have focused on Gaussian linear and multinomial directed acyclic graphs (DAGs) for continuous and discrete data, respectively. In this chapter, we cover a few generalizations of the DAG models. We consider non-Gaussian errors and nonlinear structural equation models (SEMs) for parent-child relations. An interesting observation is that many of these generalized DAG models are identifiable, unlike the simpler Gaussian or multinomial models. We also review some recent methods that learn a nonlinear DAG model with deep neural networks. Lastly, we present a generalization of DAG models to network data, in which the units in a dataset are modeled by a network.

10.1 DAG Identifiability

Recall the causal DAG model defined in Section 8.1, where the joint distribution $\mathbb{P}(X_1, \ldots, X_p)$ is defined by a DAG $\mathcal{G}$ over $V = [p]$ and a distribution $\mathbb{P}(\varepsilon) = \mathbb{P}(\varepsilon_1, \ldots, \varepsilon_p)$ for background or noise variables $\varepsilon_1, \ldots, \varepsilon_p$. Let $PA_j^{\mathcal{G}} = \{X_i : i \in \mathrm{pa}_{\mathcal{G}}(j)\}$, where $\mathrm{pa}_{\mathcal{G}}(j)$ is the parent set of node j in $\mathcal{G}$. The conditional distribution of a variable given its parents in $\mathcal{G}$ is specified through an SEM,

$$X_j = f_j(PA_j^{\mathcal{G}}, \varepsilon_j), \qquad j = 1, \ldots, p, \tag{10.1}$$

where f_j is a *deterministic* function and ε_j is a random variable. We assume the background (error) variables are jointly independent so that

$$\mathbb{P}(\varepsilon_1, \ldots, \varepsilon_p) = \prod_{j=1}^{p} \mathbb{P}(\varepsilon_j). \qquad (10.2)$$

Then, (10.1) and (10.2) define a joint distribution $\mathbb{P}(X_1, \ldots, X_p)$ that admits a recursive factorization (DF) with respect to (w.r.t.) $\mathcal{G}$, as in Equation (8.3). We say that the joint distribution $\mathbb{P}(X_1, \ldots, X_p)$ is generated by the above DAG model with the graph $\mathcal{G}$. Throughout this chapter, we assume that $\mathbb{P}(X_1, \ldots, X_p)$ does not admit a DF w.r.t. any proper subgraph of $\mathcal{G}$, which is referred to as the *causal minimality* assumption. Taking the Gaussian linear DAG model (7.4) as an example, causal minimality holds if $\beta_{ij} \neq 0$ for all j and $i \in \mathrm{pa}(j)$.

Definition 10.1 (Identifiability). Suppose a joint distribution $\mathcal{L} \equiv \mathbb{P}(X_1, \ldots, X_p)$ is generated by the model (10.1) and (10.2) with a DAG $\mathcal{G}_0$. If the distribution $\mathcal{L}$ cannot be generated by any other model, in the form of (10.1) and (10.2), with a different DAG $\mathcal{G} \neq \mathcal{G}_0$, then we say that $\mathcal{G}_0$ is identifiable from $\mathcal{L}$.

It is well known that linear Gaussian DAGs (7.4) and multinomial DAGs (7.7) are not identifiable (Heckerman *et al.*, 1995). On the other hand, the general DAG model, defined by (10.1) and (10.2) without any constraints on f_j or $\mathbb{P}(\varepsilon)$, is obviously non-identifiable.

10.1.1 *Equal error variance*

There have been some interesting identifiability results on DAG models in recent works. For example, if we assume equal error variances, then the linear Gaussian DAG model is identifiable (Peters and Bühlmann, 2014).

Theorem 10.1. *Let the distribution $\mathcal{L}$ be generated from the linear Gaussian model (7.4) with a DAG $\mathcal{G}_0$, where the error variance $\omega_j = \omega_0 > 0$ for all $j \in [p]$. Then, $\mathcal{G}_0$ is identifiable, and the coefficients β_{kj} can be reconstructed from $\mathcal{L}$ for all j and $k \in \mathrm{pa}_{\mathcal{G}_0}(j)$.*

Let Θ be the invariance covariance matrix of the joint Gaussian distribution $\mathcal{L}$ and π be a permutation of $[p]$. As shown in (7.6), the Cholesky decomposition of Θ_π is

$$\Theta_\pi = H(B_\pi, \Omega_\pi) := (I - B_\pi)\Omega_\pi^{-1}(I - B_\pi)^{\mathsf{T}}.$$

For each π, there is a pair (B_π, Ω_π) given by the above Cholesky decomposition, and the support of B_π defines a DAG $\mathcal{G}_\pi$. This shows that the linear Gaussian DAG model is *not* identifiable in general. Let Π_0 be the set of reversal orderings of the true DAG $\mathcal{G}_0$ with coefficient matrix B_0. If the true DAG model satisfies the constraint that $\Omega_0 = \omega_0^2 I_p$, then $\Omega_\pi = \omega^2 I_p$, for some ω^2, if and only if $\pi \in \Pi_0$. For each $\pi \in \Pi_0$, B_π is a permuted version of B_0 and $\Omega_\pi = \omega_0^2 I_p$. This gives an algebraic interpretation of the identifiability result in Theorem 10.1. Moreover, Ω_0 has the minimum trace among Ω_π over all permutations π (Aragam *et al.*, 2019a).

Lemma 10.1. *Suppose* $\Theta = H(B_0, \Omega_0)$ *for some* $\Omega_0 = \omega_0^2 I_p$. *Let* Π_0 *be the set of reversal orderings of the DAG defined by the support of* B_0. *Then,* $\operatorname{tr}(\Omega_\pi) > \operatorname{tr}(\Omega_0)$ *for any* $\pi \notin \Pi_0$.

In other words, the equal-variance DAG is always the unique minimum-trace DAG. Note that $\operatorname{tr}(\Omega) = \sum_j \omega_j^2$ is the sum of the error variances across all variables, suggesting least-squares estimation for a minimum-trace DAG. Aragam *et al.* (2019a) established high-dimensional consistency results for learning the structure of a minimum-trace DAG through penalized least squares. These results provide explicit, finite-sample recovery guarantees for penalized least-squares estimators without assuming faithfulness.

10.1.2 *Generalized linear DAGs*

In order to model a much larger class of random variables, Ye *et al.* (2024) proposed a generalized linear DAG model using a generalized linear model (GLM) for the conditional distribution $[X_j \mid PA_j]$. In this section, we review the details of this model and present its identifiability result established by Ye *et al.* (2024).

Denote by $x^j \in \mathbb{R}^{d_j}$ a realization of X_j, where $d_j = 1$ for a numerical X_j and $d_j = r_j - 1$ for a categorical variable X_j with r_j classes, using the one-hot encoding. Denote by pa_j a realization of PA_j. Let $\beta_{ij} \in \mathbb{R}^{d_i \times d_j}$ denote the parameters associated with the edge $i \to j$ in a DAG, and $\beta_{ij} = 0$ if $i \notin \operatorname{pa}(j)$. Put

$$\beta_j := [\beta_{0j}, \beta_{1j}, \ldots, \beta_{pj}] \in \mathbb{R}^{(d+1) \times d_j}, \tag{10.3}$$

$$x := [1, x^1, \ldots, x^p] \in \mathbb{R}^{d+1}, \tag{10.4}$$

where $\beta_{0j} \in \mathbb{R}^{1 \times d_j}$, $d = \sum_{i=1}^p d_i$, and $[x, y]$ denotes vertical concatenation of two vectors or matrices x and y. We define a *generalized linear*

DAG (GLDAG) with conditional densities given by GLMs with canonical links, i.e.,

$$p(x^j \mid pa_j, \beta_j) = c_j(x^j) \exp\left\{ \langle \beta_j^\top x, x^j \rangle - b_j(\beta_j^\top x) \right\}, \quad j \in [p], \qquad (10.5)$$

where b_j and c_j are both functions from $\mathbb{R}^{d_j}$ to $\mathbb{R}$. Note that

$$\beta_j^\top x = \beta_{0j}^\top + \sum_{i \in \mathrm{pa}(j)} \beta_{ij}^\top x^i$$

only depends on the parent set PA_j. GLDAG models allow for many common distributions via the choice of the log partition function $b_j(\cdot)$. Examples include the Bernoulli distribution for $b_j(\theta) = \log(1 + e^\theta)$, equal-variance Gaussian for $b_j(\theta) = \theta^2/2$, Poisson for $b_j(\theta) = \exp(\theta)$, Gamma for $b_j(\theta) = -\log(-\theta)$, and the multinomial for $b_j(\theta) = \log\left(1 + \sum_{k=1}^{d_j} e^{\theta_k}\right)$. In the multinomial case, $b_j(\cdot)$ is a multivariate function, operating on a vector $\theta = (\theta_k) \in \mathbb{R}^{d_j}$, in contrast to the other examples for which $b_j(\cdot)$ is a scalar function. The Bernoulli and multinomial choices above give rise to logistic and multi-logit regression models for each node.

We collect all the parameters of model (10.5) in a matrix $\beta \in \mathbb{R}^{(d+1) \times d}$, which is obtained by horizontal concatenation of $\beta_j, j = 1, \ldots, p$, each as defined in (10.3). We say that a GLDAG (10.5) is continuous if all the variables are continuous. Recall that in this case, $d_j = 1$ for all $j \in [p]$, and thus β is a $(p+1) \times p$ matrix. We rewrite the log pdf of (10.5) in the continuous case as

$$L(x; \beta) = \sum_{j=1}^{p} \left[\log c_j(x^j) + x^j \beta_j^\top x - b_j(\beta_j^\top x) \right], \qquad (10.6)$$

where $\beta_j^\top x \in \mathbb{R}$ and $\beta_{ij} \neq 0$ if and only if $i \to j$ in the DAG. Continuous GLDAG models (10.6) are identifiable under mild assumptions, as given by the following.

Theorem 10.2. *Suppose the joint distribution $\mathcal{L}(X)$ is defined by the log-pdf $L(x; \beta)$ with a DAG $\mathcal{G}_0$ according to (10.6). If $L(x; \beta)$ is second-order differentiable w.r.t. x and the first-order derivative of $b_j(\cdot)$, for all j, exists and is not constant, then $\mathcal{G}_0$ is identifiable from $\mathcal{L}(X)$.*

The linear Gaussian DAG with equal variance is a special case of the GLDAG model with $b_j(\theta) = \theta^2/2$ for all j. In this sense, the above identifiability result generalizes Theorem 10.1.

Other classes of identifiable DAGs include linear non-Gaussian DAGs and additive noise models (ANMs), which will be discussed in detail in the subsequent sections.

10.2   Linear Non-Gaussian DAGs

In this section, we assume a linear SEM, as in Equation (7.4), with non-Gaussian noise. That is, the error or background variable ε_j follows a non-Gaussian continuous distribution for each $j \in [p]$, and they are independent of each other. Such a model is called a linear non-Gaussian acyclic model (LiNGAM) by Shimizu *et al.* (2006).

10.2.1   *LiNGAM discovery algorithm*

Put $X = (X_1, \ldots, X_p)$ and $\varepsilon = (\varepsilon_1, \ldots, \varepsilon_p)$ as column vectors in $\mathbb{R}^p$, and let $B = (\beta_{ij})_{p \times p}$, where $\beta_{ij} = 0$ if and only if $i \notin \mathrm{pa}(j)$. The linear SEM (7.4) can be expressed in matrix form as

$$X = B^\mathsf{T} X + \varepsilon,$$

from which we have

$$X = A\varepsilon, \qquad A = (I_p - B^\mathsf{T})^{-1}. \tag{10.7}$$

Let $W := A^{-1} = I_p - B^\mathsf{T}$. As discussed in Section 7.4.1, B can be permuted into a strictly lower-triangular matrix $B_\pi = P_\pi B P_\pi^\mathsf{T}$ using a permutation matrix P_π, where π is the reversal of a topological ordering of the DAG associated with the SEM. In causal modeling, a topological ordering of a causal DAG represents a causal ordering among the variables.

The LiNGAM algorithm (Shimizu *et al.*, 2006) estimates a causal ordering from observed data of X. It is built upon the observation that Equation (10.7) defines the observed variables X_j as a linear mixing of independent non-Gaussian random variables ε_j. The mixing matrix A can be estimated through independent component analysis (ICA) (Comon, 1994; Hyvärinen *et al.*, 2001), up to scaling and column permutation. Note that ICA relies on non-Gaussianity to estimate the mixing matrix A and cannot recover A if the error variables are Gaussian. This explains why the Gaussian linear DAG model is not identifiable from an ICA point of view.

The first step of the LiNGAM algorithm is to estimate an $\widehat{A}$ from the observed data through ICA. Since the columns of $\widehat{A}$ correspond to the order of the error variables, which cannot be identified by ICA. Correspondingly, the rows of $\widehat{W} = \widehat{A}^{-1}$ need to be permuted to match the observed variables $X_1, \ldots, X_p$. For simplicity, assume that we are given an infinite amount of data or the distribution of X, from which $\widehat{W}$ is estimated exactly. There is only one permutation of the rows of $\widehat{W}$ such that the diagonal of the permuted $\widehat{W}$ consists of only nonzero elements, and this permutation recovers the correspondence with the observed variables. Denote by $\widetilde{W}$ the permuted $\widehat{W}$. We further normalize each row of $\widetilde{W}$ so that the diagonals are all one. Then, the estimate of B is $\widehat{B} = (I_p - \widetilde{W})^{\mathsf{T}}$, from which we can find a permutation $\widehat{\pi}$ such that $\widehat{B}_{\widehat{\pi}}$ is a strictly lower-triangular matrix. The reversal of $\widehat{\pi}$ is an estimated causal ordering. This population-level LiNGAM algorithm implies the following identifiability result.

Theorem 10.3. *Suppose the distribution $\mathcal{L}(X)$ is generated by the LiNGAM with a DAG $\mathcal{G}_0$. Let Π_0 be the set of topological sorts of $\mathcal{G}_0$. Then, both $\mathcal{G}_0$ and Π_0 are identifiable from $\mathcal{L}(X)$.*

For a finite dataset, the LiNGAM algorithm uses empirical criteria for the row permutation of $\widehat{W}$ and for learning a causal ordering of $\widehat{B}$. It finds the row permutation of $\widehat{W}$ that minimizes $\sum_i |\widetilde{w}_{ii}|^{-1}$, where $\widetilde{w}_{ij}$ is the (i, j)th element of $\widetilde{W}$. To find a causal order, the algorithm searches for $\widehat{\pi}$ such that $\widehat{B}_{\widehat{\pi}} = (\widetilde{\beta}_{ij})$ is as close as possible to a strictly lower-triangular matrix, say, by minimizing $\sum_{i \leq j} \widetilde{\beta}_{ij}^2$. In practice, this means that $\widehat{B}$ does not necessarily satisfy the acyclicity constraint. We need to threshold some elements of $\widehat{B}$ to zero to convert it to a DAG.

10.2.2 *Sequential order learning*

Motivated by the identifiability of causal orderings under the LiNGAM (Theorem 10.3), some recent methods have been developed to learn a topological order in a sequential manner (Shimizu *et al.*, 2011; Hyvärinen and Smith, 2013; Wang and Drton, 2019; Ruiz *et al.*, 2022).

Let $\mathbf{X} \in \mathbb{R}^{n \times p}$ be an observed dataset of size n. An estimated ordering $\widehat{\pi}$ is constructed by applying Algorithm 10.1 until all nodes are sorted. At step t, given the partial ordering $\mathcal{A}_{t-1} = (\widehat{\pi}(1), \ldots, \widehat{\pi}(t-1))$, the algorithm selects a node $\widehat{\pi}(t)$ to append to the estimated ordering by maximizing an appropriate score $\mathcal{S}(k, \mathcal{A}_{t-1} \mid \mathbf{X})$ over $k \notin \mathcal{A}_{t-1}$. The key is the choice of the score $\mathcal{S}(k, \mathcal{A}_{t-1} \mid \mathbf{X})$. Ruiz *et al.* (2022) proposed to use a likelihood ratio score against the Gaussian error distribution. We present this work to illustrate the idea of sequential learning.

Algorithm 10.1 (Sequential ordering learning).

1: **Input:** data $\mathbf{X}$.
2: $\mathcal{A}_0 = \varnothing$.
3: **for** $t = 1, \ldots, p$ **do**
4:     $\widehat{\pi}(t) \leftarrow \mathrm{argmax}_{k \notin \mathcal{A}_{t-1}} \mathcal{S}(k, \mathcal{A}_{t-1} \mid \mathbf{X})$.
5:     $\mathcal{A}_t \leftarrow (\mathcal{A}_{t-1}, \widehat{\pi}(t))$.
6: **end for**

Let $g(\cdot; \theta_j)$ be the probability density for the error variable ε_j in the LiNGAM, where θ_j is the parameter of a non-Gaussian distribution. We first discuss a choice of the score at the population level. At step t, define the residual variables

$$R_{kt} := \begin{cases} X_k & \text{if } |\mathcal{A}_{t-1}| = 0, \\ X_k - \beta_{kt}^{\mathsf{T}} X_{\mathcal{A}_{t-1}} & \text{otherwise,} \end{cases} \tag{10.8}$$

for $k \notin \mathcal{A}_{t-1}$, where β_{kt} is the least-squares regression coefficient vector,

$$\beta_{kt} = \left(\mathbb{E}\left[X_{\mathcal{A}_{t-1}} X_{\mathcal{A}_{t-1}}^{\mathsf{T}} \right] \right)^{-1} \mathbb{E}\left[X_{\mathcal{A}_{t-1}} X_k \right].$$

Moreover, let

$$\eta_{kt} := \mathrm{argmax}_{\eta} \, \mathbb{E}_X \left[\log g(R_{kt}; \eta) \right] \quad \text{and} \quad \sigma_{kt}^2 := \mathrm{Var}[R_{kt}],$$

where the expectations are with respect to the distribution of X defined by the LiNGAM. Denote by $\phi(\cdot; \sigma_{kt})$ the density for $\mathcal{N}\left(0, \sigma_{kt}^2\right)$, i.e., a normal distribution with a mean of zero and a variance matching that of R_{kt}. Assume that the vertex set $\mathcal{A}_{t-1}$ is ancestral (see Section 7.3.2). If $\mathcal{A}_{t-1}$ contains $\mathrm{pa}(k)$, then $R_{kt} = \varepsilon_k$ and thus $\eta_{kt} = \theta_k$. In this case, we say that k is a *valid* node to continue the partial ordering $\mathcal{A}_{t-1}$. Otherwise, it is *invalid*.

In the population version of Algorithm 10.1, we select the next node to continue our constructed topological ordering as

$$\widehat{\pi}(t) \leftarrow \mathrm{argmax}_{k \notin \mathcal{A}_{t-1}} \mathbb{E}_X \left[\log \frac{g(R_{kt}; \eta_{kt})}{\phi(R_{kt}; \sigma_{kt})} \right]. \tag{10.9}$$

The log-likelihood ratio in (10.9) can be thought of as a score that allows us to distinguish between valid and invalid nodes by telling us how "non-Gaussian" the residual R_{kt} is. If the residual is well explained by a Gaussian distribution relative to the non-Gaussian distribution in the assumed family, then we expect the log-likelihood ratio to be small. Otherwise, if the Gaussian density is not a good fit relative to $g(\cdot; \eta_{kt})$, then we have stronger evidence to believe that node k is a valid node to continue the ordering.

To establish a rigorous statement on the above likelihood ratio score in the sequential-order learning algorithm, we make a couple of assumptions for the error distributions. We restrict our class of densities $\{g(\cdot; \theta_j), j \in [p]\}$ to a scale family with a mean of zero and a scale parameter of $\theta_j > 0$, such as the Laplace family of distributions, the logistic family of distributions, or a scaled-t family with the same degree of freedom.

Assumption 10.1. Let $U \sim g(\cdot; \theta_0)$ for some $\theta_0 > 0$. For each $j = 1, \ldots, p$, the distribution of the error ε_j satisfies $[\epsilon_j] = [(\theta_k/\theta_0)U]$.

Our next assumption for the linear SEM of interest is on linear combinations of the noise terms. This condition is related to a key result about how to characterize the regression residual (10.8) as a linear combination of independent components.

Assumption 10.2. For any $j \in [p]$ and any $a \in \mathbb{R}^p$ with at least two nonzero entries, the linear combination $a^{\mathsf{T}}\varepsilon$ does not follow the same distribution as ε_j.

If ε_j are all Gaussian random variables, then Assumption 10.2 is violated. Many non-Gaussian distributions, however, satisfy this assumption, including the Laplace distribution, the logistic distribution, and the scaled-t distribution.

Theorem 10.4. *Let $X \in \mathbb{R}^p$ follow a LiNGAM with DAG $\mathcal{G}_0$. Execute Algorithm 10.1 with $\hat{\pi}(t)$ updated via (10.9) for $t = 1, \ldots, p$. If Assumptions 10.1 and 10.2 hold, then $\hat{\pi} = (\hat{\pi}(1), \ldots, \hat{\pi}(p))$ is a topological ordering of $\mathcal{G}_0$.*

See Ruiz *et al.* (2022) for a proof. Note that the maximization in (10.9) can be done easily, leading to a tractable computation of complexity $O(p^2)$. This differs from maximizing a score over a whole ordering, which is in general NP-hard and not tractable.

Assume that we have a data matrix $\mathbf{X} \in \mathbb{R}^{n \times p}$ consisting of n i.i.d. rows from a LiNGAM. Let $\mathbf{X}_j$ be the jth column and $\mathbf{X}_{\mathcal{A}_t}$ be the set of columns in $\mathcal{A}_t$. Analogous to the population version of the sorting procedure, we estimate $\widehat{\beta}_{kt}$ through least-squares regression of $\mathbf{X}_k \sim \mathbf{X}_{\mathcal{A}_{t-1}}$ at the step t in the sequential learning algorithm. Further, we find the residual vector $\widehat{R}_{kt} \in \mathbb{R}^n$ using

$$
\widehat{R}_{kt} = \begin{cases} \mathbf{X}_k & \text{if } |\mathcal{A}_{t-1}| = 0, \\ \mathbf{X}_k - \mathbf{X}_{\mathcal{A}_{t-1}}\widehat{\beta}_{kt} & \text{if } |\mathcal{A}_{t-1}| \geq 1. \end{cases}
$$

These residuals are used to estimate the pertinent scale parameters:

$$
\widehat{\eta}_{kt} = \operatorname*{argmax}_{\eta} \sum_{i=1}^{n} \log g(\widehat{R}_{i,kt}; \eta) \quad \text{and} \quad \widehat{\sigma}_{kt}^2 = \frac{1}{n}\|\widehat{R}_{kt}\|_2^2,
$$

where $\widehat{R}_{i,kt}$ is the ith component of the vector $\widehat{R}_{kt}$. Then, we select the next node to continue the ordering using the empirical analogue of the mean log-likelihood ratio in (10.9):

$$
\widehat{\pi}(t) \leftarrow \operatorname*{argmax}_{k \notin \mathcal{A}_{t-1}} \frac{1}{n} \sum_{i=1}^{n} \log \frac{g(\widehat{R}_{i,kt}; \widehat{\eta}_{kt})}{\phi(\widehat{R}_{i,kt}; \widehat{\sigma}_{kt})}. \tag{10.10}
$$

For example, if η_{kt} is the scale parameter for a Laplace distribution, it can be seen that $\widehat{\eta}_{kt} = \frac{1}{n}\|\widehat{R}_{kt}\|_1$. In this case, (10.10) is equivalent to

$$
\widehat{\pi}(t) \leftarrow \operatorname*{argmax}_{k \notin \mathcal{A}_{t-1}} \log \frac{\widehat{\sigma}_{kt}}{\widehat{\eta}_{kt}} = \operatorname*{argmax}_{k \notin \mathcal{A}_{t-1}} \frac{\|\widehat{R}_{kt}\|_2}{\|\widehat{R}_{kt}\|_1}. \tag{10.11}
$$

Remark 10.1. The Laplace update of (10.11) exemplifies how simple the maximization of the likelihood ratio score is. After the regression of each unsorted node X_k, $k \notin \mathcal{A}_{t-1}$, onto $\mathcal{A}_{t-1}$, we only need to compare the ratio between the two norms of the residual vector $\widehat{R}_{kt}$ across unsorted nodes to find $\widehat{\pi}(t)$.

If the data $\mathbf{X}$ is high-dimensional ($p > n$), we cannot perform least-squares regression to calculate the residuals. In this case, we may first estimate the Markov blanket $\widehat{N}_k$ of each node $k \in [p]$ so that $|\widehat{N}_k| \ll n$. Then, we regress $\mathbf{X}_k \sim \mathbf{X}_{\mathcal{A}_{t-1} \cap \widehat{N}_k}$ using the least squares to calculate the residuals and estimate the scale parameters.

Assume that each ε_j follows a Laplacian distribution. Let Π_0 be the set of true orderings of the underlying LiNGAM and $\widehat{\pi}$ be the ordering estimated by Algorithm 10.1 with $\widehat{\pi}(t)$ updated by (10.11). Under some regularity conditions,

$$\mathbb{P}(\widehat{\pi} \notin \Pi_0) \leq \frac{8|\Pi_0|p^2}{n^\xi},$$

where $\xi > 0$ is a constant involved in a gap condition for the scale parameters η_{kt} and σ_{kt}. Consistency of $\widehat{\pi}$ is achieved as long as $|\Pi_0|p^2 \ll n^\xi$ as $n \to \infty$. Note that ξ can be a large constant, which allows for a high-dimensional scaling, such as $n \ll p \ll n^{\xi/2}$, assuming $|\Pi_0| = O(1)$. See Theorem 3.2.2 in Ruiz (2022) for the precise statement and its proof.

10.3   Nonlinear DAG Models

Although the general SEM (10.1) is not identifiable, restrictions on the class of functions for f_j may lead to identifiable DAG models. As an example, we consider additive functions w.r.t. the error variables, which defines the additive noise model (Peters *et al.*, 2014).

10.3.1   *Additive noise models*

We focus on continuous error variables, i.e., each ε_j has a strictly positive density w.r.t. the Lebesgue measure. In an ANM, the SEM for X_j is

$$X_j = f_j(PA_j) + \varepsilon_j, \qquad j = 1, \ldots, p, \qquad (10.12)$$

where PA_j is the parent set of X_j in the true DAG. As stated in Proposition 17 in Peters *et al.* (2014), an ANM satisfies causal minimality if and only if each function f_j is not constant in any of its arguments.

In general, the true DAG for an ANM is identifiable if every f_j is nonlinear under some technical and subtle conditions. Let us start with the bivariate case ($p = 2$), in which $X = \varepsilon_1$ and $Y = f(X) + \varepsilon_2$. Zhang and Hyvärinen (2009) proved that there are only five classes of bivariate ANMs that are non-identifiable, one of them being the well-known Gaussian linear model. Peters *et al.* (2014) provide a sufficient condition on the triple $(f, \mathbb{P}(X), \mathbb{P}(\varepsilon_2))$ for the model to be identifiable; see their Condition 19. In particular, identifiability is guaranteed whenever the function f is not injective.

For the general case of $p > 2$, there is a specific identifiability result for Gaussian noise (Peters *et al.*, 2014, Corollary 31), which we present in the following theorem. We say a function $f : \mathbb{R}^m \mapsto \mathbb{R}$ is nonlinear in all components if $f(x_1, \ldots, x_{a-1}, \cdot, x_{a+1}, \ldots, x_m)$ is nonlinear, for all $a \in [m]$ and some $(x_1, \ldots, x_{a-1}, x_{a+1}, \ldots, x_m) \in \mathbb{R}^{m-1}$.

Theorem 10.5. *Let $\mathcal{L}(X)$ be generated by the SEM (10.12) with a DAG $\mathcal{G}_0$. If $\varepsilon_j \sim \mathcal{N}(0, \sigma_j^2)$ and f_j is three times differentiable and nonlinear in all components for all j, then the DAG $\mathcal{G}_0$ is identifiable from $\mathcal{L}(X)$.*

Let $\mathrm{nd}(j)$ be the non-descendants of node j. It is assumed that ε_j is independent of all variables $\{X_k : k \in \mathrm{nd}(j)\}$. In particular, for a sink node X_i, we have $\varepsilon_i \perp\!\!\!\perp X_{-i}$. Based on this observation, Peters *et al.* (2014) developed an algorithm called regression with subsequent independence test (RESIT) to sequentially estimate a topological order (or causal order). In the first phase, the algorithm repeatedly identifies a sink node and removes it from further consideration. Initially, let $S = [p]$. For each $k \in S$, we perform a nonlinear regression of X_k on $X_{S \setminus \{k\}}$ and measure the dependence ρ_k between the residual and $X_{S \setminus \{k\}}$. Let $k^* = \mathrm{argmin}_{k \in S} \rho_k$. Then, X_{k^*} is identified as a sink node among X_S, and we put k^* at the end of the ordering among S: $\widehat{\pi}(|S|) \leftarrow k^*$. We update $S \leftarrow S \setminus \{k^*\}$ to remove X_{k^*} and repeat the above process until S is empty. This process yields a topological ordering: $\widehat{\pi} = \{\widehat{\pi}(1), \ldots, \widehat{\pi}(p)\}$. Accordingly, a fully connected DAG with $\mathrm{pa}(\widehat{\pi}(j)) = \{\widehat{\pi}(1), \ldots, \widehat{\pi}(j-1)\}$ is constructed. In the second phase, edges into the node $\widehat{\pi}(j)$ are removed until the residual is no longer independent of X_i, $i \in \{\widehat{\pi}(1), \ldots, \widehat{\pi}(j-1)\}$.

The RESIT algorithm can make use of any regression and dependence measure. For example, we may apply generalized additive model (GAM) regression with the R package mgcv. One may use the Hilbert-Schmidt independence criterion (HSIC) test (Gretton *et al.*, 2007) and its p-value as a dependence measure. If the distribution $\mathcal{L}(X) = \mathbb{P}(X_1, \ldots, X_p)$ is generated by an identifiable ANM that also satisfies causal minimality, then RESIT with a consistent non-parametric regression method and an independence oracle is guaranteed to find the true DAG $\mathcal{G}_0$ from $\mathcal{L}(X)$.

A common special case of (10.12) is the additive model

$$X_j = \sum_{i \in \mathrm{pa}(j)} f_{j,i}(X_i) + \varepsilon_j, \qquad j = 1, \ldots, p, \qquad (10.13)$$

where $\varepsilon_j \sim \mathcal{N}(0, \sigma_j^2)$ and each $f_{j,i}$ is nonlinear and thrice differentiable. Under this model, Bühlmann *et al.* (2014) developed a maximum likelihood estimate for a topological order. With a little abuse of notation, let us regard X_j as a data vector of length n in an i.i.d. sample from the above model. Similarly, the function $f(X_j)$ may be interpreted as a vector in

$\mathbb{R}^n$ evaluated at the data points in X_j, depending on the context. For a permutation π, we use X_j^π to denote $X_{\pi(j)}$. For an n-vector $v = (v_1, \ldots, v_n)$, let $\|v\|_n^2 = n^{-1} \sum_{i=1}^n v_i^2$. For any π, the parameters for the sorted variables $(X_1^\pi, \ldots, X_p^\pi)$ are denoted as $f_j^\pi = (f_{j,k}^\pi : k \leq j - 1)$ and σ_j^π. Their MLEs are given by

$$\widehat{f}_j^\pi = \underset{\{g_{j,k}\}}{\operatorname{argmin}} \left\| X_j^\pi - \sum_{k=1}^{j-1} g_{j,k}(X_k^\pi) \right\|_n^2, \quad (\widehat{\sigma}_j^\pi)^2 = \left\| X_j^\pi - \sum_{k=1}^{j-1} \widehat{f}_{j,k}(X_k^\pi) \right\|_n^2.$$

The MLE of π is then defined by minimizing the negative log-likelihood

$$\widehat{\pi} \in \underset{\pi}{\operatorname{argmin}} \sum_{j=1}^p \log(\widehat{\sigma}_j^\pi).$$

For the high-dimensional case of $p > n$, a neighborhood selection is applied to learn a set $\widehat{N}_j$ for each node. Then, the MLEs $\widehat{f}_j^\pi$ and $(\widehat{\sigma}_j^\pi)^2$ are calculated via regression of $X_{\pi(j)}$ onto $\{X_k : k \in \{\pi(1), \ldots, \pi(j-1)\} \cap \widehat{N}_j\}$. Under some technical assumptions, Bühlmann *et al.* (2014) showed that $\widehat{\pi}$ is a true ordering with probability approaching one as $n \to \infty$.

It is interesting to develop a constraint-based approach under the additive nonlinear model (10.13). The first step is to learn a CPDAG using the PC algorithm (Algorithm 9.1) with a general CI test. Then, one may design a test procedure to orient an undirected edge $i - j$ in the estimated CPDAG, making use of the asymmetry between the two possible orientations. For example, if the likelihood of $X_i \to X_j$ is significantly greater than the likelihood of $X_j \to X_i$, we orient the edge into $i \to j$. Once an undirected edge is oriented, we apply Meek's rules (Figure 9.4) to further orient other undirected edges. This procedure is repeated until no more edges may be oriented. Wang and Zhou (2021) developed such an algorithm, assuming that there are both linear and nonlinear additive functions in (10.13).

10.3.2   *Learning with neural networks*

A powerful model for nonlinearity is the neural network (NN). We selectively discuss some recent works that use neural networks for structure learning of DAGs.

For easy understanding, we primarily focus on ANMs (10.12), where f_j is approximated by an NN. Consider an NN with input $u \in \mathbb{R}^p$ and L

hidden layers:

$$F(u; A^{(1)}, \ldots, A^{(L+1)}) = A^{(L+1)} g(A^{(L)} g(\cdots A^{(2)} g(A^{(1)} u))), \qquad (10.14)$$

$$A^{(k)} \in \mathbb{R}^{m_k \times m_{k-1}}, \; m_0 = p, \; m_{L+1} = 1,$$

where $g : \mathbb{R} \to \mathbb{R}$ is a nonlinearity applied element-wise. Let $\theta_j = (A_j^{(h)})$ denote all the parameters of the NN F_j for X_j. Our model for X_j can be expressed as

$$X_j = F_j(X; \theta_j) + \varepsilon_j. \qquad (10.15)$$

A key step is to enforce acyclicity for this model so that X_j depends only on variables that precede itself w.r.t. to an ordering. Let M_k be the kth column of a matrix M. It is not difficult to see that the function $F(u; A^{(1)}, \ldots, A^{(L+1)})$ in (10.14) is independent of u_k if the kth column of $A^{(1)}$ is zero, i.e., $\|(A^{(1)})_k\|_2 = 0$ (Zheng *et al.*, 2020). Put $\theta = (\theta_1, \ldots, \theta_p)$. Define a weighted adjacency matrix $W_{kj}(\theta) := \|(A_j^{(1)})_k\|_2$ such that $W_{kj} = 0$ implies that X_k is not a parent of X_j. This allows us to impose the differentiable constraint (9.27) on W so that it represents a DAG. We further regularize the ℓ_2-norm $\|(A_j^{(1)})_k\|_2$ to encourage sparsity in the DAG such that F_j only depends on a small subset of X_{-j}. This can be done by including $\|A_j^{(1)}\|_{1,2} := \sum_k \|(A_j^{(1)})_k\|_2$ as a penalty term in the loss function.

Let $\mathbf{X} = [\mathbf{x}_1 \mid \cdots \mid \mathbf{x}_p]$ be the observed data matrix. Denote by $\ell(\theta_j \mid \mathbf{X})$ the negative log-likelihood of θ_j according to (10.15). Zheng *et al.* (2020) formulated the following constrained optimization problem to learn a DAG:

$$\min_{\theta} \left\{ L(\theta) := \frac{1}{n} \sum_{j=1}^{p} \ell(\theta_j \mid \mathbf{X}) + \lambda \|A_j^{(1)}\|_{1,2} \right\} \qquad (10.16)$$

$$\text{subject to } h(W(\theta)) = 0,$$

where h is the differentiable acyclicity constraint in (9.27). Lachapelle *et al.* (2020) proposed a similar NN formulation, where the dependence of X_j on X_k is characterized through the path product of the weights in $A_j^{(h)}$, $h = 1, \ldots, L + 1$.

As suggested by Zheng *et al.* (2018), we may solve the program (10.16) approximately by the augmented Lagrangian approach, which optimizes a sequence of subproblems. The unconstrained subproblem minimizes

$$L(\theta) + \alpha_t h(W(\theta)) + \frac{\rho_t}{2} h(W(\theta))^2, \qquad (10.17)$$

where α_t and ρ_t are the dual variable and the penalty coefficient of the tth subproblem. They are updated after each subproblem is solved. Since

NN training is involved, Lachapelle *et al.* (2020) proposed to use stochastic gradient descent for the minimization of (10.17).

By using different likelihood or loss functions, we may generalize program (10.16) to other models, say, for discrete data. See Zheng *et al.* (2020) for a more detailed discussion.

10.4   Network Data

With a little abuse of notation, let $X = [X_1 \mid \cdots \mid X_p] \in \mathbb{R}^{n \times p}$ be a data matrix generated by the linear Gaussian DAG model (7.4),

$$X_j = \sum_{k \in \text{pa}(j)} \beta_{kj} X_k + \varepsilon_j, \quad \varepsilon_j = (\varepsilon_{1j}, \dots, \varepsilon_{nj}) \sim \mathcal{N}_n\left(0, \omega_j^2 I_n\right), \quad (10.18)$$

for $j = 1, \dots, p$, where $X_j \in \mathbb{R}^n$ is the jth column in X. A key assumption under (10.18) is that the rows of X are jointly independent since the covariance matrix of each ε_j is diagonal. In real applications, however, it is common for observations to be dependent, as in network data, which violates the i.i.d. assumption in (10.18). For example, when modeling the features of an individual in a social network, the observed characteristics from different individuals can be dependent because they belong to the same social group such as friends, family, and colleagues who often share some similarity. Motivated by such applications, Li *et al.* (2024) developed a Gaussian DAG model that takes the dependence between observations into account. Based on this model, they further developed an algorithm that can simultaneously infer the DAG structure and the sample dependencies given a topological ordering. When the node ordering is unknown, this algorithm can be used to estimate the correlations between samples, with which one can de-correlate the dependent data to significantly improve the performance of standard DAG learning methods for i.i.d. data. In this section, we present this work in detail.

10.4.1   *Gaussian DAG for dependent data*

We model sample dependency through an undirected graph H on n vertices, with each vertex representing a unit $x_i, i \in [n]$ in the data and the edges representing some association among the units. Each unit x_i corresponds to a row in the data matrix X. Suppose we observe not only the dependent samples $\{x_i\}_{i=1}^n$ but also the graph (network) H. We generalize the SEM in (10.18) to

$$X_j = \sum_{k \in \text{pa}(j)} \beta_{kj} X_k + \varepsilon_j, \quad \varepsilon_j = (\varepsilon_{1j}, \dots, \varepsilon_{nj}) \sim \mathcal{N}_n \left(0, \omega_j^2 \Sigma\right), \quad (10.19)$$

where $\Sigma \in \mathbb{R}^{n \times n}$ is positive definite. The support of the precision matrix $\Theta = \Sigma^{-1}$ is restricted to the edge set $E(H)$ of H, i.e., $\text{supp}(\Theta) \subseteq E(H)$. Consequently, we have the following CI among the units:

$$(i, j) \notin E(H) \Rightarrow x_i \perp\!\!\!\perp x_j \mid x_{-\{i,j\}}, \quad \forall\, i \neq j.$$

Note that when $\Sigma = I_n$, the SEM (10.19) reduces to (10.18). Hence, the classical Gaussian DAG model in (10.18) is a special case of (10.19).

The key distinction between (10.18) and (10.19) lies in the distribution for the background variables $(\varepsilon_{1j}, \dots, \varepsilon_{nj})$. The dependence among the units in (10.19) is induced by the dependence among the background variables. The variables $x_{i1}, \dots, x_{ip}$ of each unit satisfy the same conditional independence constraints defined by the underlying DAG, while the background variables $\varepsilon_{1j}, \dots, \varepsilon_{nj}$ across the n units are dependent. When estimating the DAG structure with such data, the correlations among individuals will reduce the effective sample size. Therefore, we need to take the distribution of the correlated ε_j into account.

Let $B = (\beta_{kj})$ and $\Omega = \text{diag}(\omega_j^2)$. The model (10.19) defines a matrix normal distribution for X. To see this, note that $\varepsilon = (\varepsilon_{ij})_{n \times p}$ in (10.19) follows a matrix normal distribution:

$$\varepsilon \sim \mathcal{N}_{n,p} (0, \Sigma, \Omega) \Leftrightarrow \text{vec}(\varepsilon) \sim \mathcal{N}_{np}(0, \Omega \otimes \Sigma),$$

where $\text{vec}(\cdot)$ is the vectorization operator and $\otimes$ is the Kronecker product. The density for ε is given by

$$-2 \log p(\varepsilon \mid \Sigma, \Omega) = n \log \det \Omega + p \log \det \Sigma + \text{tr}\left(\Omega^{-1} \varepsilon^{\mathsf{T}} \Sigma^{-1} \varepsilon\right), \quad (10.20)$$

up to an additive constant. Then, the random matrix

$$X \sim \mathcal{N}_{n,p}(0, \Sigma, \Psi), \quad (10.21)$$

where $\Psi = (I_p - B)^{-\mathsf{T}} \Omega (I - B)^{-1}$. We fix $\omega_1 = 1$ so that the parameters (Ω, Σ) are identifiable under model (10.19). From the properties of a matrix normal distribution, we can easily prove the following lemma, which will come in handy when estimating the row covariance matrix Σ given different orderings of nodes. For a permutation π of the set $[p]$, define P_π as the permutation matrix such that $hP_\pi = (h_{\pi(1)}, \ldots, h_{\pi(p)})$ for any row vector $h = (h_1, \ldots, h_p)$.

Lemma 10.2. *If X follows the model (10.19), then for any permutation π of $[p]$, we have*

$$XP_\pi \sim \mathcal{N}_{n,p}(0, \Sigma, P_\pi^\mathsf{T} \Psi P_\pi).$$

Let $\beta_j = (\beta_{1j}, \ldots, \beta_{pj})$ be the jth column of B. Define an $n \times n$ sample covariance matrix for $\varepsilon_1/\omega_1, \ldots, \varepsilon_p/\omega_p$ by

$$S(\Omega, B) = \frac{1}{p} \sum_{j=1}^{p} \frac{1}{\omega_j^2} (X_j - XB_j)(X_j - XB_j)^\mathsf{T} = \frac{1}{p} \varepsilon \Omega^{-1} \varepsilon^\mathsf{T}. \qquad (10.22)$$

Applying $\mathrm{tr}\left(\Omega^{-1}\varepsilon^\mathsf{T}\Sigma^{-1}\varepsilon\right) = p\,\mathrm{tr}(\Theta S(\Omega, B))$ in (10.20), the *negative* log-likelihood $\ell(B, \Omega, \Theta \mid X)$ for the model (10.19) is given by

$$2\ell(B, \Omega, \Theta \mid X) = n \log \det \Omega - p \log \det \Theta + p\,\mathrm{tr}(\Theta S(\Omega, B))$$

$$= -p \log \det \Theta + n \log \det \Omega + \sum_{j=1}^{p} \frac{1}{\omega_j^2} \|LX_j - LXB_j\|_2^2,$$

$$(10.23)$$

where L^T is the Cholesky factor of $\Theta = \Sigma^{-1}$ such that $\Theta = L^\mathsf{T}L$ and L is upper triangular.

10.4.2   *Parameter estimation*

Let us assume that a topological ordering π is given. Denote by $\mathcal{D}(\pi)$ the set of weighted adjacency matrices compatible with π such that any $B \in \mathcal{D}(\pi)$ can be sorted by π. For simplicity, assume that Ω has been pre-estimated. Fixing Ω, the function (10.23) is biconvex in (B, Θ), which can be minimized using iterative methods, such as coordinate descent. Tseng (2001) showed that the coordinate descent algorithm on a bi-convex function converges to a stationary point. Inspired by this observation, the following two-step algorithm is proposed:

(i) Pre-estimate Ω to get $\widehat{\Omega} = \mathrm{diag}(\hat{\omega}_j^2)$.

(ii) Estimate $\widehat{B}$ and $\widehat{\Theta}$ by minimizing a biconvex loss function derived from the negative log-likelihood given $\hat{\omega}_j$ through block-wise coordinate descent.

See Li *et al.* (2024) for more details on the estimation of Ω. Since one of the two subproblems in step (ii) is high-dimensional, sparse regularization is needed. Thus, they propose the following estimator:

$$
\left(\widehat{\Theta}, \widehat{B}(\pi)\right) = \operatorname*{argmin}_{\Theta \succ 0, B \in \mathcal{D}(\pi)} \left\{ -p \log \det \Theta + \sum_{j=1}^{p} \frac{1}{\hat{\omega}_j^2} \|LX_j - LX\beta_j\|_2^2 \right.
$$

$$
\left. + \sum_{j=1}^{p} \frac{\lambda_1}{\hat{\omega}_j^2} \|\beta_j\|_1 + \lambda_2 \|\Theta\|_1 \right\}. \tag{10.24}
$$

The ℓ_1 regularization on $\beta_j / \hat{\omega}_j^2$ not only helps promote sparsity in the estimated DAG but also prevents the model from overfitting due to a small error variance. The ℓ_1 regularization on Θ ensures that the estimator is unique and can improve the accuracy of $\widehat{\Theta}$ by enforcing its sparsity.

Coordinate descent for minimizing the loss in (10.24) is quite straightforward. Fixing $\Theta = \widehat{\Theta}^{(t)}$, the optimization problem in (10.24) becomes the standard lasso problem for each j:

$$
\widehat{\beta}_j^{(t+1)} = \operatorname*{argmin}_{\beta_j} \|\widehat{L}^{(t)} X_j - \widehat{L}^{(t)} X \beta_j\|_2^2 + \lambda_1 \|\beta_j\|_1, \tag{10.25}
$$

where $\widehat{\Theta}^{(t)} = (\widehat{L}^{(t)})^\mathsf{T} \widehat{L}^{(t)}$ is the Cholesky decomposition. Since an ordering π is given, we may order the columns of X according to π, so that $\widehat{\beta}_{ij}^{(t+1)} = 0$ for $i = j, \ldots, p$. Fixing $\widehat{B}^{(t+1)}$, solving for $\widehat{\Theta}^{(t+1)}$ is equivalent to a graphical lasso problem with fixed support (Ravikumar *et al.*, 2011):

$$
\widehat{\Theta}^{(t+1)} = \operatorname*{argmin}_{\operatorname{supp}(\Theta) \subseteq E(H)} -\log \det \Theta + \operatorname{tr}(\widehat{S}^{(t+1)} \Theta) + \lambda_p \|\Theta\|_1, \tag{10.26}
$$

where $\widehat{S}^{(t+1)} = S(\widehat{\Omega}, \widehat{B}^{(t+1)})$ and $\lambda_p = \lambda_2 / p$. Li *et al.* (2024) established the following convergence property of this two-block coordinate descent algorithm.

Proposition 10.1. *Let $\{(\widehat{B}^{(t)}, \widehat{\Theta}^{(t)}) : t = 1, 2, \ldots\}$ be a sequence generated by iterating between (10.25) and (10.26). Then, for any $\lambda_1, \lambda_2 > 0$ and almost all $X \in \mathbb{R}^{n \times p}$, every cluster point of $\{(\widehat{B}^{(t)}, \widehat{\Theta}^{(t)})\}$ is a stationary point of the objective function in (10.24).*

Let $\lambda_{\min}(\Theta^*)$ denote the minimum eigenvalue of the true Θ^*. If the initial value $\widehat{\Theta}^{(0)}$ is close to Θ^* in terms of operator norm,

$$
\|\widehat{\Theta}^{(0)} - \Theta^*\|_2 \leq M,
$$

where $0 < M \leq \lambda_{\min}(\Theta^*)$, then the estimate after one iteration $(\widehat{B}^{(1)}, \widehat{\Theta}^{(1)})$ becomes consistent as $n, p \to \infty$ (Li *et al.*, 2024, Theorem 4). Let m be the maximum degree of the undirected graph H^* and s be the maximum parent size of the true DAG $\mathcal{G}^*$. Assuming $p \gg n \to \infty$, Corollary 8 in Li *et al.* (2024) establishes the following convergence rates:

$$\sup_{1 \leq j \leq p} \|\widehat{\beta}_j^{(1)} - \beta_j^*\|_2 = O_p\left(\sqrt{\frac{s \log p}{n}}\right), \tag{10.27}$$

$$\|\widehat{\Theta}^{(1)} - \Theta^*\|_2 = O_p\left(ms\sqrt{\frac{(\log p)^3}{n}}\right). \tag{10.28}$$

10.4.3   *Structure learning via de-correlation*

Given any permutation π of $[p]$, the joint distribution $\mathbb{P}$ of p random variables admits a recursive factorization w.r.t. a DAG $\mathcal{G}_\pi$ of which π is a topological sort (Definition 7.6). If the true DAG $\mathcal{G}^*$ for model (10.19) is sparse, then for any random ordering π, the corresponding DAG $\mathcal{G}_\pi$ is also likely to be sparse so that the number of parents for each node is less than some constant s_π. Formally, we assume that $s_\pi \ll n/\log p$ uniformly for all permutations π. Similar assumptions have been made in other works on DAG learning from high-dimensional data, such as Condition 3.4 in van de Geer and Bühlmann (2013) and Theorem 4.1 in Aragam *et al.* (2019a).

Let us randomly pick a permutation π and apply the block-wise coordinate descent algorithm on $X_\pi := XP_\pi$, where $(XP_\pi)_{ij} = X_{i\pi(j)}$. Since the covariance Θ^* is invariant to permutations (Lemma 10.2), the resulting estimate $\widehat{\Theta} = \widehat{\Theta}^{(1)}$ under the random ordering π will be a good estimate of Θ^*, satisfying the rate in (10.28), with s_π in place of s. Given the Cholesky factor $\widehat{L}^\mathsf{T}$ of $\widehat{\Theta}$, we de-correlate the rows of X and treat

$$\widehat{X} = \widehat{L}X \tag{10.29}$$

as the new data. Because the row correlations in $\widehat{X}$ vanish, we can apply existing structure learning methods which assume independent observations to learn the underlying DAG. This de-correlation step is able to substantially improve the accuracy of structure learning by well-known, state-of-the-art methods, such as the PC algorithm (Algorithm 9.1) and GES (Algorithm 9.3), as shown by the numerical results in Li *et al.* (2024).

Chapter 11

Directed Mixed Graphs for Latent Variables

Causal graphical models with latent variables, particularly acyclic directed mixed graphs (ADMGs), offer a useful tool for describing causal relations involving latent structures that are not directly measured but affect the observed variables. ADMGs go beyond standard directed acyclic graphs by allowing for a combination of directed and bidirected edges, accommodating scenarios where variables might share latent common causes. This unique characteristic enables a more nuanced representation of causal structures, making these models particularly suitable for inferring intricate causal relationships in situations where some variables remain unobserved. As expected, learning ADMGs from observed data is more challenging than structure learning of DAGs.

11.1 Semi-Markov Causal Models

Denote by $\mathcal{G}(V)$ a graph $\mathcal{G}$ over a vertex set V. In this chapter, we encounter multiple types of edges, including directed ($\rightarrow$) and bidirected ($\leftrightarrow$) edges. There could be more than one edge between a pair of nodes. Thus, it is more convenient to define paths in terms of edges. We define a *path* as a sequence of distinct adjacent edges of any type or orientation between distinct vertices. A path $\{(a_{i-1}, a_i) : i = 1, \ldots, k\}$ is a *directed path* if all edges are directed and oriented as $a_{i-1} \rightarrow a_i$. It is a *bidirected path* if all edges are bidirected, i.e., $a_{i-1} \leftrightarrow a_i$ for all $i = 1, \ldots, k$. The vertices a_0 and

a_k are the endpoints of this path, while the other vertices $a_1, \ldots, a_{k-1}$ are called intermediate vertices.

Our starting point is a DAG $\mathcal{G}(V \cup L)$, where V is a set of observed variables and L is a set of latent variables. To represent the causal relations among the observed variables, we seek to marginalize out L from the DAG $\mathcal{G}(V \cup L)$ via the latent projection of a DAG (Tian and Pearl, 2002b).

Definition 11.1 (Latent projection). Given a DAG with latent variables $\mathcal{G}(V \cup L)$, where V is observed and L is latent, the *latent projection* $\mathcal{G}(V)$ is constructed as follows: for $a, b \in V$,

(i) $\mathcal{G}(V)$ contains an edge $a \to b$ if there is a directed path $a \to \cdots \to b$ in $\mathcal{G}(V \cup L)$ with all intermediate vertices in L;

(ii) $\mathcal{G}(V)$ contains an edge $a \leftrightarrow b$ if there is a collider-free path $a \leftarrow \cdots \to b$ with all intermediate vertices in L.

Note that step (i) adds all directed edges $a \to b$ in $\mathcal{G}(V \cup L)$ to the latent projection $\mathcal{G}(V)$, regarding the intermediate variables as $\varnothing$.

Let us illustrate latent projection with the DAG in Figure 11.1(a), where $V = \{X_1, X_2, X_3\}$ and $L = \{U_1, U_2, U_3\}$. Its latent projection $\mathcal{G}(V)$ is shown in (b). In this process, the path $X_1 \to U_3 \to X_2$ is turned into the edge $X_1 \to X_2$ by (i). The path $X_2 \leftarrow U_1 \to U_2 \to X_3$ is collider-free and thus represented by a bidirected edge $X_2 \leftrightarrow X_3$ according to (ii). In general, the output of a latent projection is an acyclic directed mixed graph (ADMG), which contains two types of edges: directed and bidirected.

Recall that a causal model is defined by an SEM associated with a DAG and a set of independent background variables (Section 8.1). Using ADMGs, we can generalize this causal model to accommodate dependent background variables. If two variables in a latent projection $\mathcal{G}(V)$ are connected by a bidirected edge, $X_i \leftrightarrow X_j$, then they share at least one latent

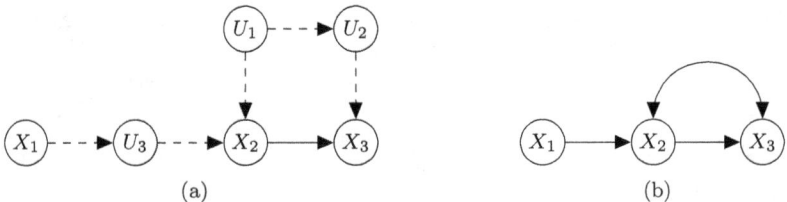

Fig. 11.1. Latent projection of a DAG. (a) A DAG over six nodes, where U_1, U_2, and U_3 are unobserved and a dashed edge incidents on a latent node. (b) The latent projection of the DAG in (a), which is an ADMG.

common ancestor. This latent dependency can be equivalently modeled using dependent background variables ε_i and ε_j.

Given an ADMG $\mathcal{G}(V)$, the distribution of $V = \{X_1, \ldots, X_p\}$ is defined by the SEM

$$X_j = f_j(PA_j, \varepsilon_j), \qquad j = 1, \ldots, p, \tag{11.1}$$

where the parent set $PA_j = \{X_i : i \to j\}$ is defined by directed edges. This is the same as the causal DAG model in Equation (8.1). The key difference lies in the distribution of the background variables $\{\varepsilon_j\}$:

$$\varepsilon_i \perp\!\!\!\perp \varepsilon_j, \quad \text{if there is no bidirected edge between } X_i \text{ and } X_j. \tag{11.2}$$

Thus, having a bidirected edge $X_i \leftrightarrow X_j$ allows for dependence between the two background variables as a result of sharing latent ancestors or latent confounders in causal inference. A causal model with dependent background variables is called a semi-Markov causal model (SMCM).

Let $Y(x)$ be the potential outcome of Y under $\text{do}(X = x)$. In other words, the distribution of the random variable $Y(x)$ is the intervention distribution $[Y \mid \text{do}(X = x)]$. Two types of restrictions on potential outcome variables are encoded by an SMCM, as implications of a missing directed edge and a missing bidirected path (Pearl, 1995):

(1) *Exclusion:* For any set of variables $S \subseteq V \setminus (PA_j \cup \{X_j\})$,

$$X_j(pa_j) = X_j(pa_j, s), \tag{11.3}$$

for any possible values of pa_j and s, where pa_j denotes the value of PA_j. Note that there is no directed edge from any variable in S to X_j. By the SEM (11.1),

$$X_j(pa_j, s) = f_j(pa_j, \varepsilon_j) = X_j(pa_j).$$

The first equality holds because interventions on S do not change the distribution of ε_j (even though there could be a bidirected edge between X_j and a variable in S).

(2) *Independence:* If there are no bidirected paths between X_j and X_i, then for any possible values of pa_i and pa_j,

$$X_j(pa_j) \perp\!\!\!\perp X_i(pa_i). \tag{11.4}$$

Under $\text{do}(PA_k = pa_k)$, the random variable $X_k(pa_k) = f_k(pa_k, \varepsilon_k)$ is a function of ε_k only for any $k \in [p]$. Since there are no bidirected paths between X_i and X_j, the background variables $\varepsilon_i \perp\!\!\!\perp \varepsilon_j$ and thus $X_j(pa_j) \perp\!\!\!\perp X_i(pa_i)$.

Consider the SMCM in Figure 11.1(b). By exclusion restrictions, we see that $X_2(x_1) = X_2(x_1, x_3)$ and $X_1 = X_1(x_2, x_3)$. The latter says that no matter how we manipulate X_2 and X_3, the distribution of X_1 remains unchanged. By independence restrictions, $X_1 \perp\!\!\!\perp \{X_2(x_1), X_3(x_2)\}$ since there are no bidirected paths from X_1 to $\{X_2, X_3\}$. Here, the intervention value x_1 can be any fixed value in the domain of X_1 and is different from the random variable X_1. However, because there is a bidirected edge $X_2 \leftrightarrow X_3$ between the two variables, $X_2(x_1)$ and $X_3(x_2)$ are in general *not* independent.

11.2 Acyclic Directed Mixed Graphs

Let $\mathcal{G}(V)$ be a directed mixed graph, i.e., a graph with two types of edges: directed ($\rightarrow$) or bidirected ($\leftrightarrow$). If $a \rightarrow b$, then a is a parent of b and b is a child of a. If there is a directed path (Section 11.1) from a to d or $a = d$, we say that a is an ancestor of d and d is a descendant of a. Non-descendants of a node a consist of every node that is not a descendant of a. If $a \leftrightarrow b$, then a is a sibling of b. We adopt the notations $\mathrm{pa}_{\mathcal{G}}(a)$, $\mathrm{ch}_{\mathcal{G}}(a)$, $\mathrm{an}_{\mathcal{G}}(a)$, $\mathrm{de}_{\mathcal{G}}(a)$, $\mathrm{nd}_{\mathcal{G}}(a)$, and $\mathrm{sib}_{\mathcal{G}}(a)$, respectively, for the sets of parents, children, ancestors, descendants, non-descendants, and siblings of node a in $\mathcal{G}$.

A *directed cycle* is a path of the form $v \rightarrow \cdots \rightarrow w$ along with an edge $w \rightarrow v$. An ADMG is a mixed graph containing no directed cycles. A topological sort of an ADMG is defined in the same way as for a DAG: a must precede b in the sort $a \prec b$ if $a \rightarrow b$.

Separation relations on an ADMG are defined via the concept of m-separation, which is a generalization of d-separation for DAGs. A vertex z is a collider on a path if the two edges incident on z in the path exhibit one of the following orientations: $\rightarrow z \leftarrow$, $\leftrightarrow z \leftrightarrow$, $\rightarrow z \leftrightarrow$, or $\leftrightarrow z \leftarrow$; otherwise, z is a non-collider. For a subset of nodes C, we define the set of ancestors $\mathrm{an}(C) := \cup_{a \in C} \mathrm{an}(a)$.

Definition 11.2 (m-connection and m-separation). A path between a and b in an ADMG $\mathcal{G}(V)$ is m-connecting given a subset of nodes C if (i) every non-collider on the path is not in C and (ii) every collider on the path is an ancestor of C. If there is no path m-connecting a and b given C, then a and b are m-separated given C. Two subsets of nodes A and B are m-separated given C if all paths from any $a \in A$ to any $b \in B$ are m-separated given C.

If $\mathcal{G}$ is a DAG, then m-separation is identical to d-separation. Moreover, m-separation in a latent projection of a DAG captures all the d-separations among observed nodes in the DAG, as summarized in the following result (Richardson *et al.*, 2023):

Proposition 11.1. *Let $\mathcal{G}(V \cup L)$ be a DAG and $\mathcal{G}(V)$ be its latent projection. For disjoint subsets $A, B, C \subseteq V$, A and B are d-separated given C in $\mathcal{G}(V \cup L)$ if and only if A and B are m-separated given C in $\mathcal{G}(V)$.*

This proposition is proved using the following fact: on every path between $a, b \in V$ in $\mathcal{G}(V \cup L)$, colliders and non-colliders in V are also, respectively, colliders and non-colliders on a path in $\mathcal{G}(V)$. Therefore, a path is m-connected given $C \subseteq V$ in $\mathcal{G}(V \cup L)$ if and only if a corresponding path in $\mathcal{G}(V)$ is m-connected by C. See Richardson *et al.* (2023) for a complete proof.

As a consequence of Proposition 11.1, m-separation in an ADMG $\mathcal{G}(V)$ encodes all conditional independence (CI) constraints among the observed variables V that are implied by d-separation in the DAG $\mathcal{G}(V \cup L)$.

11.3 Factorizations on ADMGs

In an ADMG $\mathcal{G}(V)$, replacing each bidirected edge $X_i \leftrightarrow X_j$ with a path $X_i \leftarrow U_k \rightarrow X_j$ will construct a DAG $\mathcal{G}(V \cup L)$, where $L = \{U_1, \ldots, U_d\}$ and d is the number of bidirected edges in $\mathcal{G}(V)$. We call each U_k a *latent confounder* hereafter. It is easy to see that $\mathcal{G}(V)$ is the latent projection of $\mathcal{G}(V \cup L)$. Based on this construction, the joint distribution $\mathbb{P}(V)$ defined by the ADMG $\mathcal{G}(V)$ is a marginal distribution of $\mathbb{P}(V, L)$ defined by the DAG $\mathcal{G}(V \cup L)$:

$$p(x_1, \ldots, x_p) = \sum_{u_1, \ldots, u_d} p(x_1, \ldots, x_p \mid u_1, \ldots, u_d) \prod_i p(u_i), \qquad (11.5)$$

where all the latent confounders are marginalized out.

11.3.1 *District factorization*

To work out a factorization of $\mathbb{P}(V)$ based on Equation (11.5), we make use of the concept of district in ADMG $\mathcal{G}(V)$. The *district* of vertex v, denoted $\mathrm{dis}_{\mathcal{G}}(v)$, is the set of vertices that are connected to v by a bidirected path (including v itself). If every pair of vertices in a set C is connected by at

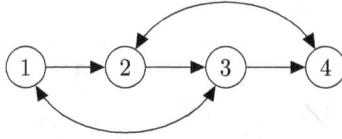

Fig. 11.2. An example ADMG over four vertices.

least one bidirected path, we say that C is a bidirected-connected set of vertices. A district of $\mathcal{G}$ is a maximal bidirected-connected set of vertices. If D is a district of $\mathcal{G}$, then $D = \text{dis}_{\mathcal{G}}(v)$ for every $v \in D$, and $D \cup \{w\}$ is not a bidirected-connected set for any $w \notin D$. A district corresponds to a confounded component (c-component) in Tian and Pearl (2002b). Districts specify variable partitions that define terms in the factorization of $\mathbb{P}(V)$. Denote the set of districts by

$$\mathcal{D}(\mathcal{G}) - \{D : D \text{ is a district of } \mathcal{G}\}.$$

Define the parent set of a district as

$$\text{pa}_{\mathcal{G}}(D) := \{\cup_{a \in D} \text{pa}_{\mathcal{G}}(a)\} \setminus D,$$

which consists of the out-of-district parents of all vertices in D.

We use a simple example in Figure 11.2 to derive the district factorization on ADMGs. There are two districts in $\mathcal{G}$: $D_1 = \{1, 3\}$ and $D_2 = \{2, 4\}$. Their parent sets are $\text{pa}_{\mathcal{G}}(D_1) = \{2\}$ and $\text{pa}_{\mathcal{G}}(D_2) = \{1, 3\}$.

Introduce U_1 as the latent confounder for X_1 and X_3, and U_2 as the latent confounder for X_2 and X_4. According to (11.5), we have

$$p(x_1, \ldots, x_4) = \left[\sum_{u_1} p(x_1 \mid u_1) p(x_3 \mid x_2, u_1) p(u_1) \right]$$

$$\times \left[\sum_{u_2} p(x_2 \mid x_1, u_2) p(x_4 \mid x_3, u_2) p(u_2) \right] \quad (11.6)$$

$$= q_{1,3}(x_1, x_3 \mid x_2) \times q_{2,4}(x_2, x_4 \mid x_1, x_3),$$

where $q_{1,3}$ and $q_{2,4}$ are defined by the two summations in the brackets. This leads to the following factorization of $p(x)$ according to the districts of this

ADMG:

$$p(x_1, \ldots, x_4) = q_{1,3}(x_1, x_3 \mid x_2) \times q_{2,4}(x_2, x_4 \mid x_1, x_3)$$
$$= q_{D_1}(x_{D_1} \mid \mathrm{pa}_{\mathcal{G}}(D_1)) \times q_{D_2}(x_{D_2} \mid \mathrm{pa}_{\mathcal{G}}(D_2)). \qquad (11.7)$$

We may easily generalize the above derivation to any ADMG $\mathcal{G}(V)$ to define the *district factorization* of $\mathbb{P}(V)$:

$$\mathbb{P}(V) = \prod_{D \in \mathcal{D}(\mathcal{G})} q_D(x_D \mid \mathrm{pa}_{\mathcal{G}}(D)). \qquad (11.8)$$

Each factor $q_Y(y \mid W)$ is a *kernel*, i.e., a probability density of Y regarding W as a parameter such that

$$\sum_y q_Y(y \mid W = w) = 1, \qquad \forall\, w.$$

In the district factorization (11.8),

$$q_Y(y \mid W = w) = \mathbb{P}(Y = y \mid \mathrm{do}(W = w)),$$

and thus, in general,

$$q_Y(y \mid W = w) \neq \mathbb{P}(Y = y \mid W = w).$$

As an example, for the ADMG in Figure 11.2, it is easy to see that

$$q_{1,3}(x_1, x_3 \mid x_2) = p(x_1, x_3 \mid \mathrm{do}(x_2)) = \sum_{u_1} p(x_1 \mid u_1) p(x_3 \mid x_2, u_1) p(u_1)$$

in the factorization (11.7).

11.3.2 *Tian factorization*

To complete the factorization, we seek to express each $q_D(x_D \mid \mathrm{pa}_{\mathcal{G}}(D))$ in (11.8) as a product of conditional densities $\prod_{i \in D} p(x_i \mid \cdots)$. To do that, we make use of the Markov blanket of a node $a \in V$ in ADMG $\mathcal{G}(V)$. Let $D = \mathrm{dis}_{\mathcal{G}}(a)$. The *Markov blanket* of a is

$$\mathrm{mb}(a, \mathcal{G}) := \mathrm{pa}_{\mathcal{G}}(D) \cup (D \setminus \{a\}).$$

By definition, we have

$$X_a \perp\!\!\!\perp \mathrm{nd}_{\mathcal{G}}(X_a) \mid \mathrm{mb}(X_a, \mathcal{G}), \qquad (11.9)$$

which generalizes the local Markov property for DAGs. In fact, the Markov blanket reduces to the parent set of a node if $\mathcal{G}$ is a DAG. Suppose that

$1 \prec \cdots \prec p = |V|$ is a topological sort of $\mathcal{G}$. Let $V_i = \{1, \ldots, i\}$ and $\mathcal{G}_i$ be the induced subgraph of $\mathcal{G}$ on V_i. Then, $i \perp\!\!\!\perp k \,|\, \mathrm{mb}(i, \mathcal{G}_i)$ for any $k < i$, from which we arrive at

$$q_D(x_D \,|\, \mathrm{pa}_{\mathcal{G}}(D)) = \prod_{i \in D} p(x_i \,|\, \mathrm{mb}(i, \mathcal{G}_i)). \tag{11.10}$$

Putting this into (11.8), we obtain the *Tian factorization* on an ADMG (Tian and Pearl, 2002b):

$$\mathbb{P}(V) = \prod_{i \in V} p(x_i \,|\, \mathrm{mb}(i, \mathcal{G}_i)). \tag{11.11}$$

Let us use the graph in Figure 11.2 to illustrate the factorization (11.11). The only topological sort of this graph is $1 \prec 2 \prec 3 \prec 4$. The involved Markov blankets according to this sort are

$$\mathrm{mb}(1, \mathcal{G}_1) = \varnothing, \qquad \mathrm{mb}(2, \mathcal{G}_2) = \{1\},$$
$$\mathrm{mb}(3, \mathcal{G}_3) = \{1, 2\}, \quad \mathrm{mb}(4, \mathcal{G}_4) = \{1, 2, 3\}.$$

Then, the two kernels in (11.7) factorize as

$$q_{1,3}(x_1, x_3 \,|\, x_2) = p(x_1)p(x_3 \,|\, x_1, x_2), \tag{11.12}$$
$$q_{2,4}(x_2, x_4 \,|\, x_1, x_3) = p(x_2 \,|\, x_1)p(x_4 \,|\, x_1, x_2, x_3), \tag{11.13}$$

which leads to the factorization of $p(x)$:

$$p(x) = p(x_1)p(x_2 \,|\, x_1)p(x_3 \,|\, x_1, x_2)p(x_4 \,|\, x_1, x_2, x_3).$$

This factorization does *not* imply any conditional independence among $X_1, \ldots, X_4$, as it corresponds to the recursive factorization of a complete DAG. In particular, the graph does not imply any CI between X_1 and X_4, even though there is no edge of either type between the two nodes. This is also seen from the fact that nodes 1 and 4 are m-connected given any $S \subseteq \{X_2, X_3\}$. Then, what is the implication of the missing edge between the two nodes? The answer to this question is provided in the following section.

11.3.3 *Nested factorization*

The missing edge between X_1 and X_4 in Figure 11.2 encodes a generalized conditional independence (GCI) constraint, also known as the Verma constraint (Verma and Pearl, 1990). A GCI constraint is an equality constraint over the functions of conditional distributions.

We use the ADMG in Figure 11.2 to demonstrate GCI constraints. Let $W = \{1, 3\}$. We represent the kernel

$$q_{2,4}(x_2, x_4 \mid x_1, x_3) = p(x_2, x_4 \mid \mathrm{do}(x_1, x_3)) \qquad (11.14)$$

by a *conditional ADMG* (CADMG) with the graph $\mathcal{G}^{|W}$ (Figure 11.3) defined by cutting all edges with an arrow into W. There are two disjoint sets of vertices in a CADMG $\mathcal{G}(V, W)$, random vertex set V and fixed vertex set W. It is required that there are no edges in $\mathcal{G}(V, W)$ with an arrow into any fixed vertex $w \in W$. That is, $\mathrm{pa}_{\mathcal{G}}(w) = \varnothing$ and $\mathrm{sib}_{\mathcal{G}}(w) = \varnothing$. The graph $\mathcal{G}(V, W)$ is, in fact, the modified graph of the ADMG $\mathcal{G}(V \cup W)$ under intervention on W. To distinguish the two types of vertices, we use square nodes for W. In Figure 11.3, $V = \{2, 4\}$ and $W = \{1, 3\}$.

The kernel $q_V(x_V \mid x_W)$ represents the distribution for V after fixing W by intervention. We may further fix other random vertices if they are *fixable* by deleting edges with an arrow into these vertices.

Definition 11.3. The set of *fixable* vertices in a CADMG $\mathcal{G}(V, W)$ is $F(\mathcal{G}) := \{v \in V : \mathrm{dis}_{\mathcal{G}}(v) \cap \mathrm{de}_{\mathcal{G}}(v) = \{v\}\}$. That is, v is fixable if none of its descendants (except v itself) is in the same district as v.

In the CADMG in Figure 11.3, both vertices 2 and 4 are fixable. If we fix X_2, we arrive at a new CADMG $\mathcal{G}(V = \{4\}, W = \{1, 2, 3\})$, shown in Figure 11.4, that represents a new kernel $q_4(x_4 \mid x_2, x_1, x_3)$. By district factorization according to $\mathcal{G}(\{4\}, \{1, 2, 3\})$, it is easy to see that

$$q_4(x_4 \mid x_2, x_1, x_3) = f_4(x_4 \mid x_3) \qquad (11.15)$$

for some kernel f_4, which is a function of x_3 and x_4 only. This is called the *nested factorization* of the kernel q_4.

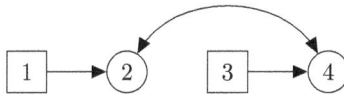

Fig. 11.3. CADMG obtained by fixing X_1 and X_3 in the ADMG in Figure 11.2.

$$\boxed{1} \quad \boxed{2} \quad \boxed{3} \longrightarrow \textcircled{4}$$

Fig. 11.4. CADMG obtained by fixing X_2 in the CADMG in Figure 11.3.

On the other hand, the kernel $q_4(x_4 \mid x_2, x_1, x_3)$ may be defined by applying the *fixing operator* on the kernel (11.14) associated with the CADMG in Figure 11.3.

Definition 11.4. Given a kernel $q_V(x_V \mid W)$ associated with a CADMG $\mathcal{G} = \mathcal{G}(V, W)$, for any fixable vertex $r \in F(\mathcal{G})$, the fixing operator ϕ_r yields a new kernel:

$$q_{V \setminus r}(x_{V \setminus r} \mid r, W) = \phi_r(q_V; \mathcal{G}) := \frac{q_V(x_V \mid W)}{q_V(x_r \mid \mathrm{mb}(r, \mathcal{G}), W)}, \qquad (11.16)$$

where $q_V(x_r \mid \mathrm{mb}(r, \mathcal{G}), W)$ is the conditional distribution $[X_r \mid \mathrm{mb}(r, \mathcal{G})]$ calculated from $q_V(x_V \mid W)$.

If r is fixable, then r can be sorted as the last vertex in its district. Denote by (r) the rank of r in this sort. Then, $\mathrm{mb}(r, \mathcal{G}_{(r)}) = \mathrm{mb}(r, \mathcal{G})$. Thus, the Tian factorization (11.11) of $q_V(x_V \mid W)$ contains the conditional density $q_V(x_r \mid \mathrm{mb}(r, \mathcal{G}), W)$ for X_r as a factor. The truncated factorization under $\mathrm{do}(X_r = x_r)$ can be obtained by removing this factor:

$$\mathbb{P}(V \setminus r \mid \mathrm{do}(x_r); \mathcal{G}) = \frac{q_V(x_V \mid W)}{q_V(x_r \mid \mathrm{mb}(r, \mathcal{G}), W)},$$

which is the kernel $\phi_r(q_V; \mathcal{G})$. This also shows that the causal effect of X_r on $V \setminus r$, defined by $\mathbb{P}(V \setminus r \mid \mathrm{do}(x_r); \mathcal{G})$, can be calculated using the fixing operator (11.16).

Let $\mathcal{G}$ be the CADMG in Figure 11.3. Applying ϕ_2 on $q_{2,4}(x_2, x_4 \mid x_1, x_3)$ associated with $\mathcal{G}$, where $\mathrm{mb}(2, \mathcal{G}) = \{1, 4, 3\}$, we arrive at

$$q_4(x_4 \mid x_2, x_1, x_3) = \phi_2(q_{2,4}; \mathcal{G}) = \frac{q_{2,4}(x_2, x_4 \mid x_1, x_3)}{q_{2,4}(x_2 \mid x_4, x_1, x_3)}$$

$$= q_{2,4}(x_4 \mid x_1, x_3)$$

$$= \sum_{x_2'} q_{2,4}(x_2', x_4 \mid x_1, x_3)$$

$$= \sum_{x_2'} p(x_2' \mid x_1) p(x_4 \mid x_1, x_2', x_3), \quad \text{(by (11.13))}$$

which expresses the kernel q_4 as a functional of conditional distributions. This expression shows that q_4 is a function of (x_1, x_3, x_4). By the nested factorization of q_4 in (11.15), however,

$$\sum_{x_2'} p(x_2' \mid x_1) p(x_4 \mid x_1, x_2', x_3) = f_4(x_4 \mid x_3),$$

and thus it does not depend on x_1. This defines a constraint on a certain function of the conditional probabilities $p(x_2 \mid x_1)$ and $p(x_4 \mid x_1, x_2, x_3)$. Such constraints are called GCI constraints.

Let us formalize nested factorization with respect to an ADMG $\mathcal{G}$. For a fixable vertex r, let $\phi_r(\mathcal{G})$ be the CADMG after fixing r in $\mathcal{G}$. A sequence of distinct vertices $w = (w_1, \ldots, w_k)$ is *valid* in $\mathcal{G}$ if $w_1 \in F(\mathcal{G})$ and $(w_2, \ldots, w_k)$ is valid in $\phi_{w_1}(\mathcal{G})$. For a valid fixing sequence $w = (w_1, \ldots, w_k)$, let $\phi_w(\mathcal{G})$ be the CADMG after fixing w sequentially and $\mathcal{D}_w = \mathcal{D}(\phi_w(\mathcal{G}))$ be the districts of the random vertices in $\phi_w(\mathcal{G})$. Theorem 11.1 follows from Proposition 29 in Richardson *et al.* (2023).

Theorem 11.1. *Suppose $p(x)$ factorizes according to a DAG $\mathcal{G}(V \cup L)$ and $\mathcal{G} = \mathcal{G}(V)$ is the ADMG defined by latent projection. For any valid fixing sequence w in $\mathcal{G}$,*

$$\phi_w(p(x_V); \mathcal{G}) = \prod_{D \in \mathcal{D}_w} f_D^w(x_D \mid pa_{\mathcal{G}}(D))$$

for some kernels $f_D^w(x_D \mid pa_{\mathcal{G}}(D))$.

11.3.4 *Finding GCI constraints*

Tian and Pearl (2002b) developed Algorithm 11.1 to systematically find CI and GCI constraints implied by an ADMG. The input is an ADMG $\mathcal{G}(V)$, assuming that $V = [p]$ is sorted such that $1 \prec \cdots \prec p$. The algorithm enumerates all CI and GCI constraints on $p(x_V)$ implied by $\mathcal{G}(V)$ by examining the constraints of the subgraph $\mathcal{G}_i$ over the vertex set $[i] = \{1, \ldots, i\}$, for $i = 1, \ldots, p$.

CI constraints involving X_i are identified on Line 3 via the local Markov property of an ADMG (11.9). The for loop from Line 5 to Line 14 identifies GCIs by recursively fixing each descendant set D in the district of i. A set D is descendant if it contains the descendants of every vertex in D.

Algorithm 11.1 (Identify CI and GCI constraints implied by an ADMG).

1: **Input:** ADMG $\mathcal{G}(V)$, $V = [p]$ is sorted such that $1 \prec \cdots \prec p$.
2: **for** $i = 1$ to p **do**
3:      Output CI $X_i \perp\!\!\!\perp X_k \,|\, \mathrm{mb}(i, \mathcal{G}_i)$ for each $k < i$, $k \notin \mathrm{mb}(i, \mathcal{G}_i)$.
4:      $S \leftarrow \mathrm{dis}_{\mathcal{G}_i}(i)$ and $G \leftarrow \phi_{[i] \setminus S}(\mathcal{G}_i)$.
5:      **for** each descendant set $D \subset S$ s.t. $i \notin D$ **do**
6:          $D' \leftarrow S \setminus D$.
7:          $\sum_{x_D} q_S = q_{D'}$ (by fixing D). [Potential GCI]
8:          $G' \leftarrow \phi_D(G)$.
9:          **if** G' has 2 or more districts **then**
10:              $E \leftarrow \mathrm{dis}_{G'}(i)$.
11:              $q_{D'} / \sum_{x_i} q_{D'}$ is a function of $E \cup \mathrm{pa}_{G'}(E)$. [Potential GCI]
12:              Go to Line 5 with $G \leftarrow \phi_{S \setminus E}(G)$ and $S \leftarrow E$.
13:          **end if**
14:      **end for**
15: **end for**

Since D is descendant in S, the kernel $q_{D'}$ yielded by fixing D can be obtained by marginalizing out x_D, hence the identity on Line 7. If the set of variables $q_{D'}$ depends on is a proper subset of those on the left-hand side, then we find a GCI constraint. Since i is the last vertex in $\mathcal{G}_i$ according to the sort,

$$\frac{q_{D'}}{\sum_{x_i'} q_{D'}} = q_{D'}(x_i \mid D'_{-i}) = q_{D'}(x_i \mid \mathrm{mb}(i, G'))$$

is a function of variables in $\{i\} \cup \mathrm{mb}(i, G') = E \cup \mathrm{pa}_{G'}(E)$ (Line 11). Therefore, a GCI constraint is identified if the left-hand side depends on some variables not in $E \cup \mathrm{pa}_{G'}(E)$.

11.4   Identification of Causal Effects

Given an ADMG $\mathcal{G}(V)$, let $k \in V$ be a single variable and $S \subset V$. The causal effect of X_k on X_S is identifiable if $\mathbb{P}(X_S \mid \mathrm{do}(X_k))$ can be computed from the joint distribution $\mathbb{P}(X_V)$. If identifiable, the causal effect can be estimated from observational data. This is a generalization of causal effect identification on DAGs.

Let us consider the causal effect of X_2 on the other three variables in Figure 11.2. Applying truncated factorization to (11.6) under $\mathrm{do}(X_2 = x_2)$,

$$p(x_1, x_3, x_4 \mid \mathrm{do}(x_2)) = \left[\sum_{u_1} p(x_1 \mid u_1)p(x_3 \mid x_2, u_1)p(u_1) \right]$$

$$\times \left[\sum_{u_2} p(x_4 \mid x_3, u_2)p(u_2) \right],$$

where the term $p(x_2 \mid x_1, u_2)$ has been removed from the kernel $q_{2,4}$. Since $\sum_{x_2'} p(x_2' \mid x_1, u_2) = 1$, we see that

$$p(x_1, x_3, x_4 \mid \mathrm{do}(x_2)) = q_{1,3}(x_1, x_3 \mid x_2) \sum_{x_2'} q_{2,4}(x_2', x_4 \mid x_1, x_3), \quad (11.17)$$

which shows that the intervention distribution on the left-hand side can be calculated from the joint distribution of $(X_1, \ldots, X_4)$. Thus, the causal effect of X_2 on any subset of the other three variables is identifiable. The key structure that ensures the identifiability in this example is that X_2 has no children in its district, and thus it only appears in the conditional distribution $p(x_2 \mid x_1, u_2)$.

Tian and Pearl (2002a) provided a general sufficient condition (Theorem 11.2) for the identifiability of causal effects on an ADMG. Recall that $\mathrm{an}(S)$ is the smallest ancestral set containing S.

Theorem 11.2. *If there is no bidirected path connecting X_k to any of its children in $\mathcal{G}_{\mathrm{an}(S)}$, then the causal effect of X_k on X_S is identifiable.*

Proof. Since only variables in $\mathrm{an}(S)$ may have a causal effect on X_S, let us assume that $V = \mathrm{an}(S)$ and $\mathcal{G} = \mathcal{G}_{\mathrm{an}(S)}$. Let $M = V \setminus \{S \cup k\}$. Then,

$$p(x_S \mid \mathrm{do}(x_k)) = \sum_{x_M} p(x_{V \setminus k} \mid \mathrm{do}(x_k)).$$

Let $\mathcal{D} = \mathcal{D}(\mathcal{G})$, $D = \mathrm{dis}_\mathcal{G}(k) \in \mathcal{D}$ and $\mathcal{D}' = \mathcal{D} \setminus \{D\}$. By district factorization,

$$p(x_V) = q_D(x_D \mid \mathrm{pa}_\mathcal{G}(D)) \prod_{D' \in \mathcal{D}'} q_{D'}(x_{D'} \mid \mathrm{pa}_\mathcal{G}(D')).$$

Since $\mathrm{ch}(k) \cap D = \varnothing$, the effect of fixing X_k on q_D is equivalent to marginalizing out X_k by summing over x_k'. Therefore,

$$p(x_{V \setminus k} \mid \mathrm{do}(x_k)) = \sum_{x_k'} q_D(x_D \mid \mathrm{pa}_\mathcal{G}(D)) \prod_{D' \in \mathcal{D}} q_{D'}(x_{D'} \mid \mathrm{pa}_\mathcal{G}(D')),$$

where the right-hand side can be calculated from the joint distribution $p(x_V)$.   □

Remark 11.1. The assumption that $\mathrm{ch}(k) \cap D = \varnothing$ is weaker than the assumption that X_k is fixable (Definition 11.3). If X_k is fixable, then we may instead apply a fixing operator to calculate

$$p(x_{V \setminus k} \mid \mathrm{do}(x_k)) = \phi_k(p(x_V); \mathcal{G}) = \frac{p(x_V)}{p(x_k \mid \mathrm{mb}(k, \mathcal{G}))}.$$

Recent results on the identification of causal effects on ADMGs include Theorem 48 in Richardson *et al.* (2023) and Corollary 16 in Bhattacharya *et al.* (2022). Richardson *et al.* (2023) provided a sufficient and necessary condition for the identifiability of $\mathbb{P}(Y \mid \mathrm{do}(A)) \equiv \mathbb{P}(X_Y \mid \mathrm{do}(X_A))$ for any two disjoint subsets $A, Y \subset V$. They also reformulated the identify algorithm by Tian and Pearl (2002a), using fixing operators, to calculate a causal effect when it is identifiable. To describe this result, we define reachable graph and intrinsic set.

Definition 11.5. A CADMG $\mathcal{G}^*(V_1, V_2)$ is a reachable graph derived from an ADMG $\mathcal{G}(V)$, $V = V_1 \cup V_2$, if there is a valid fixing sequence w of the vertices in V_2 such that $\mathcal{G}^* = \phi_w(\mathcal{G})$. A set $C \subseteq V$ is intrinsic in $\mathcal{G}(V)$ if it is a district in a reachable graph derived from $\mathcal{G}(V)$.

Let $\mathcal{G} = \mathcal{G}(V)$ and $Y^* = \mathrm{an}_{\mathcal{G}_{V \setminus A}}(Y) \supseteq Y$. By definition, there is a directed path, not intersecting A, from every $v \in Y^*$ to Y. Since $Y \subseteq Y^*$, we can find $\mathbb{P}(Y \mid \mathrm{do}(A))$ through marginalization:

$$\mathbb{P}(Y \mid \mathrm{do}(A)) = \sum_{Y^* \setminus Y} \mathbb{P}(Y^* \mid \mathrm{do}(A)).$$

Let $\mathcal{D}^* = \mathcal{D}(\mathcal{G}_{Y^*})$. Since Y^* is an ancestral set, the district factorization on $\mathcal{G}$ implies that

$$\mathbb{P}(Y^* \mid \mathrm{do}(A)) = \prod_{D \in \mathcal{D}^*} \mathbb{P}[D \mid \mathrm{do}(\mathrm{pa}_{\mathcal{G}}(D))]. \tag{11.18}$$

If every D is intrinsic in $\mathcal{G}$, then $\mathbb{P}[D \mid \mathrm{do}(\mathrm{pa}_{\mathcal{G}}(D))] = \phi_{V \setminus D}(\mathbb{P}(V); \mathcal{G})$ and

$$\mathbb{P}(Y \mid \mathrm{do}(A)) = \sum_{Y^* \setminus Y} \prod_{D \in \mathcal{D}^*} \phi_{V \setminus D}(\mathbb{P}(V); \mathcal{G}). \tag{11.19}$$

Otherwise, $\mathbb{P}[D \mid \mathrm{do}(\mathrm{pa}_{\mathcal{G}}(D))]$ is not identifiable for some D, and consequently, $\mathbb{P}(Y \mid \mathrm{do}(A))$ is not identifiable.

Let us use (11.19) to find $p(x_4 \mid \mathrm{do}(x_2))$ for the example ADMG in Figure 11.2, where $Y = \{4\}, A = \{2\}, Y^* = \{3,4\}$ and $\mathcal{D}^* = \{D_1, D_2\} = \{\{3\}, \{4\}\}$. By (11.18),

$$p(x_4 \mid \mathrm{do}(x_2)) = \sum_{x_3} p(x_3 \mid \mathrm{do}(x_2)) p(x_4 \mid \mathrm{do}(x_3)). \tag{11.20}$$

It is easy to verify that both $D_1 = \{3\}$ and $D_2 = \{4\}$ are intrinsic. Thus, to complete (11.20), it suffices to find $p(x_3 \mid \mathrm{do}(x_2))$ and $p(x_4 \mid \mathrm{do}(x_3))$ by fixing operators:

$$p(x_3 \mid \mathrm{do}(x_2)) = \phi_{4,2,1}(p(x_V); \mathcal{G}) = \phi_1(q_{1,3}(x_1, x_3 \mid x_2); \mathcal{G}^{|2,4})$$

$$= \sum_{x_1} p(x_1) p(x_3 \mid x_1, x_2),$$

$$p(x_4 \mid \mathrm{do}(x_3)) = \phi_{3,1,2}(p(x_V); \mathcal{G}) = \phi_2(q_{2,4}(x_2, x_4 \mid x_1, x_3); \mathcal{G}^{|1,3})$$

$$= \sum_{x_2'} p(x_2' \mid x_1) p(x_4 \mid x_1, x_2', x_3). \tag{11.21}$$

Note that (11.21) is independent of x_1, implying a GCI constraint. Alternatively, we can find $p(x_4 \mid \mathrm{do}(x_2))$ by summing $p(x_1, x_3, x_4 \mid \mathrm{do}(x_2))$ in (11.17) over x_1 and x_3, which leads to the same result.

11.5   Linear SEM Associated with ADMG

In this section, we consider the parameterization of an ADMG through a linear SEM with dependent Gaussian errors, which generalizes the Gaussian linear DAG model discussed in Section 7.4.1.

11.5.1   *Parameterization and identifiability*

Given an ADMG $\mathcal{G}$ with vertex set $V = [p]$, a directed edge set E_d and a bidirected edge set E_b, define a linear SEM:

$$X_j = \sum_{i \in \mathrm{pa}_{\mathcal{G}}(j)} \beta_{ij} X_i + \varepsilon_j, \quad j = 1, \ldots, p, \tag{11.22}$$

$$(\varepsilon_1, \ldots, \varepsilon_p) \sim \mathcal{N}_p(0, \Omega), \tag{11.23}$$

where $\Omega = (\omega_{ij})_{p \times p}$ such that $\omega_{ij} = 0$ if $(i, j) \notin E_b$. The key difference from the Gaussian linear DAG model (7.4) is that the error variables ε_i and ε_j may be dependent if there is a bidirected edge $i \leftrightarrow j$ between X_i and X_j.

Such dependence is introduced to model the common latent cause between the two variables.

Let $B = (\beta_{ij})_{p \times p}$. The parameter spaces for B and Ω given the graph $\mathcal{G}$ are, respectively,

$$\mathcal{B}(E_d) := \{(\beta_{ij})_{p \times p} : \beta_{ij} = 0 \text{ if } (i,j) \notin E_d\},$$
$$\mathcal{P}(E_b) := \{(\omega_{ij})_{p \times p} \succ 0 : \omega_{ij} = 0 \text{ if } (i,j) \notin E_b\}.$$

The linear SEM (11.22) defines a family of multivariate Gaussian distributions $\mathcal{N}_p(0, \Sigma)$ with

$$\Sigma = \Sigma_\mathcal{G}(B, \Omega) := (I_p - B)^{-\mathsf{T}} \Omega (I_p - B)^{-1}. \tag{11.24}$$

Definition 11.6 (Identifiability). The linear SEM for an ADMG $\mathcal{G}$ is said to be *identifiable* if $\Sigma_\mathcal{G}(B, \Omega)$ is an *injective* (one-to-one) map from $\mathcal{B}(E_d) \times \mathcal{P}(E_b)$ to the set of positive definite matrices.

The above definition is an example of parameter identifiability discussed in Remark 3.1. Given $\mathcal{G}(V)$, if the linear SEM associated with $\mathcal{G}$ is identifiable, then for any $\Sigma \succ 0$, there is a unique (B, Ω) such that (11.24) holds. As a consequence, the MLE of (B, Ω) is well defined and unique when n is large.

Next, we develop a graphical criterion for the identifiability of a linear SEM associated with an ADMG based on the concept of reachable closure (Shpitser *et al.*, 2018).

Definition 11.7. A subset $R \subseteq V$ in a CADMG $\mathcal{G}(V, W)$ is reachable if a fixing sequence exists for $V \setminus R$. A reachable subset $C \subseteq V$ is called a reachable closure for $S \subseteq C$ if the set of fixable vertices in $\phi_{V \setminus C}(\mathcal{G})$ is a subset of S.

Reachable closure is unique for any $S \subseteq V$, denoted by $\langle S \rangle$. Let $\phi_{\neg S}$ denote the fixing operator that fixes as many vertices in $V \setminus S$ as possible. Then, $\langle S \rangle$ is the set of random vertices in the CADMG $\phi_{\neg S}(\mathcal{G})$. The subgraph $\mathcal{G}_{\langle v \rangle}$ is called an arborescence converging on v if $|\langle v \rangle| \geq 2$. Theorem 2 in Drton *et al.* (2011), presented as follows, provides a sufficient and necessary condition for the identifiability of the linear SEM for an ADMG.

Theorem 11.3. *The linear SEM for an ADMG $\mathcal{G}(V)$ is identifiable if and only if $\langle v \rangle = \{v\}$ for all $v \in V$.*

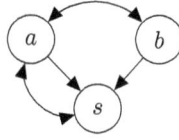

Fig. 11.5. A graph with non-identifiable SEM parameters.

In other words, the SEM parameterization is identifiable if and only if $\mathcal{G}$ does not contain any converging arborescence.

Take the ADMG in Figure 11.5 as an example. In this graph, neither vertex a nor b is fixable since they both have a child s in their district. The reachable closure for s is $\langle s \rangle = \{a, b, s\}$. The graph $\mathcal{G}$ contains a sink node s and its parents a amd b in the same district, which is an arborescence converging on s. By Theorem 11.3, the linear SEM for this graph is *not* identifiable. Indeed, the SEM has seven free parameters, $\beta_{as}, \beta_{bs}, \omega_{aa}, \omega_{bb}, \omega_{ss}, \omega_{as}$, and ω_{ab}, while the 3×3 matrix $\Sigma_{\mathcal{G}}(B, \Omega)$ in (11.24) has only six free parameters. The map $(B, \Omega) \mapsto \Sigma$ is not injective in this case.

In the graph in Figure 11.5, we see a bow pattern: a bidirected edge $a \leftrightarrow s$ connecting a parent–child pair $a \to s$. Another weaker identifiability condition requires the graph to be bow-free, i.e., there are not bow patterns in the graph (Brito and Pearl, 2002b).

Theorem 11.4. *If the ADMG $\mathcal{G}(V)$ is bow-free, then $\Sigma_{\mathcal{G}}(B, \Omega)$ is an injective map for almost all $(B, \Omega) \in \mathcal{B}(E_d) \times \mathcal{P}(E_b)$.*

11.5.2 Structure learning

There are some recent works on structure learning of ADMGs assuming the linear SEM with Gaussian noises (Nowzohour *et al.*, 2017; Frot *et al.*, 2019; Bhattacharya *et al.*, 2021). These methods all fall in the category of score-based learning, searching for graphs with a high likelihood-based score, such as the BIC score. To calculate the BIC score of a candidate graph $\mathcal{G}$, the key is to find the MLE $(\widehat{B}, \widehat{\Omega})$ under the SEM in (11.22). Although not in closed form, the MLE can be found through a residual iterative conditional fitting (RICF) algorithm (Drton *et al.*, 2009). Let

$$\beta_j = (\beta_{ij}, i \in \mathrm{pa}_{\mathcal{G}}(j)), \qquad \omega_j = (\omega_{jj}, \omega_{jk}, k \in \mathrm{sib}_{\mathcal{G}}(j)),$$

which are the parameters involved in the linear model for X_j. The RICF is a block-wise coordinate descent algorithm that cycles through the maximization of (β_j, ω_j) for $j \in V$. Given the other parameters $(\beta_{-j}, \omega_{-j})$, we first compute the residuals ε_{-j} using (11.22) and pseudo variables $Z_{-j} = (\Omega_{-j,-j})^{-1}\varepsilon_{-j}$. Then, the parameters (β_j, ω_j) are updated based on the least-squares regression $X_j \sim X_{\mathrm{pa}(j)} + Z_{\mathrm{sib}(j)}$, where $\mathrm{pa}(j) = \mathrm{pa}_{\mathcal{G}}(j)$ and $\mathrm{sib}(j) = \mathrm{sib}_{\mathcal{G}}(j)$, both specified by the candidate ADMG $\mathcal{G}$.

There are also a few methods that search for a causal ordering under specific model assumptions (Li *et al.*, 2023; Salehkaleybar *et al.*, 2020; Agrawal *et al.*, 2022). For example, the method proposed by Li *et al.* (2023) relies on normality to infer the underlying causal ordering, while the work by Agrawal *et al.* (2022) is focused on the nonlinear, pervasive confounding setting. Application of these methods has been somewhat limited by the specific and restrictive model assumptions and the challenges in the search algorithm due to the huge graph space and the combinatorial nature of the optimization problem.

Besides score-based learning, another popular approach for DAG learning is constraint-based learning, such as the PC algorithm (Spirtes and Glymour, 1991), which learns CI constraints from data to construct the Markov equivalence class of a DAG. However, there are no existing constraint-based methods for learning the structure of ADMGs. The main reason is that an ADMG not only encodes CI constraints among V but also GCI constraints, as we discussed in Section 11.3.3. In general, it is unknown how to find such constraints from data or how to use them for constraint-based learning of the underlying ADMG (Zhang *et al.*, 2020).

Nevertheless, there are efficient structure learning algorithms for some special classes of ADMGs. We review some of them in the subsequent sections.

11.6   Ancestral Graphs

An ancestral graph (Richardson and Spirtes, 2002) is a class of ADMGs that represents the conditional independence relations among the observed variables V in a DAG $\mathcal{G}(V, L)$ with latent variables L. It retains the ancestral relationships and hence causal relations among V. Like DAGs, its equivalence class can be constructed from CI relations learned from observational data. However, it does *not* preserve all confounding structures in $\mathcal{G}(V, L)$, that is, some bidirected edges in the latent projection are lost. Moreover,

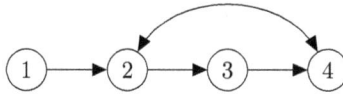

Fig. 11.6. Almost directed cycle and inducing path in ADMG.

an ancestral graph does *not* represent GCI constraints, which may result in a potential loss of efficiency in structure learning.

11.6.1 *Maximal ancestral graphs*

Let $\mathcal{G} = (V, E)$ be an ADMG. An *almost directed cycle* occurs when $a \leftrightarrow b$ and $a \in \mathrm{an}_{\mathcal{G}}(b)$, i.e., removing the arrowhead at b results in a directed cycle. Let $L \subset V$. An *inducing path relative to* L is a path on which every intermediate vertex not in L is a collider and every collider is an ancestor of an endpoint. If $L = \varnothing$, we simply call it an *inducing path*. In Figure 11.6, $2 \to 3 \to 4 \leftrightarrow 2$ is an almost directed cycle. On the path $1 \to 2 \leftrightarrow 4$, vertex 2 is a collider and an ancestor of vertex 4; therefore, this path is an inducing path. It is seen that vertices 1 and 4 are not m-separated by any subset. In general, if there is an inducing path between two vertices a and b, then a and b cannot be m-separated by any subset.

Definition 11.8 (MAG). A mixed graph is a maximal ancestral graph (MAG) if:

(i) it does not contain any directed or almost directed cycles, i.e., it is ancestral;
(ii) there is no inducing path between any two non-adjacent vertices, i.e., it is maximal.

Given a DAG $\mathcal{G} = \mathcal{G}(V, L)$ with latent variables L, the corresponding MAG $\mathcal{M} = \mathcal{M}_{\mathcal{G}}$ is constructed using a two-step procedure:

(1) For each pair $a, b \in V$, a and b are adjacent in $\mathcal{M}$ if and only if there is an inducing path between them relative to L in $\mathcal{G}$.
(2) For each adjacent pair (a, b) in $\mathcal{M}$, orient $a \to b$ in $\mathcal{M}$ if $a \in \mathrm{an}_{\mathcal{G}}(b)$, orient $b \to a$ in $\mathcal{M}$ if $b \in \mathrm{an}_{\mathcal{G}}(a)$, and orient $a \leftrightarrow b$ otherwise.

If there is an edge between a and b in $\mathcal{G}$, then the edge itself is regarded as an inducing path. Moreover, step (1) connects all pairs of vertices that have an edge in the latent projection $\mathcal{G}(V)$ (Definition 11.1), including all the edges among V in $\mathcal{G}(V \cup L)$.

(a) DAG $\mathcal{G}(V \cup L)$

(b) MAG $\mathcal{M}_\mathcal{G}$

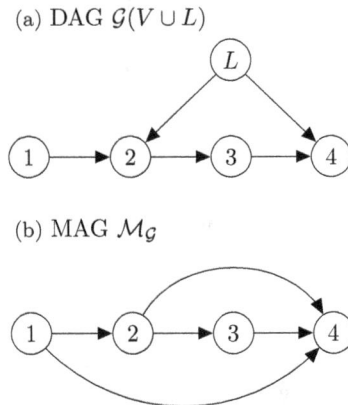

Fig. 11.7. Constructing MAG from DAG with latent variables.

One can show that the $\mathcal{M}$ constructed by the above procedure is indeed a MAG and represents, via m-separation, the CI relations among V implied by the DAG $\mathcal{G}$ (Richardson and Spirtes, 2002; Zhang, 2008a). Moreover, the orientation rules in step (2) preserve the ancestral relations among the observed variables in V, thereby retaining their causal relationships.

Figure 11.7 illustrates the construction of a MAG from a DAG with a latent variable L. Every edge among $V = \{1, 2, 3, 4\}$ in the DAG is a trivial inducing path and thus an edge in the MAG. In addition, there are two inducing paths relative to L in the DAG $\mathcal{G}(V \cup L)$: (i) $1 \to 2 \leftarrow L \to 4$, which leads to the addition of the edge $1 \to 4$ in $\mathcal{M}_\mathcal{G}$ since vertex 1 is an ancestor of vertex 4; (ii) $2 \leftarrow L \to 4$, which induces the edge $2 \to 4$ in $\mathcal{M}_\mathcal{G}$ as vertex 2 is also an ancestor of vertex 4. Compared to the latent projection $\mathcal{G}(V)$ in Figure 11.6, the MAG has an extra edge $1 \to 4$ and a different orientation for the edge $(2, 4)$. There is only one pair of non-adjacent vertices, $(1, 3)$, in the MAG $\mathcal{M}_\mathcal{G}$. The two vertices are m-separated by vertex 2, and thus $X_1 \perp\!\!\!\perp X_3 \mid X_2$, which is indeed the only CI relation among X_V.

11.6.2   *Partial ancestral graphs*

Two MAGs are Markov equivalent if they have the same set of m-separations. Consequently, equivalent MAGs encode the same set of CI constraints among the observed variables. The condition for equivalent MAGs is more complicated than that for equivalent DAGs. Besides skeleton and

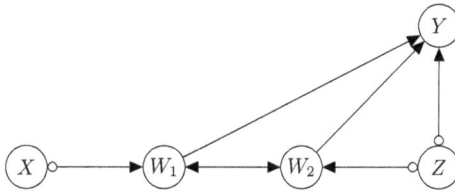

Fig. 11.8. A discriminating path for Z.

v-structure, we must also consider covered colliders on the so-called discriminating path.

Definition 11.9 (Discriminating path). In a MAG, a path $\pi = (X, \ldots, W, Z, Y)$ is a *discriminating path* for Z if:

(i) π includes at least three edges;
(ii) Z is a non-endpoint and is adjacent to Y on π;
(iii) X is not adjacent to Y, and every vertex between X and Z is a collider on π and a parent of Y.

Figure 11.8 shows a discriminating path (X, W_1, W_2, Z, Y) for Z, where the edge mark $\circ$ allows for either an arrowhead or a tail. For example, the edge $Z \circ\!\!\to Y$ may be either $Z \to Y$ or $Z \leftrightarrow Y$. If Z is a collider, i.e., the two circles at Z are replaced by arrowheads, then X and Y are m-separated given $\{W_1, W_2\}$. If Z is a non-collider, as in $W_2 \leftarrow Z \to Y$ or $W_2 \leftrightarrow Z \to Y$, then X and Y are m-connected given $\{W_1, W_2\}$. This shows that a collider at Z, though covered by the edge $W_2 \to Y$, must be preserved among equivalent MAGs.

A set of sufficient and necessary conditions for two MAGs to be equivalent is provided in the following theorem (Zhang, 2008b).

Theorem 11.5. *Two MAGs over the same vertex set are Markov equivalent if and only if the following conditions hold:*

(a) *they have the same skeleton and v-structures;*
(b) *if π is a discriminating path for a vertex Z in both graphs, then Z is a collider on π in one graph if and only if it is a collider on π in the other graph.*

Condition (a) is the same as that for DAG equivalence. However, as stated in condition (b), equivalent MAGs must also share covered colliders for which there is a discriminating path.

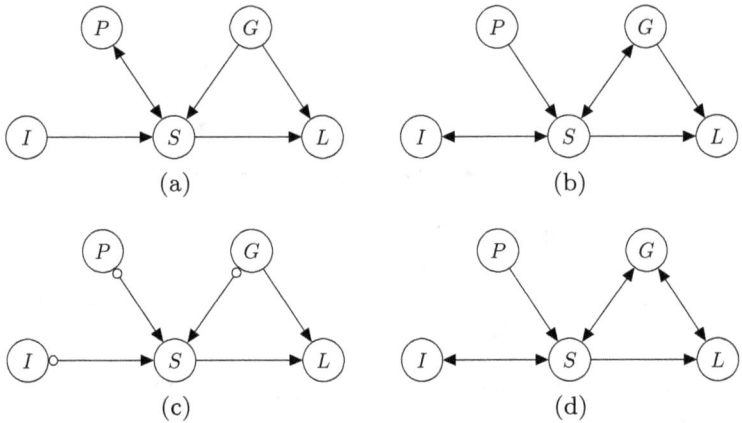

Fig. 11.9. Equivalence between MAGs. The two MAGs in (a) and (b) are in the same equivalence class represented by the PAG in (c), but the MAG in (d) is not in this equivalence class.

Figures 11.9(a) and 11.9(b) show two MAGs over five variables: income (I), smoking (S), parent smoking habits (P), genotype (G), and lung cancer (L), an example from Zhang (2008a). The path (P, S, G, L) is a discriminating path for the vertex G in the two MAGs, and G is not a collider in either of them. One can verify that these two MAGs are equivalent. If we change the edge $G \to L$ in (b) to $G \leftrightarrow L$, then G will become a collider, and thus the resulting MAG in (d) is not in this equivalence class. It is not difficult to see that in (a) and (b) P and L are m-connected given S, while in (d) P and L are m-separated given S.

The equivalence class $[\mathcal{M}]$ of a MAG $\mathcal{M}$ is represented by a *partial ancestral graph* $\mathcal{P}$ such that:

(i) $\mathcal{P}$ has the same adjacencies (skeleton) as $\mathcal{M}$;
(ii) a mark of arrowhead is in $\mathcal{P}$ if and only if it is shared by all MAGs in $[\mathcal{M}]$;
(iii) a mark of tail is in $\mathcal{P}$ if and only if it is shared by all MAGs in $[\mathcal{M}]$.

The edge marks in (ii) and (iii) are invariant across $[\mathcal{M}]$; other variable marks are represented by ∘ in $\mathcal{P}$.

Shown in Figure 11.9(c) is the PAG that represents the equivalence class of the two equivalent MAGs in (a) and (b). The edge $I \circ\!\!\to S$ means that it can be either $I \to S$ or $I \leftrightarrow S$ among the MAGs in this equivalence class, as shown in (a) and (b). Every MAG in this equivalence class must preserve the three v-structures at the uncovered collider S, and on the discriminating path (P, S, G, L), the vertex G must be a non-collider. In addition, none of the MAGs should have directed or almost directed cycles among G, S, and L. Taking into account these constraints, there are only three variant marks in the PAG (c).

11.6.3 *The FCI algorithm*

Suppose $\mathcal{M}(V)$ is a MAG constructed from a DAG $\mathcal{G}(V \cup L)$ by marginalizing out the latent variables L. We say that the random vector X_V follows a causal MAG model if the joint distribution of (X_V, X_L) is defined by the causal DAG $\mathcal{G}(V \cup L)$. Given the observed data of X_V, we consider the problem of learning the MAG $\mathcal{M}(V)$ under a similar faithfulness assumption, that is, there is a one-to-one correspondence between the CI relations in $\mathbb{P}(X_V)$ and the m-separations in $\mathcal{M}(V)$.

Constraint-based learning of MAGs can be carried out using the fast causal inference (FCI) algorithm (Spirtes *et al.*, 1999), which is similar to the PC algorithm for DAG structure learning. CI constraints learned from observational data are used to construct the equivalence class of a MAG, represented by a PAG. The two components of the FCI algorithm are to learn the skeleton of the MAG and then to identify invariant edge marks (arrowheads and tails), as outlined in Algorithm 11.2.

Algorithm 11.2 (FCI algorithm outline).

1: $E \leftarrow$ edge set of a complete graph on V; put all edge marks as ∘.
2: **for** $(i, j) \in E$ **do**
3:     Search for a subset S_{ij} such that $X_i \perp\!\!\!\perp X_j \mid S_{ij}$.
4:     If found, $E \leftarrow E \setminus \{(i, j), (j, i)\}$ and store $S_{ij} = S_{ji}$.
5: **end for**
6: Orient edges in v-structures based on E and $\{S_{ij}\}$.
7: Apply orientation rules R1 to R4 (Zhang, 2008b) until none of them applies.
8: Apply orientation rules R8 to R10 (Zhang, 2008b) until none of them applies.

The algorithm is initialized with a complete graph having an edge ∘—∘ between every pair of vertices. The for loop uses conditional independence knowledge or tests to find a subset S_{ij} that may m-separate the two vertices i and j. This is similar to the skeleton estimation of the PC algorithm, but with the subtle difference that the candidate separating set S_{ij} is not necessarily a subset of the current neighbor set of vertex i or vertex j; compare with Line 5 in Algorithm 9.1. The v-structure orientation step (Line 6) is essentially identical to that of the PC algorithm (Algorithm 9.1). For each non-adjacent pair (i, j) with a common neighbor k, orient as $i \ast\!\!\rightarrow k \leftarrow\!\ast j$ if $k \notin S_{ij}$, where the edge mark $\ast$ can be any of the three marks. Assuming a perfect CI oracle, the algorithm identifies all invariant arrowheads in the true equivalence class after completion of Line 7, but some invariant tails may not be detected at this point. The rules on Line 8 further identify the remaining invariant tails, making the algorithm complete with respect to orientation.

Suppose $\mathcal{M}$ is the true MAG. If we have a CI oracle, the FCI algorithm will correctly learn the PAG that represents the equivalence class of $\mathcal{M}$, as stated in Theorem 4 given by Zhang (2008b):

Theorem 11.6. *Suppose the joint distribution of X_V is faithful to the MAG $\mathcal{M}$. Given a perfect conditional independence oracle, the FCI algorithm returns the PAG that represents the equivalence class of $\mathcal{M}$.*

Similar to the PC algorithm, for a finite sample, we use CI tests in place of the CI oracle. If the tests are pointwise consistent (9.3), the FCI estimates the true PAG with probability approaching one as the sample size $n \to \infty$.

11.7   Latent Factor Analysis

Latent factor models are of paramount importance in biomedical, educational, and social sciences. Latent factors are used to represent hypothetical, abstract constructs which correspond to meaningful real-world concepts that are either difficult or impossible to measure directly. A great deal of literature has been dedicated to learning the structure of linear latent factor models. Typically, this problem has been divided into learning the number of factors and learning the support of the linear coefficients, separately. To choose the number of latent factors, most classical methods are related to the eigenvalues of the sample correlation matrix among the observed variables (Guttman, 1954; Kaiser, 1960; Cattell, 1966; Horn, 1965; Glorfeld, 1995). More recently, model selection and regularized approaches (Preacher *et al.*, 2013; Hirose and Yamamoto, 2014a, 2014b) have been developed to recover the support, assuming the number of factors is known.

In this section, we focus on a graphical approach to latent factor analysis, adapted from the work of Kim and Zhou (2023). In this approach, the latent factor model structure is represented by a special type of ADMGs.

11.7.1   Graphical representation

Let $X = (X_1, \ldots, X_p) \in \mathbb{R}^p$ be a vector of observed variables. The factor analysis model specifies the joint distribution of X in the form of a structural equation model

$$X = \Lambda L + \varepsilon, \tag{11.25}$$

where $L = (L_1, \ldots, L_d)$ is a vector of latent factors, $\varepsilon = (\varepsilon_1, \ldots, \varepsilon_p)$ is a vector of independent errors, and $\Lambda = (\lambda_{ij}) \in \mathbb{R}^{p \times d}$ is a matrix of coefficients, or factor loadings. For convenience, an additive mean vector μ is omitted from the model without loss of generality. We assume that $d < p$ for identifiability and dimension reduction considerations. In the context of an SEM, X_i is a function of L_j if and only if $\lambda_{ij} \neq 0$, in which case we say that L_j is a parent of X_i and X_i is a child of L_j. We assume that every X_i has at least one parent and every L_j has at least one child, i.e., there are no rows or columns of full zeroes in Λ. Here, we consider primarily the case of independent factor analysis, where $L_i, i \in [d]$ are mutually independent.

The support of Λ in the SEM (11.25) defines a DAG $\mathcal{G}(X \cup L)$ with observed variables X and latent variables L. Denote the parent set of X_i and the child set of L_k, respectively, by

$$\mathrm{pa}(i) = \{k \in [d] : \lambda_{ik} \neq 0\} \quad \text{and} \quad \mathrm{ch}(L_k) = \{i \in [p] : \lambda_{ik} \neq 0\}.$$

The latent projection $\mathcal{G}(X)$ has only bidirected edges, and there is a bidirected edge between X_i and X_j if and only if the two variables share at least one common latent parent. Correspondingly, we have the following dependence relations among X.

Lemma 11.1. *Suppose X and L are defined by (11.25), where L_i are mutually independent and $\mathrm{Var}(L_i) > 0$ for all $i \in [d]$. Then, $X_i \not\perp\!\!\!\perp X_j$ if and only if $\mathrm{pa}(i) \cap \mathrm{pa}(j) = \varnothing$, for all $i, j \in [p]$ and $i \neq j$.*

Let E be the edge set of $\mathcal{G}(X)$. This lemma shows that the edges in $\mathcal{G}(X)$ have a one-to-one correspondence with the marginal dependencies among X:

$$(i, j) \notin E \Leftrightarrow X_i \perp\!\!\!\perp X_j, \qquad \text{for all } i \neq j. \tag{11.26}$$

Thus, we call $\mathcal{G}(X)$ a marginal independence graph. The DAG for a latent factor model and its latent projection are shown in Figure 11.10, from which we see that $\{X_1, X_2\} \perp\!\!\!\perp X_4$.

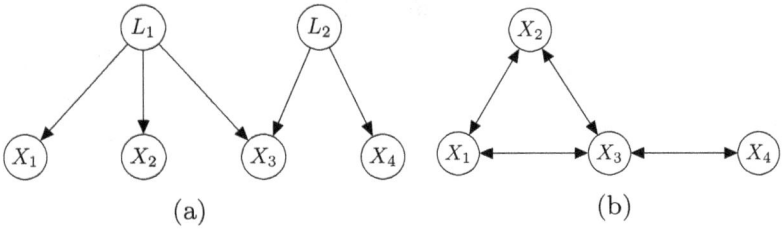

Fig. 11.10. Graphical representations for a latent factor model. (a) DAG $\mathcal{G}(X \cup L)$ with L unobserved. (b) Latent projection $\mathcal{G}(X)$, an ADMG with only bidirected edges.

11.7.2   *Structure recovery by clique search*

An interesting observation from the example in Figure 11.10 is that the maximal cliques in $\mathcal{G}(X)$, namely $\{X_1, X_2, X_3\}$ and $\{X_3, X_4\}$, have a perfect correspondence with the child sets of the latent factors, $\mathrm{ch}(L_1)$ and $\mathrm{ch}(L_2)$. This suggests that one may recover the latent factor model structure from the maximal cliques in the marginal independence graph $\mathcal{G}(X)$. To establish such a correspondence in general, we define a special class of maximal cliques.

Definition 11.10. Let $\mathcal{C} = \{C_1, \ldots, C_k\}$ be the set of all maximal cliques in a graph. Then, C_i is an *independent maximal clique* if

$$C_i \not\subseteq \bigcup_{j \neq i} C_j.$$

Essentially, an independent maximal clique is a maximal clique that contains a vertex that is not a member of any other maximal clique. We call such a vertex a *unique member* of the independent maximal clique. We use the word "independent" as an analog to the notion of linear independence in a vector space. That is, an independent maximal clique cannot be covered by the union of any subset of the other maximal cliques. It is easy to see that both maximal cliques in Figure 11.10(b) are independent maximal cliques.

A sufficient condition for recovering the structure of latent factors by independent maximal cliques is the *unique child condition*, defined as follows.

Definition 11.11. We say that X_i is a unique child of L_k if $i \in \mathrm{ch}(L_k)$ and $i \notin \bigcup_{j \neq k} \mathrm{ch}(L_j)$. If each latent factor L_k has at least one unique child, then we say that the unique child condition holds.

Given this condition, there is a bijection between the latent factors and the independent maximal cliques in $\mathcal{G}(X)$.

Theorem 11.7. *If the unique child condition holds for the SEM* (11.25), *then the set* $\{\mathrm{ch}(L_k) : k \in [d]\}$ *is identical to the set of independent maximal cliques in* $\mathcal{G}(X)$.

Proof. Recall that the edge set of $\mathcal{G}(X)$ is

$$E = \{(i,j) : \mathrm{pa}(i) \cap \mathrm{pa}(j) \neq \varnothing\}.$$

Pick any $k \in [d]$. Since every vertex $j \in \mathrm{ch}(L_k)$ shares a common parent L_k, the set $\mathrm{ch}(L_k)$ forms a clique in $\mathcal{G}(X)$. Under the unique child condition, there exists $i \in \mathrm{ch}(L_k)$ such that X_i does not have an edge connected to any node other than $\mathrm{ch}(L_k)$. Therefore, every clique that includes X_i must be a subset of $\mathrm{ch}(L_k)$. This implies that $\mathrm{ch}(L_k)$ is the only maximal clique that includes X_i, making it an independent maximal clique. The above argument shows that each $\mathrm{ch}(L_k), k \in [d]$, is an independent maximal clique. Since $\cup_k \mathrm{ch}(L_k) = [p]$, other maximal cliques, if any, cannot be independent. This shows that $\{\mathrm{ch}(L_k) : k \in [d]\}$ is the set of independent maximal cliques in $\mathcal{G}(X)$. $\qquad\square$

Observe that the child sets $\{\mathrm{ch}(L_k) : k \in [d]\}$ completely encode the structure of the latent factor model, including both the dimension of L and the support of Λ, i.e., $(d, \mathrm{supp}(\Lambda))$. Therefore, this theorem shows that the structure of the latent factor model (11.25) can be *perfectly* recovered from the independent maximal cliques in the marginal independence graph $\mathcal{G}(X)$.

All independent maximal cliques in a graph can be found quickly using the following result.

Lemma 11.2. *Given a graph* $\mathcal{G}(X)$, *let* $\bar{N}(X_i)$ *be the set of vertices that contains* X_i *and every node that shares an edge with* X_i.

(i) *If* $\bar{N}(X_i)$ *is a clique, then* $\bar{N}(X_i)$ *is also an independent maximal clique and* X_i *is a unique member of this clique.*

(ii) *If* C *is an independent maximal clique, then* $C = \bar{N}(X_i)$ *for any unique member* $X_i \in C$.

See Kim and Zhou (2023) for a proof. In the worst-case scenario, all independent maximal cliques can be found by checking whether $\bar{N}(X_i)$ is a clique for every node X_i. The computational cost for checking if $\bar{N}(X_i)$ is a clique has a brute force complexity of $O(k^2)$, assuming a maximum neighborhood size of k. Thus, the total computational cost on all p nodes is no greater than and, usually, well below $O(k^2 p)$, which is very efficient, even for large graphs. This is in sharp contrast to the exponential complexity in enumerating all maximal cliques in a graph (Eppstein *et al.*, 2010).

According to Theorem 11.7, we design Algorithm 11.3 to recover the latent factor structure through clique search on $\mathcal{G}(X)$. The first step relies on an independence check to construct the graph $\mathcal{G}(X)$. For a finite sample, this may be implemented through independence tests or thresholding sample correlation. Kim and Zhou (2023) developed a correlation thresholding (CT) algorithm to construct the edge sets $E_m = \{(i,j) : |r_{ij}| > \tau_m\}$, where r_{ij} is the sample correlation between X_i and X_j, for a sequence of thresholding values $\{\tau_m : m \in M\}$. A candidate model structure, $(\widehat{d}, \{\widehat{\mathrm{ch}}(L_k)\})$, is estimated through clique search on each graph $\mathcal{G}(X, E_m)$. Then, a model selection criterion is utilized to pick an estimated structure. As shown through a series of simulation studies, the CT algorithm is an accurate method for learning the structure of factor analysis models and is robust to violations of its assumptions. The algorithm scales well up to thousands of variables, while competing methods are not applicable in a reasonable amount of running time.

Algorithm 11.3 (Latent factor recovery via clique search).

1: $E \leftarrow \{(i,j) \in [p] \times [p] : X_i \not\perp\!\!\!\perp X_j, \ i \neq j\}$.
2: $\mathcal{G} \leftarrow \mathcal{G}(X, E)$.
3: $\{\mathcal{C}_1, \ldots, \mathcal{C}_q\} \leftarrow$ independent maximal cliques of $\mathcal{G}$.
4: $\widehat{d} \leftarrow q$, $\widehat{\mathrm{ch}}(L_k) \leftarrow \mathcal{C}_k$ for $k = 1, \ldots, q$.

Other methods have been designed to infer latent factors in hierarchically structured graphs; see Xie *et al.* (2020), Chen *et al.* (2022), Huang *et al.* (2022), and Xie *et al.* (2022) for recent examples. There are also methods that leverage algebraic constraints on the sample covariance matrix of observed variables to infer latent factor clusters under certain structural assumptions (Kummerfeld and Ramsey, 2016; Squires *et al.*, 2022).

Chapter 12

Partitioned, Federated, and Active Learning

This chapter explores a diverse array of recent subjects within the realm of DAG learning. The initial focus is placed on the challenges associated with learning large DAGs encompassing numerous variables, employing effective divide-and-conquer strategies. The discussion then shifts to the task of learning a DAG when confronted with distributed data dispersed across multiple local clients. A noteworthy aspect is the simultaneous consideration of protecting data privacy while making full use of all local data in a coordinated fashion. Lastly, we introduce the intriguing problem of sequential design of interventions to optimize rewards within a decision-making process. This exploration is conducted under the assumption of an underlying causal graphical model, offering insights into the intersection of active structure learning and dynamic decision optimization.

12.1   Divide and Conquer

We have discussed various methods for structure learning of DAGs in Chapters 9 and 10. Despite these great efforts, learning big DAGs over a large number of nodes remains challenging. The DAG space grows super-exponentially in the number of nodes p (Robinson, 1977), and learning DAGs has been shown to be NP-hard in general (Chickering *et al.*, 2004). As p increases, many structure learning methods slow down dramatically and become much less accurate, making them incompetent for large datasets. This motivated the development of a divide-and-conquer approach by Gu and Zhou (2020), which consists of three steps, namely partition, estimation, and fusion (PEF for short):

(a) *P-step*: partition the nodes into clusters based on some clustering algorithm.
(b) *E-step*: apply an existing structure learning algorithm to estimate a subgraph on each cluster of nodes.
(c) *F-step*: merge the estimated subgraphs into one graph over all nodes.

Note that the number of nodes in a cluster is usually much smaller than p. This greatly speeds up structure learning in the estimation step, as most algorithms scale at least as $O(p^k)$ for some $k \geq 2$, e.g., Kalisch and Bühlmann (2007). Moreover, this step can be easily parallelized, leading to further improvements in computational efficiency.

Following the PEF framework, Huang and Zhou (2022) developed a partitioned PC (pPC) algorithm to improve the computational efficiency of the PC algorithm while retaining its attractive theoretical guarantees. Since the computational cost of edge orientation is negligible compared to skeleton estimation, the PEF approach is applied to the skeleton learning of the PC algorithm. In what follows, we elaborate on the main components of the pPC algorithm.

12.1.1 *Partition and estimation*

Suppose our input dataset is an $n \times p$ matrix $X = [X_1 \mid \cdots \mid X_p]$, where each column X_j consists of the observed values of the jth variable (node) in a DAG across n units. In the P-step, we cluster the p variables into K clusters, according to a chosen distance matrix $D = (d_{ij})_{p \times p}$, using a modified hierarchical clustering algorithm that chooses a cut to determine the number of clusters (Gu and Zhou, 2020). Succinctly described, we choose the highest cut such that the result consists of the greatest number of large clusters of size $\geq 0.05p$. We then merge clusters of size less than $0.05p$ with other small clusters or into large clusters sequentially, ordered by average linkage, until every cluster is a large cluster. For further details regarding the algorithm, we refer to the original paper by Gu and Zhou (2020). The clustering step partitions the p nodes into K clusters, returning the cluster labels $\mathbf{c} = \{c_1, \ldots, c_p\}$, with $c_i \in \{1, \ldots, K\}$ denoting the cluster label of node i.

We now apply the skeleton learning phase of the PC algorithm to estimate K disconnected undirected subgraphs according to the partition obtained in the clustering step. Practically, independently applying the PC algorithm to each node cluster benefits from at most K processors if distributed as such for parallel processing. Furthermore, the speedup is limited by the longest estimation runtime, usually corresponding to the largest node cluster. In contrast, the design of the PC-stable implementation (Algorithm 12.1) by Colombo and Maathuis (2014) allows for parallelizing the computation (Line 5) for all adjacent node pairs by deferring the update of the estimated skeleton E' to a synchronization step (Line 9) between the iterations of ℓ. In the original PC algorithm (Algorithm 9.1), the edge set E of the estimated skeleton is updated immediately after an edge $i - j$ is deleted, which is the key difference.

Algorithm 12.1 (PC-stable algorithm).

1: $E \leftarrow$ edge set of complete undirected graph on V.
2: $\ell \leftarrow 0$, $E' \leftarrow E$.
3: **for** $\ell \leq \ell_{\max}$ **do**
4:     **for all** unordered pair $(i,j) \in E'$ **do**
5:         Search for size-ℓ set $S \subseteq N_i(E') \setminus \{j\}$ or $S \subseteq N_j(E') \setminus \{i\}$
        such that $X_i \perp\!\!\!\perp X_j \mid S$.
6:         If found, $E \leftarrow E \setminus \{(i,j),(j,i)\}$ and store $S_{ij} = S_{ji} \leftarrow S$.
7:     **end for**
8:     $\ell = \ell + 1$.
9:     $E' \leftarrow E$ (synchronized update).
10: **end for**

Several parallel implementations exist for the PC-stable approach, which addresses the case of a large p problem (Kalisch *et al.*, 2012; Scutari, 2017; Zarebavani *et al.*, 2020). In these parallelization paradigms, due to the large number of distributed tasks, the number of utilizable computing processors is not practically limited, and the computational load is reasonably expected to be evenly distributed. To take advantage of these developments in parallelizing the PC algorithm, we estimate skeletons for the K node clusters by executing Algorithm 12.1, initializing with K complete undirected graphs, each over a cluster of nodes. The result is an undirected graph $\mathcal{G}$ consisting of K disconnected subgraphs, each corresponding to the estimated skeleton for a cluster of nodes.

12.1.2   *Fusion*

At this stage, given a good partition, we expect the node adjacencies to be relatively well estimated, with the exception of extraneous edge connections within a cluster due to the violation of causal sufficiency and missing edge connections between clusters that are disconnected by the partition. The fusion of the K subgraphs consists of two steps. First, we apply a screening process to add edges between clusters. Then, we prune existing edges by identifying candidate separating sets that have not been investigated.

Let $\mathcal{G} = (V, E)$ be the current skeleton and $N_{i\backslash j}^E := N_i(E) \backslash \{j\}$. The between-cluster screening process consists of the following operations:

$$E' \leftarrow E, (\text{to fix adjacency})$$

$$E \leftarrow E \cup \{(i,j) : c_i \neq c_j, X_i \not\perp\!\!\!\perp X_j \text{ and } X_i \not\perp\!\!\!\perp X_j \mid N_i(E') \cup N_j(E')\}, \tag{12.1}$$

$$E' \leftarrow E,$$

$$E \leftarrow E \backslash \left\{ (i,j) \in E : c_i \neq c_j, X_i \perp\!\!\!\perp X_j \mid N_{i\backslash j}^{E'} \text{ or } X_i \perp\!\!\!\perp X_j \mid N_{j\backslash i}^{E'} \right\}, \tag{12.2}$$

where c_i is the cluster label of X_i and (i,j) is understood as an unordered pair. The first screen in (12.1) constructively connects between-cluster pairs that are dependent marginally as well as conditioned on the union of the neighbor sets. With the addition of edges between clusters, the second screen in (12.2) disconnects pairs that are separated by the newly updated adjacencies. Note that every between-cluster edge that is present in the underlying DAG will be connected by (12.1), and operation (12.2) can only prune false positives. Thus, after this step, every node pair will have been considered and, in the population setting, every true edge in the underlying DAG will be connected in $\mathcal{G}$.

However, it is important to note that all possible conditioning sets have not yet been fully investigated for every node pair. The tests for edges within a cluster only considered conditioning sets consisting of nodes within the same cluster, and tests for edges between clusters were limited to the sets in (12.1) and (12.2). In particular, for each remaining adjacent node pair $(i, j) \in E$, the possible separation sets that have not been evaluated are limited to either of the following cases, if any:

(a) $c_i = c_j$ (within a cluster), sets S where $\exists\, k \in S$ such that $c_k \neq c_i$;
(b) $c_i \neq c_j$ (between clusters), sets S not used in (12.1) or (12.2).

The most straightforward continuation to achieve completeness in our partitioned skeleton learning process would be to exhaustively evaluate the dependence of the remaining connected node pairs in $\mathcal{G}$ conditioned on the remaining conditioning sets. We accomplish this by restarting the PC algorithm on the current skeleton, evaluating independence conditioned on sets restricted to criteria (a) and (b), before finally orienting the resulting skeleton into a CPDAG to complete structure learning.

It is not difficult to show that, under faithfulness, the pPC algorithm is sound and complete given a conditional independence oracle; see Theorem 2 in Huang and Zhou (2022) for the precise statement. The numerical results in Section 5.2 of that paper demonstrate that pPC can achieve comparable or better accuracy than PC with substantial computational savings.

12.2   Federated Learning

Federated learning is a machine learning approach that enables training models across decentralized devices or clients while keeping the data localized. Instead of collecting all data on a central server, federated learning allows model training to occur on the devices or local clients where the data are generated. This approach empowers local entities to collaboratively learn from decentralized data without the need for direct data sharing. As a result, federated learning not only addresses privacy and logistical concerns but also facilitates the development of robust and generalized models by learning from a diverse range of data sources.

In this section, we focus on some recent work on federated DAG learning. A unique challenge in structure learning of DAGs is to enforce the acyclicity constraint. Averaging estimated graphs from local clients is likely to violate this constraint. To overcome this difficulty, Ye *et al.* (2024) proposed an order-based learning method in which the objective function is equivalent to a regularized log-likelihood of the overall data. The central server proposes a candidate ordering π, where the score of π is evaluated via distributed optimization with multiple local clients. Then, the candidate sort π is selected by simulated annealing. Because every DAG has at least one ordering, searching over the space of orderings ensures that the acyclicity constraint is always satisfied. Another critical contribution of this work is the convergence results of the federated estimator to the *oracle* estimate constructed assuming access to all data across local clients. The convergence rate is $O(\log(n)/\sqrt{m})$, where n is the total sample size across all local clients and m is the smallest local sample size. Therefore, even for a finite sample, as long as $\log(n)/\sqrt{m}$ is small, the federated estimate will be essentially identical to the oracle solution, achieving the ideal efficiency while preserving data privacy. To the best of our knowledge, the work of Ye *et al.* (2024) was the first federated causal discovery method with such a nice theoretical guarantee. In the sequel, we present this method and its theoretical guarantees. See Gao *et al.* (2023) and Ng and Zhang (2022) for other recent federated DAG learning methods.

12.2.1 *Global objective function*

We assume the generalized linear DAG model introduced in Section 10.1.2 for this work, where each variable $X_j \in \mathbb{R}^{d_j}$. Here, $d_j = 1$ if X_j is continuous and $d_j = r_j - 1$ if X_j is discrete with r_j possible levels. Let $\{x_h\}_{h=1}^n$ be an i.i.d. sample of size n from this model. We also let x_h^j represent the observed value of the jth variable (X_j) in the hth data point. Consider a subset $\mathcal{I} \subseteq [n]$. The normalized negative log-likelihood of the subsample $\{x_h\}_{h \in \mathcal{I}}$ is given, up to an additive constant, by

$$\ell_{\mathcal{I}}(\beta) := \frac{1}{|\mathcal{I}|} \sum_{h \in \mathcal{I}} \sum_{j=1}^p [b_j(\beta_j^\top x_h) - \langle \beta_j^\top x_h, x_h^j \rangle], \qquad (12.3)$$

where β_j is given in (10.3) and $\beta = [\beta_1, \ldots, \beta_p]$ is arranged into a $(d+1) \times d$ matrix.

Suppose there are K different local clients $\mathcal{M}_1, \ldots, \mathcal{M}_K$. Each $\mathcal{M}_k$ holds its private data $\{x_h\}_{h \in \mathcal{I}_k}$ and communicates with a central server $\mathcal{C}$. Let $n_k = |\mathcal{I}_k|$ be the sample size in $\mathcal{M}_k$ so that $\sum_{k=1}^K n_k = n$. The normalized negative log-likelihood based on the entire data can be decomposed as $\ell_{[n]}(\beta) = \sum_{k=1}^K \frac{n_k}{n} \ell_{\mathcal{I}_k}(\beta)$. Let $\mathcal{P}$ be the set of all permutations on $[p]$ and $\mathcal{D}(\pi) \subset \mathbb{R}^{(d+1) \times d}$ be the set of DAGs whose topological sorts are compatible with a permutation $\pi \in \mathcal{P}$. Note that $\mathcal{D}(\pi)$ is a linear subspace of $\mathbb{R}^{(d+1) \times d}$. We would ideally like to estimate β by minimizing a regularized loss function of the form

$$\min_{\pi \in \mathcal{P}} \left[f(\pi) := \min_{\beta \in \mathcal{D}(\pi)} \sum_{k=1}^K \frac{n_k}{n} \ell_{\mathcal{I}_k}(\beta) + \rho(\beta) \right], \qquad (12.4)$$

where $\rho(\cdot)$ is an appropriate regularizer to promote sparsity in β. We call $f(\pi)$ the global objective function, since it is defined using all data across local clients.

As defined in Section 10.1.2, β_{ij} is a parameter matrix associated with the edge $i \to j$ and $\beta_{ij} \neq 0$ if and only if $i \in \text{pa}(j)$. To learn a sparse DAG, we apply group regularization of the form $\rho(\beta) = \lambda \sum_{i,j} \rho_g(\beta_{ij})$, where $\rho_g(\cdot)$ is a nonnegative and nondecreasing group regularizer and $\lambda > 0$ is a tuning parameter. Restricted to $\mathcal{D}(\pi)$, the regularizer can be further simplified to

$$\rho(\beta) = \lambda \sum_{j} \sum_{i \prec_\pi j} \rho_g(\beta_{ij}), \tag{12.5}$$

such as the group ℓ_2 penalty with the choice $\rho_g(\beta_{ij}) = \|\beta_{ij}\|_F$, where $\|\beta_{ij}\|_F$ is the Frobenius norm of the matrix β_{ij}.

To search over $\pi \in \mathcal{P}$ and $\beta \in \mathcal{D}(\pi)$ with distributed data, as in (12.4), Ye *et al.* (2024) developed the *distributed annealing on regularized likelihood score* (DARLS) algorithm, which applies simulated annealing to search over the permutation space, coupled with a distributed optimization method. Such a manner of joint optimization over the topological sort space and the DAG space has demonstrated great effectiveness in structure learning (Larranaga *et al.*, 1996; Scanagatta *et al.*, 2015; Ye *et al.*, 2021). At each annealing iteration, a permutation, π^+, is proposed based on current $\hat{\pi}$ and is accepted with probability according to simulated annealing, given a decreasing temperature schedule. A *distributed optimization* approach, discussed in Section 12.2.2, is used to compute the score $f(\pi)$ of the optimal DAG structure for a given permutation π. This approach allows multiple rounds of communications between local clients and the central server to update and synthesize information.

12.2.2 *Distributed optimization*

For any fixed π, we use distributed computing to evaluate $f(\pi)$, as the samples $\mathcal{I}_k, k \in [K]$ are not shared among the local clients. We rely on local versions of the objective function,

$$f_k(\pi) := \min_{\beta \in \mathcal{D}(\pi)} F_k(\beta), \quad \text{where } F_k(\beta) := \ell_{\mathcal{I}_k}(\beta) + \rho(\beta),$$

to guide a distributed algorithm that divides the task of computing $f(\pi)$ among the K local clients. The global version (12.4) can be rewritten as $f(\pi) = \min_{\beta \in \mathcal{D}(\pi)} F(\beta)$, where $F(\beta) := \ell_{[n]}(\beta) + \rho(\beta)$. Typically, each of F and F_k is nonsmooth due to the presence of the regularizer ρ, but the difference $h_k := F_k - F = \ell_{\mathcal{I}_k} - \ell_{[n]}$ is often smooth. The gradient of h_k

5

is used to guide iterations in each local client. That is, given the current (global) estimate β, the local client $\mathcal{M}_k$ performs the update

$$\varphi_{k,\pi}(\beta) := \underset{\xi \in \mathcal{D}(\pi)}{\operatorname{argmin}} \left[\widetilde{F}_k(\xi) := F_k(\xi) - \langle \nabla h_k(\beta), \xi \rangle \right.$$

$$\left. = F_k(\xi) - \langle \nabla \ell_{\mathcal{I}_k}(\beta) - \nabla \ell_{[n]}(\beta), \xi \rangle \right]. \quad (12.6)$$

The local regularized loss F_k guided by ∇h_k, denoted by $\widetilde{F}_k$, is a first-order approximation to the global regularized loss F, up to an additive constant. Let $\beta_\pi^{(t)}$ be the global estimate of the algorithm at iteration t. At the next iteration, $t+1$, we obtain local estimates $\beta_{k,\pi}^{(t+1)} = \varphi_{k,\pi}(\beta_\pi^{(t)})$ for $k = 1, \ldots, K$. These local estimates are then passed to the central server $\mathcal{C}$ to compute the global estimate by averaging, i.e., $\beta_\pi^{(t+1)} = \sum_{k=1}^K \frac{n_k}{n} \beta_{k,\pi}^{(t+1)}$, for the next iteration.

The main steps of this distributed optimization method are outlined in Algorithm 12.2. This approach is essentially a version of the distributed approximate Newton (DANE) algorithm (Shamir *et al.*, 2014; Zhang *et al.*, 2013). Note that to calculate local updates $\beta_{k,\pi}^{(t+1)}$ (Line 5), only the current global estimate $\beta_\pi^{(t)}$ and the global gradient $\nabla \ell_{[n]}(\beta_\pi^{(t)})$ need to be communicated to each local client. Thus, DARLS can be applied to any objective function as long as the gradient w.r.t. β has a closed-form expression. We use the proximal gradient algorithm to compute the local updates on Line 5; see Ye *et al.* (2024) for the technical details.

Algorithm 12.2 (Distributed optimization for $f(\pi)$).

1: Server $\mathcal{C}$ broadcasts π to local clients $\{\mathcal{M}_k\}_{k=1}^K$.
2: **for** $t = 0, 1, \ldots, T-1$ **do**
3:     Each $\mathcal{M}_k$ computes $\nabla \ell_{\mathcal{I}_k}(\beta_\pi^{(t)})$ and sends it to $\mathcal{C}$.
4:     $\mathcal{C}$ computes $\nabla \ell_{[n]}(\beta_\pi^{(t)}) = \frac{1}{n} \sum_k n_k \nabla \ell_{\mathcal{I}_k}(\beta_\pi^{(t)})$ and broadcasts it to local clients.
5:     Each $\mathcal{M}_k$ computes $\beta_{k,\pi}^{(t+1)} = \varphi_{k,\pi}(\beta_\pi^{(t)})$ via (12.6) and sends it to $\mathcal{C}$.
6:     $\mathcal{C}$ computes $\beta_\pi^{(t+1)} = \frac{1}{n} \sum_k n_k \beta_{k,\pi}^{(t+1)}$ and broadcasts it to local clients.
7: **end for**
8: Each $\mathcal{M}_k$ reports $F_k^{(T)} := n_k F_k(\beta_\pi^{(T)})$ to $\mathcal{C}$, and $\mathcal{C}$ sets $\widehat{\beta}_\pi \leftarrow \beta_\pi^{(T)}$ and $f(\pi) \leftarrow \frac{1}{n} \sum_k F_k^{(T)}$.

12.2.3 *Theoretical guarantees*

Let $\widehat{\beta}_\pi$ be any global minimizer of the global objective function, i.e.,

$$\widehat{\beta}_\pi \in \operatorname*{argmin}_{\xi \in \mathcal{D}(\pi)} \ell_{[n]}(\xi) + \rho(\xi),$$

where $\rho(\xi)$ is a regularizer. In the distributed data setting, $\widehat{\beta}_\pi$ is an oracle estimate with access to all data across multiple local clients. Theorem 2 in Ye *et al.* (2024) establishes that, for any π, the global estimator $\widehat{\beta}_\pi$ is consistent for the minimizer of the population loss

$$\beta_\pi^* := \operatorname*{argmin}_{\xi \in \mathcal{D}(\pi)} \mathbb{E}[\ell_{[n]}(\xi)].$$

Under some technical conditions, $\{\beta_\pi^{(t)}\}_{t \geq 0}$ produced by Algorithm 12.2 converges to a global minimizer $\widehat{\beta}_\pi$ (Ye *et al.*, 2024, Theorem 1). The overall iteration function for Algorithm 12.2 can be written as

$$\Phi_\pi(\cdot) := \sum_k \frac{n_k}{n} \varphi_{k,\pi}(\cdot).$$

For a matrix β, let $\mathbb{B}_F(\beta; r)$ denote the Frobenius ball of radius $r > 0$ centered at β. Under some conditions, there are positive constants c_1 and C such that

$$\|\Phi_\pi(\beta) - \widehat{\beta}_\pi\|_F \leq C\zeta_n(r)\|\beta - \widehat{\beta}_\pi\|_F, \quad \text{for all } \beta \in \mathbb{B}_F(\widehat{\beta}_\pi, r),$$

with probability at least

$$1 - 3(np)^{-c_1} - \mathbb{P}(\|\widehat{\beta}_\pi - \beta_\pi^*\|_F > r),$$

where $\zeta_n(r) = O(\log(n)/\sqrt{m})$ assuming p is fixed. The consistency of $\widehat{\beta}_\pi$ to β_π^* for any π implies that $\mathbb{P}(\|\widehat{\beta}_\pi - \beta_\pi^*\|_F > r)$ goes to zero as n grows. Thus, with high probability, the iteration operator $\Phi_\pi(\cdot)$ will be a contraction: the sequence $\{\beta_\pi^{(t)}\}_{t \geq 0}$ produced by the distributed algorithm converges geometrically to the oracle estimator $\widehat{\beta}_\pi$ if $C\zeta_n(r) < 1$. Let $m := \min_k n_k$ be the minimum sample size among the local clients. For fixed p and for sufficiently large r such that $\beta_\pi^{(0)} \in \mathbb{B}_F(\widehat{\beta}_\pi, r)$, one can always satisfy the condition of $C\zeta_n(r) < 1$ by taking m large enough.

12.3   Causal Bandit via Sequential Interventions

In previous chapters, we embarked on a journey through the rich literature on causal structure learning after observational or experimental data are collected, sometimes called *passive learning*. There are also growing efforts in intervention design, which generates experimental data for *active causal learning* (Ghassami *et al.*, 2019; Hauser and Bühlmann, 2014; Hu *et al.*, 2014; Hyttinen *et al.*, 2013; Kocaoglu *et al.*, 2017). All of these methods are non-adaptive in nature; they involve designing experiments before collecting any interventional data. As such, they assume perfect conditional independence oracles in structure learning, equivalent to assuming an infinite sample size of experimental data.

A different and more practical approach is to develop adaptive methods that iterate between intervention design, data generation, and active causal structure learning. It is usually formulated within the framework of sequential decision processes. We first review the multi-armed bandit (MAB) problem and then discuss its generalization to causal graphical models.

12.3.1   *Multi-armed bandit*

Sequential decision processes are a fundamental concept in the fields of artificial intelligence and reinforcement learning. These processes model decision-making situations where an agent interacts with an environment over a series of discrete time steps. At each step, the agent makes a decision based on its current knowledge or data, leading to a corresponding reward. The key challenge in sequential decision processes is devising a strategy or policy that maximizes the cumulative reward over time. This involves balancing the exploration of different actions and exploiting the knowledge gained to make optimal decisions.

The MAB problem is a well-known sequential allocation framework for experimental investigations (Berry and Fristedt, 1985). Classically, the MAB problem formulation features an action set $\mathcal{A}$ consisting of $|\mathcal{A}| = K$ actions, also called arms, typically corresponding to interventions. Each arm $a \in \mathcal{A}$ defines a real-valued distribution for the reward signal, with expected reward μ_a. The objective of an allocation policy is to sequentially pick arms $a_t \in \mathcal{A}$, $t = 1, \ldots, T$ in a manner that maintains a balance between exploration and exploitation in the interest of identifying and obtaining the greatest reward. This overall goal is quantified as minimizing the cumulative regret

$$R_T := \sum_{t=1}^{T} \mathbb{E}(\mu^* - \mu_{a_t}) = T\mu^* - \mathbb{E}\left[\sum_{t=1}^{T} \mu_{a_t}\right], \tag{12.7}$$

where $\mu^* = \max_a \mu_a$ is the optimal reward. Maximally and effectively utilizing all available information is imperative, especially when investigating interventions that are resource-demanding or time-consuming.

A well-known class of algorithms used to learn an optimal decision for the MAB problem is the upper confidence bound (UCB) (Agrawal, 1995; Auer *et al.*, 2002), under the principle of optimism in the face of uncertainty (Lai and Robbins, 1985). Let Y_t be the reward variable generated when arm a_t is pulled at time t. Denote by $n_a(t)$ the number of times arm $a \in \mathcal{A}$ has been pulled up to time t, and let

$$\mathcal{D}_a[t] = \{Y_s : a_s = a, s \leq t\}.$$

To consider uncertainty and allow for exploration, we calculate a confidence upper bound for μ_a:

$$U_a(t) = \bar{Y}_{a,t} + c\sqrt{\log(t)/n_a(t)} \quad \text{for all } a \in \mathcal{A}, \tag{12.8}$$

where $\bar{Y}_{a,t}$ is the sample average of $\mathcal{D}_a[t]$ and $c > 0$ is a chosen constant. Then, the arm with the greatest U_a is pulled for the next time step:

$$a_{t+1} = \underset{a \in \mathcal{A}}{\operatorname{argmax}} U_a(t). \tag{12.9}$$

Thompson sampling (Russo *et al.*, 2018; Thompson, 1933) is a Bayesian approach to the MAB problem. Instead of an upper bound, we draw $U_a(t)$ from a posterior distribution of μ_a given the data $\mathcal{D}_a[t]$:

$$U_a(t) \sim p(\mu_a \mid \mathcal{D}_a[t]), \tag{12.10}$$

where the sample $U_a(t)$ naturally incorporates the uncertainty of μ_a in a Bayesian way. Then, we choose the arm with the greatest $U_a(t)$ for time $(t+1)$, as in (12.9).

For both the UCB and Thompson sampling, the cumulative regret

$$R_T = O(\sqrt{KT \log T}) \tag{12.11}$$

for any finite T; see, for example, Russo and Van Roy (2014).

12.3.2   *Causal bandit*

Lattimore *et al.* (2016) proposed the causal bandit (CB) problem, wherein a causal graphical model (Section 8.1) is assumed to govern the distribution of the reward variable and its covariates. The addition of causal assumptions introduces avenues by which interventional distributions may be inferred from observational data and information may be shared between arms. Most works addressing the CB problem exploit strong assumptions regarding prior knowledge of the underlying causal model to achieve improvements over standard MAB algorithms. In a recent paper, Huang and Zhou (2023) developed a Bayesian CB framework that does not require prior knowledge of the underlying causal structure, but instead efficiently utilizes available observational data and sequentially acquired interventional data to inform exploitation and guide exploration in a decision-making process.

For illustrative purposes, we borrow and adapt the farming example described by Lattimore *et al.* (2016) as an illuminating and motivating example of the problem setting of interest and the surrounding challenges. Suppose a farmer wishes to optimize the yield (Y) of a certain crop, which she knows is only dependent on temperature (H), a particular soil nutrient (N), and moisture level (M). While she understands that crop yield is somehow affected by these factors, the underlying causality governing this system of four variables (including crop yield) is unknown to her. The farmer's resource limitations restrict her to intervening on at most one factor in each crop season by either adjusting the temperature or regulating the moisture level. These experimental interventions are costly to perform, and each realization of the interventional data can only be observed once a season. Hence, it is in the farmer's best interest to leverage her historical logs containing observational data accrued from previous seasons where no interventions were performed but rather the variables were passively observed as they naturally varied from season to season.

The causal diagram among the four variables, $X = \{H, N, M, Y\}$, is shown in Figure 12.1. Suppose the experiments are designed to either increase or decrease one unit, relative to the mean, of the variable under intervention. Using do-operators, they are expressed as $\mathrm{do}(X_i = 1)$ or $\mathrm{do}(X_i = -1)$, assuming $\mathbb{E}(X_i) = 0$, for $X_i \in \{H, M\}$. Thus, the action space $\mathcal{A}$ consists of these four do-operations. The reward variable is Y, and the reward for an arm a is $\mu_a = \mathbb{E}(Y \mid \mathrm{do}(X_i = x))$, the expectation of Y under one of the four do-operations. In a bandit process, at each time

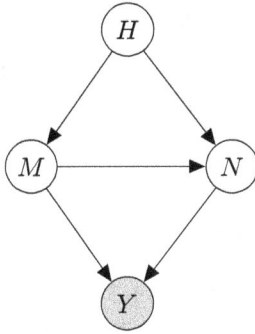

Fig. 12.1. An example for causal bandit, where Y (shaded) is the reward variable.

step, we execute an intervention, say $\operatorname{do}(H = -1)$, and measure all the variables in X. Based on the accumulated data, we hope to better estimate the rewards μ_a and choose the optimal arm to minimize the cumulative regret.

For a general causal bandit problem on p variables, $X = \{X_1, \ldots, X_p\}$, the action set $\mathcal{A}$ consists of K arms that correspond to interventions on variables in $X \setminus Y$, where $Y = X_p$ is the reward variable. In particular, let arm $a \in \mathcal{A}$ correspond to the intervention $\operatorname{do}(X_{\langle a \rangle} = x_a)$, fixing $X_{\langle a \rangle}$ to some value x_a, where $\langle a \rangle \in [p]$ is the node corresponding to the intervened variable. The expected reward is given by $\mu_a := \mathbb{E}[Y \mid \operatorname{do}(X_{\langle a \rangle} = x_a)]$, where $\mathbb{E}[\cdot]$ takes expectation with respect to the causal DAG model. There is an optimal arm a^* corresponding to the optimal reward $\mu^* := \mu_{a^*}$. Given a horizon of time steps T, let $a_t \in \mathcal{A}$ be the arm pulled by an algorithm at time step $t \in \{1, \ldots, T\}$. The objective of the algorithm is to pull arms over T time steps with a balance between exploring different arms and exploiting the reward signal to minimize the cumulative regret R_T (12.7).

For a causal bandit problem, all rewards μ_a are functions of the DAG model parameters. If the causal DAG structure is known, one may use historical observational data to estimate the DAG parameters, based on which all the rewards can be estimated. However, for many practical scenarios, the underlying causality is unknown or partially unknown, and therefore, one needs to take into account the uncertainty in the DAG structure in a bandit algorithm. The approach adopted by Huang and Zhou (2023) tackles this challenge by learning a posterior distribution of the graphical structure while exploiting backdoor adjustment to optimize the reward. For this reason, this method is called Bayesian back-door bandit (BBB).

12.3.3   *Bayesian backdoor bandit*

We first describe a Bayesian approach to the general bandit problem, with some notation adapted from Kaufmann *et al.* (2012). A set of parameters $\Theta_{\mathcal{A}} = (\theta_a)_{a \in \mathcal{A}}$ is assumed to define the corresponding marginal reward distributions:

$$p_{\theta_a}(y) := \mathbb{P}[Y = y \mid do(X_{\langle a \rangle} = x_a)].$$

We put a modular prior distribution $\Pi^0(\Theta_{\mathcal{A}}) = \prod_{a \in \mathcal{A}} \pi_a^0(\theta_a)$ over these parameters. Typically, $\{\pi_a^0 : a \in \mathcal{A}\}$ are chosen to be all equal and uninformative. When arm $a_t \in \mathcal{A}$ is pulled at time step t and a realization $y_t \sim [Y \mid do(X_{\langle a_t \rangle} = x_{a_t})]$ is observed, the posterior Π^t is computed by updating according to

$$\pi_{a_t}^t(\theta_{a_t}) \propto p_{\theta_{a_t}}(y_t)\, \pi_{a_t}^{t-1}(\theta_{a_t}),$$

while $\pi_a^t = \pi_a^{t-1}$ for $a \neq a_t$. For each arm $a \in \mathcal{A}$, the posterior π_a^t induces a posterior distribution for the expected reward μ_a, which is simply a marginal or transformation of π_a^t, since μ_a is a function of θ_a. These posteriors are utilized by Bayesian MAB algorithms, such as Thompson sampling, to draw μ_a.

In our problem formulation, we assume possession of observational data $\mathcal{D}_0$ of sample size n_0 prior to investigating the arms. We denote by $\mathcal{D}^{(t)}$ the interventional data acquired by pulling arm a_t at time t and by

$$\mathcal{D}_a[t] = \left\{ \mathcal{D}^{(s)} : a_s = a, s \leq t \right\}$$

the accumulated interventional data from arm a through time t. The combined observational and interventional data accrued through time t is

$$\mathcal{D}[t] = \mathcal{D}_0 \cup \{\mathcal{D}_a[t], a \in \mathcal{A}\},$$

which we refer to as ensemble data.

If a set of variables $Z \subset X$ satisfies the backdoor criterion relative to $(X_{\langle a \rangle}, Y)$, then the interventional distribution $[Y \mid do(X_{\langle a \rangle} = x_a)]$ may be expressed in terms of the joint distribution of $\{X_{\langle a \rangle}, Y\} \cup Z$ via the backdoor adjustment (Theorem 8.2),

$$\mathbb{P}[Y = y \mid do(X_{\langle a \rangle} = x_a)] = \sum_z \mathbb{P}(Y = y \mid X_{\langle a \rangle} = x_a, Z = z)\mathbb{P}(Z = z).$$

$$(12.12)$$

Equation (12.12) provides an avenue through which an estimator $\widehat{\mu}_{a,\text{bda}}(Z)$ for μ_a using observational data $\mathcal{D}_0$ may be derived, conditional on Z satisfying the backdoor criterion. A convenient choice is $Z = \text{pa}(X_{\langle a \rangle})$, the parent set of $X_{\langle a \rangle}$. Its uncertainty is quantified by the posterior probability $\mathbb{P}(\text{pa}(X_{\langle a \rangle}) = Z \mid \mathcal{D}[t])$ given the ensemble data at time t.

We construct a conditional prior $\pi^0_{a \mid Z}(\theta_a \mid Z)$ for θ_a, given $\text{pa}(X_{\langle a \rangle}) = Z$, such that the induced prior distribution of μ_a satisfies

$$\mathbb{E}[\mu_a \mid Z] = \widehat{\mu}_{a,\text{bda}}(Z), \quad \text{Var}[\mu_a \mid Z] = \widehat{\text{SE}}^2[\widehat{\mu}_{a,\text{bda}}(Z)]. \tag{12.13}$$

Here, $\mathbb{E}[\cdot \mid Z]$ and $\text{Var}[\cdot \mid Z]$ are, respectively, the expectation and variance with respect to $\pi^0_{a \mid Z}$, and $\widehat{\text{SE}}^2[\widehat{\mu}_{a,\text{bda}}(Z)]$ is the estimated sample variance of the estimator $\widehat{\mu}_{a,\text{bda}}(Z)$. The matching of the prior variance with the sampling variance of the backdoor adjustment estimator endeavors to assign the appropriate prior effective sample size. When arm $a_t = a$ is pulled at time step t and a realization of the reward $y_t \sim [Y \mid \text{do}(X_{\langle a \rangle} = x_a)]$ is observed in $\mathcal{D}^{(t)}$, the conditional posterior $\pi^t_{a \mid Z}$ is updated to

$$\pi^t_{a \mid Z}(\theta_a) \propto p_{\theta_a}(y_t)\, \pi^{t-1}_{a \mid Z}(\theta_a), \qquad t \geq 1.$$

Accordingly, the marginal posterior of θ_a is determined by averaging over all possible parent sets for $X_{\langle a \rangle}$:

$$\pi^t_a(\theta_a) = \sum_{Z \subseteq X \setminus X_{\langle a \rangle}} \pi^t_{a \mid Z}(\theta_a \mid Z)\mathbb{P}(\text{pa}(X_{\langle a \rangle}) = Z \mid \mathcal{D}[t]), \tag{12.14}$$

which is the key posterior distribution to be updated at each time step t in the Bayesian CB problem. See Huang and Zhou (2023) for the calculation of the posterior probability $\mathbb{P}(\text{pa}(X_{\langle a \rangle}) = Z \mid \mathcal{D}[t])$, which is similar to that used in the Bayesian Dirichlet score discussed in Section 9.4.1.

Algorithm 12.3 shows how to implement Thompson sampling under the BBB framework. We initialize the priors using observational data $\mathcal{D}_0$ (Line 2). In Thompson sampling (Line 6), we first draw θ_a from the posterior distribution $\pi^{t-1}_a(\theta_a)$ (12.14) and then calculate μ_a as a function of θ_a. The arm with the greatest sample of reward $U_a(t)$ is pulled, and experimental data $\mathcal{D}^{(t)}$ is generated by the underlying DAG model under $\text{do}(X_{\langle k \rangle}) = x_k$, where $k = a_t$. Lastly, we update the conditional posterior $\pi^t_{k \mid Z}(\theta_k)$ and the posterior probabilities of the parent sets.

Algorithm 12.3 (Bayesian backdoor bandit algorithm).

1: **for all** $a \in \mathcal{A}$ and $Z \subseteq X \setminus X_{\langle a \rangle}$ **do**
2:      Compute $\pi^0_{a|Z}$ and $\mathbb{P}(\mathrm{pa}(X_{\langle a \rangle}) = Z \mid \mathcal{D}_0)$.
3: **end for**
4: **for** $t = 1, \ldots, T$ **do**
5:      **for all** $a \in \mathcal{A}$ **do**
6:          Sample $U_a(t) \sim \pi^{t-1}_a(\mu_a)$.
7:      **end for**
8:      Pull arm $a_t \in \mathrm{argmax}_{a \in \mathcal{A}} U_a(t)$ and generate $\mathcal{D}^{(t)}$.
9:      $k \leftarrow a_t$.
10:      **for all** $Z \subseteq X \setminus X_{\langle k \rangle}$ **do**
11:          Update $\pi^t_{k|Z}(\theta_k) \propto p_{\theta_k}(y_t) \, \pi^{t-1}_{k|Z}(\theta_k)$.
12:      **end for**
13:      $\pi^t_{a|Z} \leftarrow \pi^{t-1}_{a|Z}$ for all $a \neq k$.
14:      Update $\mathbb{P}(\mathrm{pa}(X_{\langle a \rangle}) = Z \mid \mathcal{D}[t])$ for all a and Z.
15: **end for**

To use UCB in the BBB algorithm, we simply need to replace the Thompson sampling step on Line 6 with an upper bound for μ_a. We estimate the expected reward using the posterior mean $\mathbb{E}_{\pi^{t-1}_a}[\mu_a]$ with respect to π^{t-1}_a, and we replace $1/n_a(t)$ in (12.8) with the posterior variance $\mathrm{Var}_{\pi^{t-1}_a}[\mu_a]$. For each arm $a \in \mathcal{A}$, we compute the upper bound

$$U_a(t) = \mathbb{E}_{\pi^{t-1}_a}[\mu_a] + c\sqrt{\mathrm{Var}_{\pi^{t-1}_a}[\mu_a]\log(t)}. \tag{12.15}$$

One may also implement a Bayes-UCB instead, computing an upper quantile of μ_a based on its posterior distribution:

$$U_a(t) = Q\left(1 - \frac{1}{t(\log T)^c}, \pi^{t-1}_a\right), \tag{12.16}$$

where $Q(r, \pi)$ is the r-quantile of the probability distribution π and c is a constant for computing the quantile used in the theoretical analysis of Bayes-UCB, with $c = 0$ empirically preferred (Kaufmann *et al.*, 2012).

Under a Bayesian approach, the cumulative regret defined in (12.7) is conditioned on a fixed parameter $\Theta_{\mathcal{A}}$, rewritten as

$$R_T(\Theta_{\mathcal{A}}) = \mathbb{E}\left[\sum_{t=1}^{T}(\mu^* - \mu_{a_t}) \,\middle|\, \Theta_{\mathcal{A}}\right].$$

We specify a prior $P(\mathcal{G})$ over the DAG $\mathcal{G}$ and conditional priors $\pi^0_{a|Z}$ for the parameters of the reward distribution under interventions $a \in \mathcal{A}$, which defines a prior Π^0 over $\Theta_\mathcal{A}$. The Bayesian regret averages the regret $R_T(\Theta_\mathcal{A})$ over the prior distribution Π^0:

$$R_T^B = \mathbb{E}_{\Pi^0}\left[R_T(\Theta_\mathcal{A})\right] = \sum_{t=1}^{T} \mathbb{E}\left[\mu^* - \mu_{a_t}\right],$$

where the second expectation is taken with respect to the joint distribution over the parameters $(\mathcal{G}, \Theta_\mathcal{A})$, the data $\mathcal{D}[T]$, and the actions $\{a_t\}$. Note that $\mu_a = \mu_a(\Theta_\mathcal{A})$ is a random variable under the Bayesian setting. If $R_T^B = O(g(T))$, then $R_T(\Theta_\mathcal{A}) = O_p(g(T))$ with respect to the prior distribution of $(\mathcal{G}, \Theta_\mathcal{A})$.

Under some assumptions over the posterior distributions, Proposition 1 in Huang and Zhou (2023) provides an upper bound:

$$R_T^B \leq C\sqrt{\log T}\left[\sqrt{KT + K^2 n_0} - \sqrt{K^2 n_0} + O(\sqrt{T})\right], \qquad (12.17)$$

where $C > 0$ is a constant. The benefit of the backdoor adjustment prior is seen when n_0 is large relative to T. If $T/n_0 \leq M$, where M is a constant, then

$$\sqrt{K^2 n_0 + KT} - \sqrt{K^2 n_0} = \sqrt{K^2 n_0}\left[\left\{1 + \frac{T}{K n_0}\right\}^{1/2} - 1\right]$$

$$\leq \frac{T}{2\sqrt{n_0}} \leq \frac{\sqrt{MT}}{2},$$

where the first inequality is due to $(1 + x)^{1/2} \leq 1 + x/2$ for $x \geq 0$. In this case, we obtain a regret bound $R_T^B = O(\sqrt{T \log T})$ independent of K. This confirms the advantage of using observational data to *simultaneously* estimate all rewards $\mu_a, a \in \mathcal{A}$ through backdoor adjustment, which largely relaxes the dependence of the regret on the number of actions. The numerical results in the work of Huang and Zhou (2023) provide additional empirical evidence for the effectiveness of the BBB approach.

The structure learning component of Algorithm 12.3 is embodied in the update of the posterior probabilities $\mathbb{P}(\mathrm{pa}(X_{\langle a \rangle}) = Z \mid \mathcal{D}[t])$ for all $X_{\langle a \rangle}$. In general, this is expected to accurately estimate the local graphical structure of variables under sufficient interventions. After a bandit algorithm converges to the optimal arm a^*, the experimental data are generated under

$\text{do}(X_{\langle a^* \rangle} = x_{a^*})$ with high probability. When T is large, we should be able to learn the parent set of $X_{\langle a^* \rangle}$ and the interventional equivalence class (Hauser and Bühlmann, 2015) of the true graph under $\text{do}(X_{\langle a^* \rangle})$. Rigorous theoretical analysis of structure learning in a causal bandit process could be a fruitful future research area.

Bibliography

Abbe, E. (2018). Community detection and stochastic block models: Recent developments, *Journal of Machine Learning Research*, **18**, 1–86.

Agrawal, R. (1995). Sample mean based index policies by $O(\log n)$ regret for the multi-armed bandit problem, *Advances in Applied Probability*, **27**, 1054–1078.

Agrawal, R., Squires, C., Prasad, N., and Uhler, C. (2022). The DeCAMFounder: Non-linear causal discovery in the presence of hidden variables. *arXiv Preprint* 2102.07921v2.

Albert, R. and Barabási, A.-L. (2002). Statistical mechanics of complex networks, *Reviews of Modern Physics*, **74**, 47–97.

Aldous, D. (1981). Representations for partially exchangeable arrays of random variables, *Journal of Multivariate Analysis*, **11**, 581–598.

Amini, A., Aragam, B., and Zhou, Q. (2022a). On perfectness in Gaussian graphical models, in *Proceedings of Machine Learning Research (AISTATS)*, Vol. 151, pp. 7505–7517.

Amini, A. A., Aragam, B., and Zhou, Q. (2022b). A non-graphical representation of conditional independence via the neighborhood lattice, 1–30. *arXiv Preprint* 2206.05829v1.

Andersson, S., Madigan, D., and Perlman, M. D. (1997). A characterization of Markov equivalence classes for acyclic digraphs, *Annals of Statistics*, **25**, 505–542.

Aragam, B., Amini, A., and Zhou, Q. (2019a). Globally optimal score-based learning of directed acyclic graphs in high-dimensions, in *Advances in Neural Information Processing Systems (NeurIPS)*, Vol. 32, pp. 4452–4464.

Aragam, B., Gu, J., and Zhou, Q. (2019b). Learning large-scale Bayesian networks with the sparsebn package, *Journal of Statistical Software*, **91**(11), 1–38.

Aragam, B. and Zhou, Q. (2015). Concave penalized estimation of sparse Gaussian Bayesian networks, *Journal of Machine Learning Research*, **16**, 2273–2328.

Auer, P., Cesa-Bianchi, N., and Fischer, P. (2002). Finite-time analysis of the multiarmed bandit problem, *Machine Learning*, **47**, 235–256.

Banerjee, O., El Ghaoui, L., and d'Aspremont, A. (2008). Model selection through sparse maximum likelihood estimation for multivariate Gaussian or binary data, *Journal of Machine Learning Research*, **9**, 485–516.

Baum, L. (1972). An equality and associated maximization technique in statistical estimation for probabilistic functions of Markov processes, *Inequalities*, **3**, 1–8.

Baum, L., Petrie, T., Soules, G., and Weiss, N. (1970). A maximization technique occuring in the statistical analysis of probabilistic functions of Markov processes, *Annals of Mathematical Statistics*, **41**, 164–171.

Bento, J. and Montanari, A. (2009). Which graphical models are difficult to learn? in *Advances in Neural Information Processing Systems (NIPS)*, Vol. 22.

Berry, D. A. and Fristedt, B. (1985). *Bandit problems: Sequential allocation of experiments*, in Monographs on Statistics and Applied Probability, Vol. 5. London: Chapman and Hall, p. 7.

Bhattacharya, R., Nabi, R., and Shpitser, I. (2022). Semiparametric inference for causal effects in graphical models with hidden variables, *Journal of Machine Learning Research*, **23**, 1–76.

Bhattacharya, R., Nagarajan, T., Malinsky, D., and Shpitser, I. (2021). Differentiable causal discovery under unmeasured confounding, in *Proceedings of the 24th International Conference on AI and STATS, PMLR*, Vol. 130.

Bickel, P., Choi, D., Chang, X., and Zhang, H. (2013). Asymptotic normality of maximum likelihood and its variational approximation for stochastic block-models, *Annals of Statistics*, **41**, 1922–1943.

Blei, D. M., Kucukelbir, A., and McAuliffe, J. D. (2017). Variational inference: A review for statisticians, *Journal of the American Statistical Association*, **112**, 859–877.

Bouckaert, R. R. (1993). Probabilistic network construction using the minimum description length principle, in *Symbolic and Quantitative Approaches to Reasoning and Uncertainty: European Conference ECSQARU'93*. Lecture Notes in Computer Science, Vol. 747. Springer-Verlag, pp. 41–48.

Bowden, R. and Day, N. (1984). *Instrumental Variables*. Cambridge University Press, Cambridge, UK.

Brito, C. and Pearl, J. (2002a). Generalized instrumental variables, in *Proceedings of the 18th Conferences on Uncertainty in Artificial Intelligence*, Vol. 18, pp. 85–93.

Brito, C. and Pearl, J. (2002b). A new identification condition for recursive models with correlated errors, *Structural Equation Modeling*, **94**, 459–474.

Bühlmann, P., Peters, J., and Ernest, J. (2014). CAM: Causal Additive Models, high-dimensional order search and penalized regression, *Annals of Statistics*, **42**, 2526–2556.

Buntine, W. (1991). Theory refinement on Bayesian networks, in *Proceedings of the Seventh Annual Conference on Uncertainty in Artificial Intelligence*. Morgan Kaufmann, pp. 52–60.

Cattell, R. B. (1966). The Scree test for the number of factors, *Multivariate Behavioral Research*, **1**, 245–276.

Celisse, A., Daudin, J., and Pierre, L. (2012). Consistency of maximum-likelihood and variational estimators in the stochastic block model, *Electronic Journal of Statistics*, **6**, 1847–1899.

Chen, Z., Xie, F., Qiao, J., Hao, Z., Zhang, K., and Cai, R. (2022). Identification of linear latent variable model with arbitrary distribution, in *Proceedings of the 36th AAAI Conference on Artificial Intelligence*, Vol. 36, pp. 6350–6357.

Chickering, D. M. (1995). A transformational characterization of Bayesian network structures, in Hanks, S. and Besnard, P. (eds.), *Proceedings of the 18th Conferences on Uncertainty in Artificial Intelligence*. Morgan Kaufmann, pp. 87–98.

Chickering, D. M. (1996). Learning Bayesian networks is NP-complete, in *Learning From Data*. Springer, New York, USA, pp. 121–130.

Chickering, D. M. (2002). Optimal structure identification with greedy search, *Journal of Machine Learning Research*, **3**, 507–554.

Chickering, D. M. and Heckerman, D. (1997). Efficient approximations for the marginal likelihood of Bayesian networks with hidden variables, *Machine Learning*, **29**, 181–212.

Chickering, D. M., Heckerman, D., and Meek, C. (2004). Large-sample learning of Bayesian networks is NP-hard, *Journal of Machine Learning Research*, **5**, 1287–1330.

Colombo, D. and Maathuis, M. H. (2014). Order-independent constraint-based causal structure learning, *Journal of Machine Learning Research*, **15**(116), 3921–3962.

Comon, P. (1994). Independent component analysis – a new concept? *Signal Processing*, **36**, 287–314.

Cooper, G. F. and Herskovits, E. (1992). A Bayesian method for the induction of probabilistic networks from data, *Machine Learning*, **9**, 309–347.

Daudin, J., Picard, F., and Robin, S. (2008). A mixture model for random graphs, *Statistics and Computing*, **18**, 173–183.

Dawid, A. (1979). Conditional independence in statistical theory, *Journal of the Royal Statistical Society: Series B (Statistical Methodology)*, **41**, 1–31.

de Leeuw, J. (1994). Block relaxation algorithms in statistics, in Bock, H., Lenski, W., and Richter, M. (eds.), *Information Systems and Data Analysis*. Springer Verlag, New York, USA, pp. 308–324.

Dempster, A. P. (1972). Covariance selection, *Biometrics*, **28**, 157–175.

Dempster, A., Laird, N., and Rubin, D. (1977). Maximum likelihood estimation from incomplete data via the EM algorithm (with discussion), *Journal of the Royal Statistical Society Series B*, **39**, 1–38.

Drton, M., Eichler, M., and Richardson, T. S. (2009). Computing maximum likelihood estimates in recursive linear models with correlated errors, *Journal of Machine Learning Research*, **10**, 2329–2348.

Drton, M., Foygel, R., and Sullivant, S. (2011). Global identifiability of linear structural equation models, *The Annals of Statistics*, **39**(2), 865–886.

Ellis, B. and Wong, W. H. (2008). Learning causal Bayesian network structures from experimental data, *Journal of the American Statistical Association*, **103**, 778–789.

Eppstein, D., Löffler, M., and Strash, D. (2010). Listing all maximal cliques in sparse graphs in near-optimal time, in *International Symposium on Algorithms and Computation*, pp. 403–414.

Fienberg, S. (2012). A brief history of statistical models for network analysis and open challenges, *Journal of Computational and Graphical Statistics*, **21**, 825–839.

Friedman, J., Hastie, T., and Tibshirani, R. (2008). Sparse inverse covariance estimation with the Graphical Lasso, *Biostatistics*, **9**(3), 432–441.

Frot, B., Nandy, P., and Maathuis, M. H. (2019). Robust causal structure learning with some hidden variables, *Journal of the Royal Statistical Society: Series B (Statistical Methodology)*, **81**, 459–487.

Fu, F. and Zhou, Q. (2013). Learning sparse causal Gaussian networks with experimental intervention: Regularization and coordinate descent, *Journal of the American Statistical Association*, **108**, 288–300.

Gámez, J. A., Mateo, J. L., and Puerta, J. M. (2011). Learning Bayesian networks by hill climbing: Efficient methods based on progressive restriction of the neighborhood, *Data Mining and Knowledge Discovery*, **22**, 106–148.

Gao, E., Chen, J., Shen, L., Liu, T., Gong, M., and Bondell, H. (2023). FedDAG: Federated DAG structure learning, *Transactions of Machine Learning Research*, **01/2023**, 1–36.

Geman, S. and Geman, D. (1984). Stochastic relaxation, Gibbs distributions, and the Bayesian restoration of images, *IEEE Transactions on Pattern Analysis and Machine Intelligence*, **6**, 721–741.

Ghassami, A., Salehkaleybar, S., and Kiyavash, N. (2019). Interventional experimental design for causal structure learning. *arXiv Preprint* 1910.05651v1.

Glorfeld, L. W. (1995). An improvement on Horn's parallel analysis methodology for selecting the correct number of factors to retain, *Educational and Psychological Measurement*, **55**(3), 377–393.

Gretton, A., Fukumizu, K., Teo, C., Song, L., Schölkopf, B., and Smola, A. (2007). A kernel statistical test of independence, in *NIPS 2007*, Vol. 20, pp. 585–592.

Gu, J., Fu, F., and Zhou, Q. (2019). Penalized estimation of directed acyclic graphs from discrete data, *Statistics and Computing*, **29**, 161–176.

Gu, J. and Zhou, Q. (2020). Learning big Gaussian Bayesian networks: Partition, estimation, and fusion, *Journal of Machine Learning Research*, **21**(158), 1–31.

Guttman, L. (1954). Some necessary conditions for common-factor analysis, *Psychometrika*, **19**(2), 149–161.

Haughton, D. M. (1988). On the choice of a model to fit data from an exponential family, *Annals of Statistics*, **16**, 342–355.

Hauser, A. and Bühlmann, P. (2014). Two optimal strategies for active learning of causal models from interventional data, *International Journal of Approximate Reasoning*, **55**, 926–939.

Hauser, A. and Bühlmann, P. (2015). Jointly interventional and observational data: Estimation of interventional Markov equivalence classes of directed acyclic graphs, *Journal of the Royal Statistical Society: Series B (Statistical Methodology)*, **77**, 291–318.

Heckerman, D., Geiger, D., and Chickering, D. M. (1995). Learning Bayesian networks: The combination of knowledge and statistical data, *Machine Learning*, **20**(3), 197–243.

Herskovits, E. and Cooper, G. (1990). Kutató: An entropy-driven system for construction of probabilistic expert systems from databases, in *Proceedings of the Sixth Annual Conference on Uncertainty in Artificial Intelligence*, pp. 54–62.

Hirose, K. and Yamamoto, M. (2014a). Estimation of an oblique structure via penalized likelihood factor analysis, *Computational Statistics and Data Analysis*, **79**, 120–132.

Hirose, K. and Yamamoto, M. (2014b). Sparse estimation via nonconcave penalized likelihood in factor analysis model, *Statistics and Computing*, **25**(5), 863–875.

Hoff, P., Raftery, A., and Handcock, M. (2002). Latent space approaches to social network analysis, *Journal of the American Statistical Association*, **97**, 1090–1098.

Horn, J. L. (1965). A rationale and test for the number of factors in factor analysis, *Psychometrika*, **30**(2), 179–185.

Hu, H., Li, Z., and Vetta, A. (2014). Randomized experimental design for causal graph discovery, in *Advances in Neural Information Processing Systems*, Vol. 27.

Huang, B., Low, C., Xie, F., Glymour, C., and Zhang, K. (2022). Latent hierarchical causal structure discovery with rank constraints, in *Procceedings of the 36th Conference on Neural Information Processing Systems (NeurIPS)*, Vol. 36.

Huang, J. and Zhou, Q. (2022). Partitioned hybrid learning of Bayesian network structures, *Machine Learning*, **111**, 1695–1738.

Huang, J. and Zhou, Q. (2023). Bayesian causal bandits with backdoor adjustment prior, *Transactions on Machine Learning Research*, **01/2023**, 1–37.

Hunter, D. and Lange, K. (2004). A tutorial on MM algorithms, *The American Statistician*, **58**, 30–37.

Hyttinen, A., Eberhardt, F., and Hoyer, P. O. (2013). Experimental selection for causal discovery, *Journal of Machine Learning Research*, **14**, 3041–3071.

Hyvärinen, A., Karhunen, J., and Oja, E. (2001). *Independent Component Analysis*. Wiley Interscience, New Jersey, USA.

Hyvärinen, A. and Smith, S. (2013). Pairwise likelihood ratios for estimation of non-Gaussian structural equation models, *Journal of Machine Learning Research*, **14**, 111–152.

Jensen, S., Liu, X., Zhou, Q., and Liu, J. (2004). Computational discovery of gene regulatory binding motifs: A Bayesian perspective, *Statistical Science*, **19**, 188–204.

Jordan, M., Ghahramani, Z., Jaakkola, T., and Saul, L. (1999). Introduction to variational methods for graphical models, *Machine Learning*, **37**, 183–233.

Kaiser, H. F. (1960). The application of electronic computers to factor analysis, *Educational and Psychological Measurement*, **20**(1), 141–151.

Kalisch, M. and Bühlmann, P. (2007). Estimating high-dimensional directed acyclic graphs with the PC-algorithm, *Journal of Machine Learning Research*, **8**, 613–636.

Kalisch, M., Mächler, M., Colombo, D., Maathuis, M. H., and Bühlmann, P. (2012). Causal inference using graphical models with the R package pcalg, *Journal of Statistical Software*, **47**(11), 1–26.

Kalman, R. (1960). A new approach to linear filtering and prediction problems, *Transactions of the ASME–Journal of Basic Engineering*, **82**, 35–45.

Kaufmann, E., Cappe, O., and Garivier, A. (2012). On Bayesian upper confidence bounds for bandit problems, in *Proceedings of Machine Learning Research (AISTATS)*, Vol. 22, pp. 592–600.

Kim, D. S. and Zhou, Q. (2023). Structure learning of latent factors via clique search on correlation thresholded graphs, in *Proceedings of Machine Learning Research (ICML)*, Vol. 202, pp. 16978–16996.

Kocaoglu, M., Dimakis, A., and Vishwanath, S. (2017). Cost-optimal learning of causal graphs, in *Proceedings of the 34th International Conference on Machine Learning, PMLR*, Vol. 70.

Krogh, A., Brown, M., Mian, I., Sjolander, K., and Haussler, D. (1994). Hidden Markov models in computational biology: Applications to protein modeling, *Journal of Molecular Biology*, **235**, 1501–1531.

Kummerfeld, E. and Ramsey, J. (2016). Causal clustering for 1-factor measurement models, in *Proceedings of the 22nd ACM SIGKDD International Conference on Knowledge Discovery and Data Mining*, Vol. 22, pp. 1655–1664.

Lachapelle, S., Brouillard, P., Deleu, T., and Lacoste-Julien, S. (2020). Gradient-based neural DAG learning, in *Proceedings of the International Conferences on Learning Representations (ICLR)*, Vol. 8.

Lai, T. L. and Robbins, H. (1985). Asymptotically efficient adaptive allocation rules, *Advances in Applied Mathematics*, **6**, 4–22.

Lam, W. and Bacchus, F. (1994). Learning Bayesian belief networks: An approach based on the MDL principle, *Computational Intelligence*, **10**, 269–293.

Larranaga, P., Poza, M., Yurramendi, Y., Murga, R., and Kuijpers, C. (1996). Structure learning of Bayesian networks by genetic algorithms: A performance analysis of control parameters, *IEEE Transactions on Pattern Analysis and Machine Intelligence*, **18**, 912–926.

Lattimore, F., Lattimore, T., and Reid, M. D. (2016). Causal bandits: Learning good interventions via causal inference, in *Advances in Neural Information Processing Systems*, Vol. 29.

Lauritzen, S. L. (1996). *Graphical Models*. Oxford University Press, Oxford, UK.

Lauritzen, S. L. and Richardson, T. S. (2002). Chain graph models and their causal interpretations, *Journal of the Royal Statistical Society: Series B (Statistical Methodology)*, **64**, 312–361.

Lauritzen, S. L. and Sadeghi, K. (2018). Unifying Markov properties for graphical models, *Annals of Statistics*, **46**, 2251–2278.

Lawrence, C., Altschul, S., Boguski, M., Liu, J., Neuwald, A., and Wootton, J. (1993). Detecting subtle sequence signals: A Gibbs sampling strategy for multiple alignment, *Science*, **262**, 208–214.

Lawrence, C. and Reilly, A. (1990). An expectation-maximization (EM) algorithm for the identification and characterization of common sites in unaligned biopolymer sequences, *Proteins*, **7**, 41–51.

Lee, C. and Wilkinson, D. (2019). A review of stochastic block models and extensions for graph clustering, *Applied Network Science*, **4**(122), 1–50.

Levitz, M., Perlman, M. D., and Madigan, D. (2001). Separation and completeness properties for AMP chain graph Markov models, *Annals of Statistics*, **29**, 1751–1784.

Li, C., Shen, X., and Pan, W. (2023). Nonlinear causal discovery with confounders, *Journal of the American Statistical Association* (Early Access), 1–10.

Li, H., Madrid Padilla, O., and Zhou, Q. (2024). Learning Gaussian DAGs from network data. *Journal of Machine Learning Research*, **25**(377), 1–52.

Liu, J., Neuwald, A., and Lawrence, C. (1995). Bayesian models for multiple local sequence alignment and Gibbs sampling strategies, *Journal of the American Statistical Association*, **90**, 1156–1170.

Lovasz, L. and Szegedy, B. (2006). Limits of dense graph sequences, *Journal of Combinatorial Theory, Series B*, **96**, 933–957.

Margaritis, D. and Thrun, S. (1999). Bayesian network induction via local neighborhoods, in *Advances in Neural Information Processing Systems (NIPS)*, Vol. 12, pp. 505–511.

Matias, C. and Robin, S. (2014). Modeling heterogeneity in random graphs through latent space models: A selective review, *ESAIM: Proceedings and Surveys*, **47**, 55–74.

Meek, C. (1995). Causal inference and causal explanation with background knowledge, *Uncertainty in Artificial Intelligence*, **11**, 403–410.

Meinshausen, N. and Bühlmann, P. (2006). High-dimensional graphs and variable selection with the Lasso, *The Annals of Statistics*, **34**(3), 1436–1462.

Mohan, K. and Pearl, J. (2021). Graphical models for processing missing data, *Journal of the American Statistical Association*, **116**, 1023–1037.

Ng, I. and Zhang, K. (2022). Towards federated Bayesian network structure learning with continuous optimization, *International Conference on Artificial Intelligence and Statistics*, *PMLR*, Vol. 151, pp. 8085–8111.

Niinimäki, T., Parviainen, P., and Koivisto, M. (2016). Structure discovery in Bayesian networks by sampling partial orders, *Journal of Machine Learning Research*, **17**(1), 2002–2048.

Nowzohour, C., Maathuis, M. H., and Evans, R. (2017). Distributional equivalence and structure learning for bow-free acyclic path diagrams, *Electronic Journal of Statistics*, **11**, 5342–5374.

Orbanz, P. and Roy, D. (2015). Bayesian models of graphs, arrays and other exchangeable random structures, *IEEE Transactions on Pattern Analysis and Machine Intelligence*, **37**, 437–461.

Orchard, T. and Woodbury, M. (1972). A missing information principle: Theory and applications, in *Proceedings of the 6th Berkeley Symposium on Mathematical Statistics*, Vol. 1, pp. 697–715.

Pearl, J. (1988). *Probabilistic Reasoning in Intelligent Systems: Networks of Plausible Inference*. Morgan Kaufmann, California, USA.

Pearl, J. (1995). Causal diagrams for empirical research, *Biometrika*, **82**, 669–710.

Pearl, J. (2000). *Causality: Models, Reasoning and Inference*. Cambridge University Press, Cambridge, UK.

Peña, J. M. (2011). Faithfulness in chain graphs: the Gaussian case, in *International Conference on Artificial Intelligence and Statistics*, pp. 588–599.

Peng, Z. and Zhou, Q. (2022). An empirical Bayes approach to stochastic block-models and graphons: Shrinkage estimation and model selection, *PeerJ Computer Science*, **8**(e1006), 1–29.

Perrier, E., Imoto, S., and Miyano, S. (2008). Finding optimal Bayesian network given a super-structure, *Journal of Machine Learning Research*, **9**, 2251–2286.

Peters, J. and Bühlmann, P. (2014). Identifiability of Gaussian structural equation models with equal error variances, *Biometrika*, **101**, 219–228.

Peters, J., Mooij, J., Janzing, D., and Schölkopf, B. (2014). Causal discovery with continuous additive noise models, *Journal of Machine Learning Research*, **15**, 2009–2053.

Pfister, N., Buehlmann, P., Schölkopf, B., and Peters, J. (2018). Kernel-based tests for joint independence, *Journal of the Royal Statistical Society: Series B (Statistical Methodology)*, **80**, 5–31.

Preacher, K. J., Zhang, G., Kim, C., and Mels, G. (2013). Choosing the optimal number of factors in exploratory factor analysis: A model selection perspective, *Multivariate Behavioral Research*, **48**(1), 28–56.

Rabiner, L. (1989). A tutorial on hidden Markov models and selected applications in speech recognition, *Proceedings of the IEEE*, **77**, 257–286.

Ravikumar, P., Wainwright, M. J., and Lafferty, J. (2010). High-dimensional Ising model selection using ℓ_1-regularized logistic regression, *Annals of Statistics*, **38**, 1287–1319.

Ravikumar, P., Wainwright, M., Raskutti, G., and Yu, B. (2011). High-dimensional covariance estimation by minimizing ℓ_1-penalized log-determinant divergence, *Electronic Journal of Statistics*, **5**, 935–980.

Richardson, T., Evans, R., Robins, J., and Shpitser, I. (2023). Nested Markov properties for acyclic directed mixed graphs, *Annals of Statistics*, **51**, 334–361.

Richardson, T. S. and Spirtes, P. (2002). Ancestral graph Markov models, *Annals of Statistics*, **30**, 962–1030.

Robins, J. (1986). A new approach to causal inference in mortality studies with a sustained exposure period–applications to control of the healthy worker survivor effect, *Mathematical Modeling*, **7**, 1393–1512.

Robinson, R. W. (1977). Counting unlabeled acyclic digraphs, in *Combinatorial Mathematics V*. Springer, New York, USA, pp. 28–43.

Rohe, K., Chatterjee, S., and Yu, B. (2011). Spectral clustering and the high-dimensional stochastic block model, *Annals of Statistics*, **39**, 1878–1915.

Rosenbaum, P. and Rubin, D. (1983). The central role of propensity score in observational studies for causal effects, *Biometrika*, **70**, 41–55.

Rubin, D. (1990). Formal models of statistical inference for causal effects, *Journal of Statistical Planning and Inference*, **25**, 279–292.

Ruiz, G. (2022). Plug-in estimation approaches to causal inference and discovery. Ph.D. Dissertation, Department of Statistics, UCLA.

Ruiz, G., Madrid Padilla, O., and Zhou, Q. (2022). Sequentially learning the topological ordering of causal directed acyclic graphs with likelihood ratio scores, *Transactions on Machine Learning Research*, **12/2022**, 1–30.

Russo, D. and Van Roy, B. (2014). Learning to optimize via posterior sampling, *Mathematics of Operations Research*, **39**, 1221–1243.

Russo, D. J., Van Roy, B., Kazerouni, A., Osband, I., and Wen, Z. (2018). A tutorial on Thompson sampling, *Foundations and Trends in Machine Learning*, **11**(1), 1–96.

Salehkaleybar, S., Ghassami, A., Kiyavash, N., and Zhang, K. (2020). Learning linear non-Gaussian models in the presence of latent variables, *Journal of Machine Learning Research*, **21**, 1–24.

Scanagatta, M., de Campos, C. P., Corani, G., and Zaffalon, M. (2015). Learning Bayesian networks with thousands of variables, in *Advances in Neural Information Processing Systems*, Vol. 28, pp. 1864–1872.

Schafer, J. (1997). *Analysis of Incomplete Multivariate Data*, 1st edn. Chapman & Hall/CRC, Florida, USA.

Schneider, T. and Stephens, R. (1990). Sequence logos: A new way to display consensus sequences, *Nucleic Acids Research*, **18**, 6097–6100.

Scutari, M. (2017). Bayesian network constraint-based structure learning algorithms: Parallel and optimized implementations in the bnlearn R package, *Journal of Statistical Software*, **77**(2), 1–20.

Shah, R. and Peters, J. (2020). The hardness of conditional independence testing and the generalized covariance measure, *Annals of Statistics*, **48**, 1514–1538.

Shamir, O., Srebro, N., and Zhang, T. (2014). Communication-efficient distributed optimization using an approximate Newton-type method, in *Proceedings of the 31st International Conference on International Conference on Machine Learnings*, Vol. 32(2), pp. 1000–1008.

Shimizu, S., Hoyer, P. O., Hyvärinen, A., and Kerminen, A. (2006). A linear non-Gaussian acyclic model for causal discovery, *Journal of Machine Learning Research*, **7**, 2003–2030.

Shimizu, S., Inazumi, T., Sogawa, Y., Hyvärinen, A., Kawahara, Y., Washio, T., Hoyer, P. O., and Bollen, K. (2011). DirectLiNGAM: A direct method for learning a linear non-Gaussian structural equation model, *Journal of Machine Learning Research*, **12**, 1225–1248.

Shpitser, I., Evans, R., and Richardson, T. S. (2018). Acyclic linear SEMs obey the nested Markov property, in *Proceedings of the 34th Conference on Uncertainty in Artificial Intelligence*, Vol. 34.

Spirtes, P. and Glymour, C. (1991). An algorithm for fast recovery of sparse causal graphs, *Social Science Computer Review*, **9**(1), 62–72.

Spirtes, P., Glymour, C., and Scheines, R. (2000). *Causation, Prediction, and Search*, 2nd edn. The MIT Press, Massachusetts, USA.

Spirtes, P., Meek, C., and Richardson, T. S. (1999). An algorithm for causal inference in the presence of latent variables and selection bias, in Glymour,

C. and Cooper, G. F. (eds.), *Computation, Causation, and Discovery*. MIT Press, pp. 211–252, Massachusetts, USA.

Squires, C., Yun, A., Nichani, E., Agrawal, R., and Uhler, C. (2022). Causal structure discovery between clusters of nodes induced by latent factors, in *Proceedings of Machine Learning Research*, Vol. 140, pp. 1–19.

Städler, N., Bühlmann, P., and Van De Geer, S. (2010). ℓ_1-penalization for mixture regression models, *Test*, **19**(2), 209–256.

Suzuki, J. (1993). A construction of Bayesian networks from databases based on an MDL principle, in *Proceedings of the Ninth Annual Conference on Uncertainty in Artificial Intelligence*, pp. 266–273.

Tanner, M. and Wong, W. H. (1987). The calculation of posterior distributions by data augmentation, *Journal of the American Statistical Association*, **82**, 528–540.

Teyssier, M. and Koller, D. (2005). Ordering-based search: A simple and effective algorithm for learning Bayesian networks, in *Proceedings of the 21st Conferences on Uncertainty in Artificial Intelligence*, Vol. 21, pp. 584–590.

Thompson, W. R. (1933). On the likelihood that one unknown probability exceeds another in view of the evidence of two samples, *Biometrika*, **25**, 285–294.

Tian, J. and Pearl, J. (2002a). A general identification condition for causal effects, in *Proceedings of the AAAI*, pp. 567–573.

Tian, J. and Pearl, J. (2002b). On the testable implications of causal models with hidden variables, in *Proceedings of the 18th Conference on Uncertainty in Artificial Intelligence*, Vol. 18, pp. 519–527.

Tibshirani, R. (1996). Regression shrinkage and selection via the lasso, *Journal of the Royal Statistical Society Series B*, **58**(1), 267–288.

Tsamardinos, I., Brown, L. E., and Aliferis, C. F. (2006). The max-min hill-climbing Bayesian network structure learning algorithm, *Machine Learning*, **65**(1), 31–78.

Tseng, P. (2001). Convergence of a block coordinate descent method for nondifferentiable minimization, *Journal of Optimization Theory and Applications*, **109**, 475–494.

van de Geer, S. and Bühlmann, P. (2013). ℓ_0-penalized maximum likelihood for sparse directed acyclic graphs, *The Annals of Statistics*, **41**(2), 536–567.

Verma, T. (1991). Graphical aspects of causal models. Technical Report R-191, UCLA Computer Science Department, Cognitive Systems Lab.

Verma, T. and Pearl, J. (1990). Equivalence and synthesis of causal models, in *Proceedings of the Sixth Annual Conference on Uncertainty in Artificial Intelligence*, pp. 220–227.

Viterbi, A. (1967). Error bounds for convolutional codes and an asymptotically optimum decoding algorithm, *IEEE Transactions on Information Theory*, **13**, 260–269.

von Luxburg, U. (2007). A tutorial on spectral clustering, *Statistics and Computing*, **17**, 395–416.

Wang, B. and Zhou, Q. (2021). Causal network learning with non-invertible functional relationships, *Computational Statistics and Data Analysis*, **156**(107141), 1–18.

Wang, Y. and Drton, M. (2019). High dimensional causal discovery under non-Gaussianity, *Biometrika*, **107**, 41–59.

Wright, S. (1934). The method of path coefficients, *Annals of Mathematical Statistics*, **5**, 161–215.

Wu, C. (1983). On the convergence properties of the EM algorithm, *Annals of Statistics*, **11**, 95–103.

Xie, F., Cai, R., Huang, B., Glymour, C., Hao, Z., and Zhang, K. (2020). Generalized independent noise condition for estimating latent variable causal graphs, in *Advances in Neural Information Processing Systems (NeurIPS 2020)*, Vol. 33.

Xie, F., Huang, B., Chen, Z., He, Y., Geng, Z., and Zhang, K. (2022). Identification of linear non-Gaussian latent hierarchical structure, in *Proceedings of Machine Learning Research (ICML)*, Vol. 162.

Ye, Q., Amini, A., and Zhou, Q. (2021). Optimizing regularized Cholesky score for order-based learning of Bayesian networks, *IEEE Transactions on Pattern Analysis and Machine Intelligence*, **43**, 3555–3572.

Ye, Q., Amini, A., and Zhou, Q. (2024). Federated learning of generalized linear causal networks, *IEEE Transactions on Pattern Analysis and Machine Intelligence*, **46**, 6623–6636.

Yuan, M. and Lin, Y. (2006). Model selection and estimation in regression with grouped variables, *Journal of the Royal Statistical Society. Series B*, **68**, 49–67.

Yuan, M. and Lin, Y. (2007). Model selection and estimation in the Gaussian graphical model, *Biometrika*, **94**(1), 19–35.

Zarebavani, B., Jafarinejad, F., Hashemi, M., and Salehkaleybar, S. (2020). cuPC: CUDA-based parallel PC algorithm for causal structure learning on GPU, *IEEE Transactions on Parallel and Distributed Systems*, **31**(3), 530–542.

Zhang, C., Chen, B., and Pearl, J. (2020). A simultaneous discover-identify approach to causal inference in linear models, in *Proceedings of the 34th AAAI Conference on Artificial Intelligence (AAAI-20)*, pp. 10318–10325.

Zhang, C.-H. (2010). Nearly unbiased variable selection under minimax concave penalty, *The Annals of Statistics*, **38**(2), 894–942.

Zhang, J. (2008a). Causal reasoning with ancestral graphs, *Journal of Machine Learning Research*, **9**, 1437–1474.

Zhang, J. (2008b). On the completeness of orientation rules for causal discovery in the presence of latent confounders and selection bias, *Artificial Intelligence*, **172**, 1873–1896.

Zhang, K. and Hyvärinen, A. (2009). On the identifiability of the post-nonlinear causal model, in *Proceedings of the 25th Annual Conference on Uncertainty in Artificial Intelligence*, pp. 647–655.

Zhang, Y., Duchi, J. C., and Wainwright, M. J. (2013). Communication-efficient algorithms for statistical optimization, *Journal of Machine Learning Research*, **14**, 3321–3363.

Zheng, X., Aragam, B., Ravikumar, P., and Xing, E. (2018). DAGs with no tears: Smooth optimization for structure learning, in *Advances in Neural Information Processing Systems*, Vol. 31.

Zheng, X., Dan, C., Aragam, B., Ravikumar, P., and Xing, E. P. (2020). Learning sparse nonparametric DAGs, in *Proceedings of the 23rd International Conferences on Artificial Intelligence and Statistics, PMLR*, Vol. 108, pp. 3414–3425.

Zhou, Q. (2011). Multi-domain dampling with applications to structural inference of Bayesian networks, *Journal of the American Statistical Association*, **106**, 1317–1330.

Zhou, Q. and Wong, W. H. (2004). CisModule: *De novo* discovery of cis-regulatory modules by hierarchical mixture modeling, *Proceedings of the National Academy of Sciences of USA*, **101**, 12114–12119.

Subject Index

A

active causal learning, 240
additive noise model, 194
acyclic directed mixed graphs
  (ADMGs), 206
  district factorization, 209
  nested factorization, 213
  Tian factorization, 210
almost directed cycle, 221
ancestor, 114, 127
ancestral graph, 220
ancestral set, 119, 127
average treatment effect, 150

B

back-door adjustment, 143
Baum–Welch algorithm, 60, 64
Bayesian back-door bandit,
  243
Bayesian Dirichlet score,
  167
Bayesian inference, 2
  Bayesian estimate, 3
  conjugate prior, 7
  credible interval, 3
  posterior distribution, 3
  prior distribution, 2
Bayesian network, 122
BIC score, 166
bidirected path, 203

blockwise coordinate descent, 178
  bi-convex function, 200
boundary, 127
bow-free, 219

C

causal bandit, 242
causal discovery, 153
causal effect, 134
  identifiable, 142
  linear SEM, 140
  on ADMG, 215
causal minimality, 186
causal model, 133
causal sufficiency, 155
chain (in a DAG), 115
chain component, 126
chain graph, 126
  moral graph of, 127
child, 114
clique, 97
  independent maximal, 228
  maximal, 97
clustering, 46
  EM clustering, 47
  k-means clustering, 48
  maximum *a posteriori*, 47
collider, 115
  covered, 160
  in ADMG, 206
  uncovered, 120, 156

community detection, 73, 81
compelled edge, 121, 156
complete lattice, 109
complete-data likelihood, 18
completed partially DAG, 156
  characterization, 157
conditional ADMG, 211
conditional ignorability, 150
conditional independence, 10
  axioms, 91
  definition, 10
conditional independence graph,
  99
conditional independence test, 93
  G^2 test, 94
  partial correlation test, 93
consistent score, 169
continuous relaxation, 175
converging arborescence, 218
correlation thresholding algorithm,
  230
counterfactual, 149
covariance selection, 101
cycle, 113

D

d-connection, 115
d-separation, 115
DAG for dependent data, 199
DAG identifiability, 186
  additive noise model, 195
  equal error variance, 186
  experimental data, 175
  generalized linear DAG, 188
  LiNGAM, 190
DANE algorithm, 238
data augmentation, 36
de-correlation, 202
decomposable score, 167
descendant, 114, 127
descendant set, 213
differentiable DAG constraint, 181
directed acyclic graph, 113
  equivalence class, 156
directed cycle, 113, 206

directed path, 126, 203
directed walk, 181
Dirichlet distribution, 7
discrete Bayesian network, 124
  multi-logit regression, 125
discriminating path, 223
distinctness of parameters, 17
district, 207
do-calculus, 147
do-operator, 134
dynamic programming, 66

E

EM algorithm, 18
  ascent property, 22
  convergence, 23
  for exponential family, 28
  for mixture model, 43
  for multivariate normal data, 33
  MM view of, 21
essential graph, 156
evidence lower bound, 76
exchangeable random array, 85
exchangeable random graph, 85
experimental data, 155, 172
exponential family, 26

F

faithfulness, 105, 129
FCI algorithm, 225
federated learning, 235
fixable vertex, 211
fixing operator, 212
forward-backward summation,
  63
fraction of the missing information,
  24
front-door adjustment, 146

G

g-formula, 137
Gaussian Bayesian network, 122
Gaussian DAG, 122, 140
Gaussian graphical model, 100
generalized linear DAG, 188

Gibbs sampler
  for missing data, 35
  Gibbs motif sampler, 54
graph, 95
  ADMG, 206
  chain graph, 126
  chordal or triangulated, 157
  connectivity component, 126
  directed acyclic graph, 113
  maximal ancestral graph, 221
  partial ancestral graph, 224
  undirected graph, 95
graph Laplacian, 81
graphical lasso, 102
graphical model, 96
graphoid, 92
  axioms, 92
  compositional graphoid, 93
  semi-graphoid, 92
graphon, 84
greedy equivalence search, 171
greedy hill-climbing, 170
grow-shrink algorithm, 107

H

hidden Markov model
  conditional independence, 59
  continuous observation, 66
  elements, 58
  factorization, 58
  graphical representation, 59
  observed-data likelihood, 64
  online prediction, 67
  sufficient statistic, 61

I

identify algorithm, 216
ignorability, 16
independence model, 108
inducing path, 221
infinitely exchangeable array, 85
instrumental variable, 147
  conditional, 149
  formula, 148
intervention, 134

intrinsic set, 216
Ising model, 103

K

Kalman filter, 67
kernel, 209
Kronecker product, 199

L

latent confounder, 142
latent factor analysis, 227
latent projection, 204
latent space model, 72
linear structure equation model, 139
  associated with ADMG, 217
LiNGAM algorithm, 189

M

m-connection, 206
m-separation, 206
manipulation theorem, 137
marginal independence graph, 227
Markov blanket, 107
  in ADMG, 209
Markov boundary, 107
Markov equivalence, 120
Markov property
  chain graph
    global, 128
    local, 128
  directed
    global, 116
    local, 116
    pairwise, 117
    recursive factorization, 116
  undirected
    factorization, 97
    global, 96
    local, 96
    pairwise, 96
matrix normal distribution, 199
maximal ancestral graph, 221
Meek's rules, 160
minimum-trace DAG, 187
missing at random, 16

missing completely at random,
  16
missing information principle, 24
missing not at random, 16
mixture model
  definition, 41
  MLE by EM algorithm, 43
MM algorithm, 21
moral graph, 118, 127
motif discovery, 49
  cis-regulatory module, 54
  EM algorithm, 52
  Gibbs motif sampler, 54
multi-armed bandit, 240
multi-domain sampler, 171
multivariate normal distribution,
  30

N

neighbor, 95
neighborhood lattice, 109
  decomposition, 110
neighborhood regression, 103, 105
neural network, 196
non-collider, 115, 206
non-descendant, 114, 127

O

observational data, 155
observed-data likelihood, 17
observed-data posterior, 18

P

parameter identifiability, 42
  linear regression, 12
  linear SEM for ADMG, 218
  mixture model, 42
parent, 114
parent set adjustment, 138
  for linear SEM, 141
partial ancestral graph, 224
partially DAG, 156
partitioned PC algorithm, 232
path, 95, 113, 203
PATH algorithm, 164

PC algorithm, 159
  consistency, 162
  PC-stable, 233
point-wise consistent test, 162
potential outcome, 149

R

random graph, 71
  binary graph, 71
  exchangeable, 85
  weighted graph, 71
reachable closure, 218
reachable graph, 216
reachable set, 218
regression with subsequent
  independence test (RESIT), 195
reversible edge, 121, 156
RICF algorithm, 219

S

score-equivalence, 169
semi-Markov causal model, 205
separation (undirected graph), 95
sequential decision process, 240
sibling, 206
skeleton, 120
sparse Cholesky factorization, 180
sparse regularization
  concave penalty, 177
  lasso or ℓ_1 penalty, 177
spectral clustering, 81
stochastic block model, 73
  asymptotic normality, 80
  definition, 73
  degree-corrected, 83
  maximum likelihood estimate
    (MLE), 74, 80
  modeling covariates, 83
  variational EM for, 77
  variational estimator, 80
strongly protected arrow, 157
structural equation model, 122, 125,
  133
structure learning
  constraint-based, 154, 158

directed acyclic graph, 153
federated DAG learning, 235
Gaussian graphical model, 101
hybrid method, 154
Ising model, 105
latent factor model, 226
score-based, 154, 166

T

Thompson sampling, 241
topological sort, 114, 206
truncated factorization, 137

U

upper confidence bound, 241

V

v-structure, 120
variational distribution, 76
variational EM, 75
  for SBM, 77
variational inference, 75
Verma constraint, 211
Viterbi algorithm, 66

Author Index

A

Abbe, E., 73
Agrawal, R., 241
Aldous, D., 85
Amini, A., 109, 123, 172, 180, 187, 235
Andersson, S., 157
Aragam, B., 109, 123, 176, 181, 187, 197
Auer, P., 241

B

Banerjee, O., 102
Baum, L., 59
Bento, J., 105
Berry, D., 240
Bickel, P., 80
Blei, D., 75
Bowden, R., 147
Brito, C., 149, 219
Brouillard, P., 197
Bühlmann, P., 103, 123, 176, 186, 195, 202

C

Cappe, O., 244
Celisse, A., 81
Cesa-Bianchi, N., 241
Chang, X., 80

Chatterjee, S., 81
Chickering, D., 167, 169, 171, 175
Choi, D., 80
Colombo, D., 233

D

d'Aspremont, A., 102
Dan, C., 197
Daudin, J., 77, 81
Dawid, A., 11
Day, N., 147
de Leeuw, J., 21
Deleu, T., 197
Dempster, A., 18, 101
Drton, M., 218, 219
Duchi, J., 238

E

Eichler, M., 219
El Ghaoui, L., 102
Ernest, J., 195
Evans, R., 207, 213, 216, 218

F

Fienberg, S., 71
Fischer, P., 241
Foygel, R., 218
Friedman, J., 102

Fristedt, B., 240
Fu, F., 125, 175, 180

G

Gámez, J., 170
Garivier, A., 244
Geiger, D., 167, 170
Geman, D., 35
Geman, S., 35
Ghahramani, Z., 75
Glymour, C., 137, 158
Gu, J., 125, 175, 180, 232

H

Handcock, M., 72
Hastie, T., 102
Haughton, D., 169
Heckerman, D., 167, 170
Hoff, P., 72
Hoyer, P., 189
Huang, J., 143, 163, 232, 243
Hunter, D., 21
Hyvärinen, A., 189, 194

J

Jaakkola, T., 75
Janzing, D., 194
Jensen, S., 54
Jordan, M., 75

K

Kalman, R., 67
Kaufmann, E., 244
Kazerouni, A., 241
Kerminen, A., 189
Kim, D., 226
Koller, D., 172
Krogh, A., 57
Kucukelbir, A., 75
Kuijpers, C., 172

L

Lachapelle, S., 197
Lacoste-Julien, S., 197

Lafferty, J., 105
Lai, T., 241
Laird, N., 18
Lange, K., 21
Larranaga, P., 172
Lattimore, F., 242
Lattimore, T., 242
Lauritzen, S., 10, 92, 96, 108, 117,
  120, 126, 128
Lawrence, C., 50, 54
Lee, C., 73
Li, H., 198
Lin, Y., 102
Liu, J., 54
Liu, X., 54
Lovasz, L., 84

M

Maathuis, M., 233
Madigan, D., 157
Madrid Padilla, O., 191, 198
Margaritis, D., 107
Mateo, J., 170
Matias, C., 72, 84
McAuliffe, J., 75
Meek, C., 160, 225
Meinshausen, N., 103
Mohan, K., 16
Montanari, A., 105
Mooij, J., 194
Murga, R., 172

N

Neuwald, A., 54

O

Orbanz, P., 86
Orchard, T., 24
Osband, I., 241

P

Pearl, J., 16, 91, 114, 120, 133, 137,
  143, 146, 149, 204, 210–211, 215,
  219
Peng, Z., 80

Perlman, M., 157
Peters, J., 186, 194–195
Picard, F., 77
Pierre, L., 81
Poza, M., 172
Puerta, J., 170

R

Rabiner, L., 57
Raftery, A., 72
Raskutti, G., 102
Ravikumar, P., 102, 105, 181, 197
Reid, M., 242
Reilly, A., 50
Richardson, T., 126, 207, 213, 216,
    218–220, 225
Robbins, H., 241
Robin, S., 72, 77, 84
Robins, J., 137, 207, 213, 216
Rohe, K., 81
Rosenbaum, P., 150
Roy, D., 86
Rubin, D., 18, 149–150
Ruiz, G., 191
Russo, D., 241

S

Sadeghi, K., 108
Saul, L., 75
Schafer, J., 16
Scheines, R., 137, 158
Schölkopf, B., 194
Shamir, O., 238
Shimizu, S., 189
Shpitser, I., 207, 213, 216, 218
Spirtes, P., 137, 158, 220, 225
Srebro, N., 238
Städler, N., 176
Sullivant, S., 218
Szegedy, B., 84

T

Tanner, A., 36
Teyssier, M., 172

Thompson, W., 241
Thrun, S., 107
Tian, J., 204, 210, 215
Tibshirani, R., 102, 177
Tseng, P., 200

V

van de Geer, S., 123, 176, 202
Van Roy, B., 241
Verma, T., 120, 211
Viterbi, A., 65
von Luxburg, U., 81

W

Wainwright, M., 102, 105, 238
Wang, B., 196
Wen, Z., 241
Wilkinson, D., 73
Wong, W., 36, 54
Woodbury, M., 24
Wright, S., 140
Wu, C., 23

X

Xing, E., 181, 197

Y

Ye, Q., 123, 172, 180, 187, 235
Yu, B., 81, 102
Yuan, M., 102
Yurramendi, Y., 172

Z

Zhang, C., 177
Zhang, H., 80
Zhang, J., 222–223
Zhang, K., 194
Zhang, T., 238
Zhang, Y., 238
Zheng, X., 181, 197
Zhou, Q., 54, 80, 109, 123, 125, 143,
    163, 171, 175–176, 180, 187, 191,
    196, 198, 226, 232, 235, 243

www.ingramcontent.com/pod-product-compliance
Lightning Source LLC
Chambersburg PA
CBHW050547190326
41458CB00007B/1954